BEAUTIFUL MACHINE

BEAUTIFUL MACHINE

Rivers and the Republican Plan
1755–1825

JOHN SEELYE

New York Oxford
OXFORD UNIVERSITY PRESS
1991

Oxford University Press

Oxford New York Toronto
Delhi Bombay Calcutta Madras Karachi
Petaling Jaya Singapore Hong Kong Tokyo
Nairobi Dar es Salaam Cape Town
Melbourne Auckland

and associated companies in
Berlin Ibadan

Published by Oxford University Press, Inc.,
200 Madison Avenue, New York, New York 10016

Library of Congress Cataloging-in-Publication Data
Seelye, John D.
Beautiful machine : rivers and the Republican plan, 1755–1825 / John Seelye.
p. cm. Includes bibliographical references and index.
ISBN 0–19–504551–3
1. American literature—Colonial period, ca. 1600–1775—History and criticism.
2. Rivers in literature.
3. American literature—Revolutionary period, 1775–1783—History and criticism.
4. American literature—1783–1850—History and criticism.
5. River engineering—United States—History.
6. United States—Civilization—1783–1865.
I. Title.
PS195.R55S39 1991 810.9'321693—dc20
90–46000

1 3 5 7 9 8 6 4 2

Printed in the United States of America
on acid-free paper

C. M. S.

1968–1981

T. D. L.

Acknowledgements

Until, through the generous terms of the chair I now hold at the University of Florida, I was enabled to structure a schedule enabling research and writing, as with all other scholars my time free from teaching was contingent upon the generosity of sponsoring foundations. Let me begin therefore by declaring my indebtedness to the National Endowment for the Humanities, for two fellowships that allowed me to undertake the research and much of the writing for the initial stage of this study; as well as to the Mellon Foundation, which, through the agency of the National Humanities Center, Research Triangle Park, North Carolina, provided a year in residence at that sanctuary, which I used to recast the manuscript. This occurred while I was still on the faculty of the University of North Carolina, Chapel Hill, which generously made up the difference between my salary and the stipend provided by the Center.

A number of the chapters that follow have already been published, in expanded, partial, or altered form, as independent essays, and I list them and the places and dates of their publication in the order in which they appear in the following pages, stating my gratitude to the editors of the various journals and books for allowing their resurrection here: Chapter IV was first published as "Flashing Eyes and Floating Hair: The Visionary Mode in Early American Poetry," in the *Virginia Quarterly Review,* v, 65 (1989); Chapter V, as "Beauty Bare: William Bartram's Triangulated Wilderness," in *Prospects 6* (1981); Chapter VI, as "The Jacobin Mode in Early American Fiction: Gilbert Imlay's *The Emigrants,* in *Early American Literature,* v, 22 (1987), and "Charles Brockden Brown and Early American Fiction," in *Columbia Literary History of the United States* (New York: Columbia University Press, 1988); Chapter VII, as "Beyond the Shining Mountains: The Lewis and Clark Expedition as Enlightenment Epic," in the *Virginia*

Quarterly Review, v, 63 (1987); Chapters XIV and XV, as "Rational Exultation: The Erie Canal Celebration," in *Proceedings of the American Antiquarian Society*, v, 94 (1984); and Chapter XVI, as "Root and Branch: Washington Irving and American Humor," in *Nineteenth-Century Fiction*, v, 38 (1984).

Let me now move to personal indebtedness, commencing with the many conversations I have had with Professor Arthur Marks of the Art Department at the University of North Carolina, Chapel Hill, whose wisdom regarding the subtleties of American iconography is only superficially reflected here. While I was in residence at the National Humanities Center, my colleague for the year, Professor Timothy H. Breen of Northwestern University, graciously took the time necessary to read an overlong manuscript and provide trenchant and useful advice concerning details and overview, which dictated the shape this book now assumes. And at the far end, my colleague in the English Department here, David Leverenz, gave the finished manuscript a detailed scrutiny; had I been able to follow all of his advice, it would have been a much different and better text, but readers will be grateful for the significant cuts in size that his recommendations occasioned. In this respect, also, I am in debt to the editorial consultants engaged by Oxford University Press, whose suggestions were extremely helpful in reducing an 800-page manuscript to reasonable dimensions. To Sheldon Meyer, Senior Vice-President at Oxford, I am grateful for patient forbearance during the long, often troubled period through which this study dragged its unwieldy length; and my gratitude to James Raimes, now at Columbia University Press, for his initial enthusiasm for this project, which resulted in the Oxford connection, remains in place in full.

The process of revision was made easier by the bank of word processors (persons, not machines) made available at the Humanities Center, and by those patient people in the Department of English here, Marie Nelson and Muriel Burks, who rendered further changes into readable copy—to whom let me here in print state my thanks. My gratitude also goes to Professor Melvyn New and Dean Charles Sidman, who prevailed upon me to join the faculty at Gainesville, providing the Graduate Research Professorship that made the final impetus for completing this book possible.

The rest goes to Alice, as all who know us will understand.

Illustrations

Contents

EPILOGUE CEASELESS TURMOIL

BEAUTIFUL MACHINE

And so no force, however great,
Can stretch a cord, however fine,
Into a horizontal line
That shall be absolutely straight.

Originally written as prose in
Wm. Whewell's *Elementary Treatise on Mechanics* (1819).

Prologue

The Shadow of the Dome

Our sensibilities having been shaped by romantic attitudes, we tend to think of primitive people as living in harmony with the natural world, an organic closure best signified by the notion of a terrestrial paradise. But if the world of nature, like so much romantic painting, may be figured as having a river running through it, then the fragments of ancient myths that have survived as scriptures suggest an antagonistic relationship, a constant struggle between men and their riverine gods. Jacob wrestling with his angel, Hercules with Proteus, Ulysses crossing the reluctant Scamander, the Lord heaping up Jordan's waters so that Joshua and his people might cross dry-shod—these images suggest that the river is never an easy ally, but must be forced into an accommodating posture or shape, made to conform with human desires. At the fragmentary start of American literature, also, we find a similar situation, for in the earliest records of encounter, European explorers uniformly tell of meetings with Indians at the river entrances to the continent, wild men who held the knowledge needed by the agents of civilization but who yielded it reluctantly, often refusing to the point of fiercely resisting entry. Even the most obliging Indians proved equivocal agents, becoming shifting shapes equivalent to Proteus.

This antagonistic pattern of colonial encounter is consistent, a continuity that may be traced from the accounts of Cartier exploring the St. Lawrence, Captain John Smith the James, or Captain Miles Standish leading the Pilgrims in their wintry search for an adequate river upon which to plant their hopes. Almost two centuries later to the year, Captains Lewis and Clark repeated the experience of John Smith as they sought to establish a river passage from the Mississippi Valley to the Pacific Ocean. Once again, American rivers and the red men who lived on their banks were wily and reluctant servants to civilized progress. And yet, by 1803, the historical record had given the ultimate victory to civilization, which over the intervening two centuries had imposed the rudiments of order upon eastern rivers, giving shape to the emerging republic. It was a neoclassical design, over all, even then being given symbolic expression by the plan for the new nation's capital to be laid out at the headwaters of navigation on the Potomac.

At the start, the Anglo-American experience in North America was inherently neoclassical, in that the military men who led the expeditions of entry were indebted to Roman models both for their tactics and for their heroic ambitions. By the opening years of the nineteenth century the neoclassical pattern had hardened into stone, an architectonics of stability implying permanence. For quite different reasons both the French Revolution and the emerging American republic generated Greek-and-Roman influenced architecture and icons, the French in reaction to the rococo superabundance of monarchical decadence, the Americans employing Palladian designs in order to lend their nascent nation the authority of Roman example. The Forefathers were either soldiers or lawyers for the most part and understood the sanctity of uniformity and precedent, and the greatest forefather of them all was not only a soldier but a surveyor, who appreciated the power of the Line.

That the Father of His Country was also a patron saint of civil engineering in America is not well remembered, but not because there is a lack of contemporary celebration of the idea. Washington's favorite scheme was a system of canals that would open the Potomac for navigation to its head, but he was also the champion of a plan earlier proposed by William Byrd of Westover for a canal that would trans-navigate and drain the Great Dismal Swamp. Drainage was ever a Roman ideal, or the Cloaca Maxima was dug in vain, and connecting lines of communication are essential to imperial design, the kind of closure promoting consolidation while permitting further expansion, which in turn justified further extension of the diagram.

The most memorable image of Washington is Leutze's depiction of the general crossing the Delaware River, an updating of a literary motif introduced by Joel Barlow, who thereby gave ideal form to the New England tradition of the Joshuan contingency, for in that region rivers were generally figured by the bridges necessary to transcolonial communication. But in Virginia the pattern was otherwise, for rivers there served as corridors of expansion, and promoters of empire sought to improve them for navigation, not cross them by bridges. The idea was to free rivers of impediments, thus increasing the flow, both of water and commerce. Because our national myths were mostly of New England manufacture, we have no suitable contemporary iconograph that best expresses Washington's role as the premier river man of his

William Rush (1756–1833). *The Schuylkill Chained* (*above*) and *The Schuylkill Freed* (1825). On deposit to the Philadelphia Museum of Art from the Commissioners of Fairmount Park. (Courtesy of the Philadelphia Museum of Art)

Carved as pedimental figures for the Fairmount Waterworks in Philadelphia, these allegorical representations have given modern commentators interpretive difficulties. The "chains" borne by the river god (traditionally an old man in classical iconography) have been read as symbolic of the waterworks, when, more likely, they are signifiers of the impediments to motion in the Schuylkill, the rocks and rapids upon which the god reclines. Likewise, one historian has suggested that the figure of the woman with her waterwheel should be entitled "The Schuylkill Harnessed," when Rush's intention was precisely what his original titles indicate.

age. Perhaps the Pennsylvania poet and novelist Hugh Henry Brackenridge came close when, in 1784, during Washington's stopover in the fledgling spa called Bath in the foothills of Virginia, he greeted the hero of the late Revolution with a masque dominated by subservient river gods.

Washington's subsequent experiences (indeed his prior ordeals) associated with the Potomac suggest something other than riverine subordination. But they provide a pattern dominated by a neoclassical ideal; reticulated devices of closure and consolidation figured more often than not by a canal. The implications of this linear figure may best be understood by resorting to etymology: "Canal" is from the Latin *canallis,* meaning "a pipe," suggesting the convenient passage of water by means of an enclosed conduit. In this root there lies a certain irony, water being made to flow more freely by enclosing it, a notion of liberty completely in accord with neoclassical ideals of aesthetic order. As Governor John Winthrop observed, it is laws that make men free, whether figured as necessary constraints or mutual consent to be ruled. Something of this meaning may be found in William Rush's designs for symbolic statuary decorating the Schuylkill waterworks in 1825.

Rush carved two figures, one a river god (an old, bearded man) lying fastened to a bed of rocks as behind him an American eagle rises, the other a seated woman, in classical dress, tending a conduit and waterwheel. The first figure was entitled "The River Enchained," the second, "The River Freed," and the implication is clear. It is the rapids that "enchain" the river, while a dam and its conduit will "free" the water to run smoothly and easily as it serves people's needs. The motto of John Quincy Adams, who was President of the United States when Rush carved his statues, "Liberty Is Power," suggests the full implication of the allegory, for the Schuylkill is most at liberty when enabled to produce the most power. The "proper" channel for a river is not necessarily the one it has carved for itself: By means of canals and locks it can be guided by men along a straight and level line, thereby improving upon natural design.

Across the breadth of the continental interior there lay a great diagram of natural waterways, a ragged Z provided by the confluence of the Ohio, Mississippi, and Missouri rivers, but this imperial pattern was blocked both in the east and the west from access to oceans by barrier ranges of mountains, visible impediments to perfect national union. Moreover, movement on those rivers was impeded by falls and retarded by the swift currents of both the Missouri and the Mississippi, necessitating arduous and dangerous labor in order to effect passage westward. If canals were the answer to the problems of barrier ranges and rapids, then the steamboat provided the solution to the problem of swift currents; Fulton was but one of many inventors who recognized the primacy of steam power as an antidote to river obduracy. Together, the canal and the steamboat were seen as marvels of engineering that made a more perfect Union possible, coefficients of the Constitution, the benevolent mechanism that was perhaps the most important contribution of the Enlightenment to the American experiment.

The great trinity of rivers, canals, and steamboats acted together not only to realize Washington's ideal of a more perfect Union but to exemplify the Horatian aesthetic that balances natural beauty with considerations of use: those features of the landscape that serve mankind are more admirable than those that do not. If the Romantic spirit as exemplified in America by Emerson prized vistas of natural scenery as seen by a naked eyeball, the neoclassical sensibility preferred to impose a grid of utilitarian considerations upon the landscape, best signified perhaps by the view through a surveyor's transit. For if rivers are naturally lovely, their meanders providing what Hogarth called the "Line of Beauty," in the neoclassical view they were most attractive when they yielded most to humanity's needs, whether as mechanisms of transportation or as sites for nascent towns.

During the colonial period, the neoclassical penchant for level and line in the service of utility gained further power through association with the millennial ideal, reinforced by evidences of providential design. The workings of Providence were intimate with the idea of progress in North America from colonial beginnings: God, in the Puritan phrase, "opened a way" for his chosen people, and if that way did not have a natural avenue, then God put into the hands of his people the instruments necessary for making an artificial passage in the providentially determined direction. If God failed to "heap up" waters, then the Puritans heaped up stones in the shape of dams, effecting the same (and a more permanent) effect. If Revelations promised a future moment when valleys would be raised, mountains lowered, and the crooked made straight, the Puritans saw no harm in hastening forward the ideal of a leveled and linear landscape.

The framers of the Constitution were careful to leave God out of the process of government, but not so the projectors of canals and proponents of improvements of river navigation. They characteristically evoked the deity as a transcendent form of landscape engineer who had designed the North American continent in a shape predisposed to accommodate man's propositions. It were a sin, went the reasoning, to neglect the intention of God so clearly marked out on the map. It would be a supreme virtue, likewise, to complete the work necessary to realize the divine scheme of natural waterways by means of artificial canals. Thus Duty, that beau ideal of neoclassical morality, was given a sacred aura, and if there was a living personification of duty on American soil during the early years of the republic, then George Washington was he, a correspondence incarnate of divine and human disposition that helps to explain the worshipfulness of his contemporaries, so mysterious to modern sensibilities.

During his lifetime, Washington was celebrated as the American Cincinnatus, the farmer-soldier who was a Roman and a republican ideal of unflinching dutifulness, but that is only half of the symbolic picture. Washington was also an American Hercules, type of all epic heroes, in whom endeavor is figured as Labors, which we may Americanize as Works, giving the ideals of the Protestant ethic an epical shape, a wrestling with natural forces until they submit and answer to the uses of man. In the story of Hercules, the river Alphaeus plays such a part, serving man by having its channel turned to cleanse the Augean stables. (It is in quite another myth that its waters rise in a fountain associated with poetic inspiration.) Washington's Alph, once again, was the Potomac, which he viewed from the front porch of his home and he necessarily saw as the premier avenue of empire for the emerging United States. Championing canals that would bypass its falls, backing inventors of steamboats that would accelerate its upstream navigation, and lending his name to the emerging capital on its banks, Washington emerges supreme at the end of the eighteenth century as the Man of Improvements, even to his symbolic presence at the Constitutional Convention, from which emerged the most beautiful machine of all.

But as in the story of Hercules and the Alphaeus there is another dimension to the technological epic of wrestling nature into a posture of accommodation. The Constitution as an Enlightenment machine was a creation of compromise that like the divine plan for the American continent left a certain amount of business untransacted, handing its completion down to a later generation, the kind of national debt not envisioned by Alexander Hamilton. In borrowing their aesthetic from Greek and Roman ideals and their millennialism from the Bible, the founders of the republic also borrowed from the ancient world the institution of slavery. Freeing rivers by restraining them within artificial channels, the founders of the republic promoted a paradoxical notion of liberty in human terms also, by dividing mankind into two separate groups, those who could perfect themselves by the instruments of reason, and those who, though contributing to that process, were kept from sharing in its benefits. Today the nineteenth-century image of a man in chains is less often associated with unimproved streams than with chattel slavery, an institution inseparable from the Constitution's bedrock foundation on the sacredness of property, yet a painful reminder of the inadequacy of compromise as a mechanism for realizing

Enlightenment ideals. In drafting the Declaration of Independence, Jefferson equated happiness with the pursuit of property, but the property Virginians increasingly found themselves pursuing had two legs and thought of happiness in terms of life beyond the Ohio River, their temporal Jordan.

Washington died in 1799, at the threshold to a new century, by which time the Ohio had become a symbolic line of demarcation between two ways of life, the one established in the Western Reserve by migrating Yankees, and the other in Kentucky, originally part of Virginia. Few travelers along the Ohio had much good to say about life in Kentucky, which was customarily described as a barbarous zone, whose inhabitants were given to horse-racing and hard drink when they were not fighting one another and flogging slaves. Ohio, by contrast, put forth the image of orderliness and industriousness, as symbolized by white-painted, clapboard architecture, the sort of rectilinearity that is to rectitude close-allied. The New England connection was further strengthened in 1825, the year in which the Erie Canal was finished, and by 1832 the Ohio Canal had completed a design that George Washington anticipated but was antithetical to the one he had favored, for it ensured that the great wealth of the interior would be drawn down the Hudson, not the Potomac River. Thus the Ohio, gateway to the West, increasingly became a symbol of division, not unity, separating the South from the rest of the expanding Union.

The metamorphosis of the Ohio River in fact and symbol demonstrates that within the schemes for national union and expansion promoted by General Washington were the fulcrum and lever that would forever displace the Virginian hegemony. We may trace, therefore, a fatal dimension in the events that gave geopolitical shape to an emerging nation, commencing with the first shots fired in the French and Indian War and ending with the cannonade that signaled the completion of the Erie Canal. In an earlier volume I attempted to show the extent to which the explorations and schemes for expansion drafted by the earliest colonists in Virginia and New England laid down the base lines for a disjunctive imperial plan, defined by the rivers along and across which the bifurcated diagram was laid. In the present volume I hope to delineate the further dimensions of that divided plan, which took additional meaning from the optimism regarding human progress generated by the Enlightenment.

During the preceding epoch, from Raleigh's first outpost of English empire in America on Roanoke Island, to William Byrd's survey line that separated North Carolina from Virginia and pointed the way to a further West, the differences between North and South were matters of regional distinction keyed by the intentions of the original settlers. But after 1755, with the outbreak of the French and Indian War, these differences became increasingly identified with sectional interests, hence with forces inimical to the republican ideal of national unity. Geopoliticians in Virginia and New England were unanimous in promoting schemes that would give the new nation as a work of art a coupleted closure and a neoclassical plan, but the design that emerged worked to opposite ends, promoting apocalyptic, not millennial, conclusions.

As we shall see, the forces of division and disunion may be traced by means both of events as literature and of literature as event, whether as Washington's western adventures or as the gothic novels of Charles Brockden Brown. Art likewise shall be considered within the Enlightenment purview, which saw no difference between aquatints and aqueducts in aesthetic terms. In these writings and actions and artifacts we shall see demonstrated the propensity of manmade schemes to go awry because of the tendency of events to elude the control of those who seek to direct them. Moreover, if in Washington's adventures and the geopolitical literature promoting territorial expansion we can detect a clearly neoclassical line, evinced also in the architectonics of the age, in the fiction of Brockden Brown we will find an antithetical aesthetic, an incipient romanticism that calls into question the primacy of neoclassical closure, most particularly when figured as the works of men.

I

Ancestral Voices

i

In 1745 the Reverend Peter Fontaine of Westover Parish, on the James River in Virginia, was attempting, despite the infirmities of old age, to resume contact through letters with his brothers in England, whom he had not seen for thirty-five years. The Fontaines were a divided family, for though John and Moses had remained in the Old World, Peter, his other brothers, James and Francis, and his sister, Mary Ann (Maury), had continued their parents' Huguenot exodus all the way to America. Peter's letters home to England are for the most part an account of the generational growth and spread of the Fontaine family in Virginia. So numerous were the progeny that he had no doubt that "when I am gone, there will not be wanting some to brighten the chain between us here and you in England, many years to come, an Indian but very significant expression, signifying to renew the affection or alliance between people of different nations, or friends, at a distance one from another" (Fontaine: 344).* After Peter's death the chain was kept bright for a time by his son and namesake, and also by the Reverend James Maury, son of Mary Ann, but in the aftermath of the French and Indian War the chain seems to have been broken not only by the disruptions of death but by the increasing tensions between the colony and the mother country.

As a family chronicle, the Fontaine correspondence evinces considerable human interest, and yet its chief value lies in the extent to which the affairs of its members parallel (much as the letters comment upon) contemporary events. It is therefore doubly significant as a record of an important period in America's history, a time when colonists were assuming a proto-national identity that would more clearly emerge in

* Throughout, abbreviated references in parentheses are keys to entries in the Bibliography (p. 397)—which see.

the years following the French and Indian War. Where the letters of Peter Senior are fairly ordinary, filled with family news and detailed accounts of his physical ills, the letters sent to England by both his son and his nephew reflect the expansive mood of Virginia at mid-century. It was a mood that contributed to the hostilities that would result in the impending conflict and afterwards would inform the ever-growing shape and imperial ambitions of what became the United States. Even at the start the Fontaine family was intimate with certain strategic moments in Virginia's ceaseless westward growth: though he returned to England, John Fontaine came to the New World in 1715 to buy land for his brothers and sister, a circumstance that resulted in his being among the group of merry adventurers that accompanied Governor Spotswood when he made his epochal crossing of the Blue Ridge into the Shenandoah Valley.

John kept a journal during his trip to America, which contains the only extended record we have of Spotswood's excursion, and though his brother, Peter, was not a diarist, the rector of Westover parish frequently came into the orbit of an indefatigable journal keeper, William Byrd, in whose vast catalogue of minutiae he occasionally appears. A greater measure of immortality, albeit dubious, was given to Peter when he accompanied Byrd as chaplain to the great Dividing Line survey party of 1728—an expedition equivalent to Governor Spotswood's—for in Byrd's "secret history" of the event, the Reverend Mr. Fontaine is satirized as Parson Humdrum, a Hudibrastic caricature of punctilious piety. Whether as recording witness or as participant, the brothers Fontaine played roles similar to the "heroes" in the Waverley Novels, being not major movers of great occasions—albeit on a provincial scale—but fellow travelers. Likewise, where the fortunes of the Byrd family in America provide a saga of dynastic rise and fall, the Fontaines were consensus colonials, who followed a middling path of modest but steady and respectable success, with the occasional reversals and disappointments to which families are heir and with which heirs are familiar.

In laying the dividing line between North Carolina and Virginia, William Byrd prophesied that settlement would follow, and by 1752, the son and namesake of Byrd's chaplain had settled near the forks of the Roanoke formed by the Stanton and the Dan—in the very heart of the frontier territory given rudimentary order by Byrd's great Line. Like his cousin James—who had in the words of his uncle left the tidewater region for a remote parish "amongst the mountains, and is concerned in the Ohio Company, who have an entry on Halifax, beginning on the other side, or properly, west side of the great mountains, upon the line between North Carolina and Virginia" (Fontaine: 342)—young Peter had followed his fortunes westward, but as a surveyor, not a minister. Writing to his Uncle John in England, Peter sent along a sketch map of the area in which he lived, thinking it "might be of some entertainment" to the man who had "rambled" with Governor Spotswood "over a good deal of the southern part of Virginia" (357).

For the work of a "surveyor," Peter Fontaine's map is a curiosity, for he has scrambled cardinal directions: north is at the bottom, west to the right-hand side. The effect, however, is to emphasize the long arm of the Stanton River, which, running off at an angle to the dividing line and cradling Halifax County, reaches into the Allegheny mountains towards "a tract of land to the westward of the Blue Ridge" and towards "PARTS UNKNOWN" (356). In the upper right-hand (northwest) corner Peter drew a short meander called "New River," which he labeled "a branch of Mississippi River," leaving a short space between it and the headwaters of the tributary Virginian streams. The iconography as well as the cartography is familiar: from the days of Captain John Smith, the symbolic lay of the land in Virginia tended to put forth the Alleghenies as a pyramidal Great Divide, where the headwaters of eastern-running streams reached out as with extended fingers toward the headwaters of western-running streams. In time, the symbolic topography would be, in part, corroborated, but as Captain Smith's hope that the western rivers ran into the Pacific Ocean had been disproved by the end of the seventeenth century, so Peter Fontaine's information concerning the New River would also prove false.

But false hopes often express true feelings, and it is clear that by mid-century, Virginians were looking longingly over the Alleghenies even as they surveyed and parceled out the upriver regions. Still, Peter inherited something of his father's sedentary disposition, and his title of "Surveyor" was mostly honorific, for Peter suffered from a "weakly constitution," and did "not go at all in the woods," leaving the actual work with chain and rod to his assistants (358). After the initial hostilities of the French and Indian War broke out, he was forced to abandon his "Clapboard Castle" in Halifax County, retreating to his "Rock Castle" in Hanover County, close to the tidewater line, where he consoled himself with the Franklinesque wisdom "that a middle station [is] the happiest" (356, 370). Peter's advance into the wilderness, his eventual retreat, and his final stand in the middle landscape would be an experience shared by many Americans in years to come, yet other Virginians would continue to push ever farther west.

What was a permanent retreat for Peter Fontaine, Junior, was only a momentary setback for the forces of Virginian expansion, and the foremost champion of that imperial advance, another surveyor, was already in the field. In June 1754, Peter Junior sent news to England of a

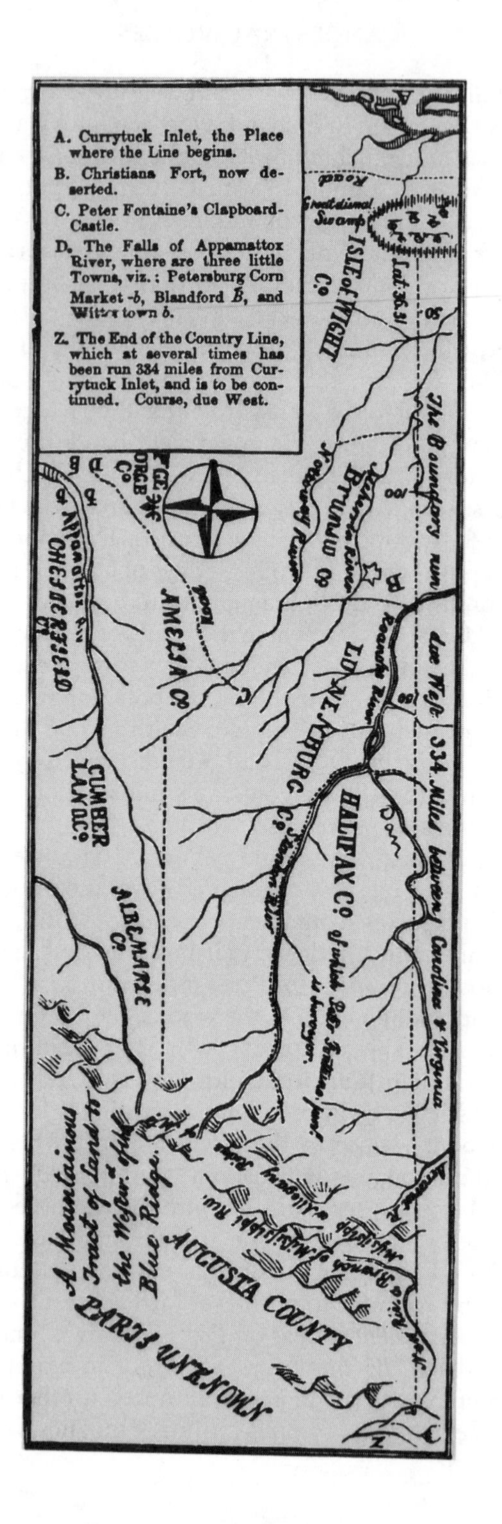
A. Currytuck Inlet, the Place where the Line begins.
B. Christians Fort, now deserted.
C. Peter Fontaine's Clapboard-Castle.
D. The Falls of Appamattox River, where are three little Towns, viz.: Petersburg Corn Market -*b*, Blandford *B̄*, and town *b*.
Z. The End of the Country Line, which at several times has been run 334 miles from Currytuck Inlet, and is to be continued. Course, due West.
ISLE of WIGHT Co
The Boundary run due West 334 Miles between Carolina & Virginia
AMELIA Co
HALIFAX Co
CUMBERLAND Co
ALBEMARLE Co
AUGUSTA COUNTY
PARTS UNKNOWN
A Mountainous Tract of Land to the Westw.d of the Blue Ridge.

victory by Virginia militia over a detachment of French soldiers "on the Ohio River": the French were hiding "in ambush in the woods. . . . Each party fired, and it has pleased God that we have killed or taken them all" (361). Actually it was the French who were ambushed; nor were all killed or taken, and Peter's jubilation was short-lived. Still, though "Colonel Washington, the commander of our men from Virginia," would soon suffer an ignominious defeat, he would rise again, a phoenixlike transformation in which the youthful representative of colonial expansion would become the mature champion of a young republic.

In a poetical retrospective of the French and Indian War, the New England almanac maker Nathaniel Ames in 1763 made one of the earliest contributions to the glorification of George Washington, who, according to Ames, first entered "the list of fame" by his "journey to Lake Erie" and his "defeat" of a "French detached Band" (Ames: 339). Washington's subsequent victories during the War for Independence would considerably expand his fame, but in terms of geopolitics, his presence and performance at the start of the great struggle to secure the Ohio Valley for Great Britain and Virginia is not only an indisputable historical fact but an indispensable symbol. For Washington of Westmoreland County is the metaphorical successor to William Byrd of Westover, the earlier Anglo-American planter-statesman giving way to a rustic yet classical Virginian type. A representative squirearch, Washington provides a live (if stiffly articulated) model for the emerging republican ideal of militant agrarianism, the alternative symbols of which were Cincinnatus, the Minuteman, and the Frontier Rifleman.

Washington had none of Byrd's cosmopolitan polish, and his youthful attempts at evoking the muse and imitating polite manners were stilted. Still, this rough edge, which with time would assume a finer line because of an enlarged perspective, was essential to the character of the emerging American hero, whose costume, like his customs, would be homespun. He would also be in manners manly, much as "Manly" was the name Royall Tyler gave to the first authentic American hero in our early republican literature—a character inspired in large part by Washington's example—who is given a dandified American cosmopolite as his satiric counterpart. The Doric was ever an Enlightened aesthetic ideal.

(*Facing*) Peter Fontaine, Junior. *Map of the Virginia and North Carolina Dividing Line* (1752). From *Memoirs of a Huguenot Family* (New York: 1852).

William Byrd's Dividing Line, originally regarded as a stabilizing factor, becomes here a mark of expansion, bisecting transmontane rivers running west.

Byrd's London wit and cavalier airs were part and parcel of Tidewater gentility early in the century. Though his purposes were often serious and his ambitions undeniable, Byrd affected in his writings a negligent and even indifferent air, making over his great western adventure into a bawdy, picaresque comedy. By contrast, Washington's genuine respect for belles lettres is reflected only indirectly through his writings, and his affability is likewise attested to chiefly through the testimony of his contemporaries. Though capable of a wry, even sardonic, humor in his reports and letters, Washington never essayed an extended literary composition, satiric or otherwise. And where Byrd's diary often reveals the personal, even very private, side of plantation life in Virginia during the first half of the eighteenth century, Washington's is literally matter of fact, a farmer's impersonal catalogue of weather, planting, hunts, harvests, visits.

Upon occasion, however, Washington's journals expand into a full-fledged narrative, and no one can read his accounts of his adventures in the Ohio Valley without acknowledging their worth beyond matters of record. If they are less artful than Byrd's twin histories of the Dividing Line, they put forth a consistently heroic face and have an epic élan given force by the author's spare prose style. In these writings, young Washington leaves behind strained efforts at gallantry, and the native Virginian takes on the outlines of a truly western man, yet Roman withal, an American metamorphosis anticipated by that earlier soldier-surveyor, Captain John Smith, for whom the Virgin Land was also Virgilian territory. For Washington, as for Captain John Smith, the epic zone is defined by rivers, a century and a half of Virginian enterprise having carried empire from its earliest probings of the entrance to the Potomac through the rift in the Blue Ridge where it is joined by the Shenandoah, an upriver progress that would, with Washington's descent into the Ohio Valley along the Monongahela, begin to run in a much more rapid fashion toward the West.

ii

Peter Fontaine, Junior, may have known who "Colonel Washington" was, but it is doubtful that his uncles in England had any idea. And yet, for several easily defined reasons, the young officer quickly emerged as a major, even international, figure. At the start of his public career, Washington was an anonymous member of up-country Virginia society, and even as the Reverend Peter Fontaine was brightening the metaphorical chain between the New World and the Old, the future father of his country was carrying a surveyor's chain into the western regions beyond the Blue Ridge. Save for the trajectory that initial sortie established, it was a routine affair, and only gains meaning in retrospect, for the young Washington of transit and line makes straight the way for the imperial presence that followed.

The surveying party that Washington accompanied as a student of the art crossed over into the Shenandoah Valley early in 1748. A routine occasion, it was a novelty for the young Washington, and resulted in the first of his diaries still extant. The record is a slim account of places visited, meals eaten, body lice encountered, and the like details of backwoods accommodations and discomforts, the last of which were increased by rain and spring freshets. Both journey and journal are important only because the author was destined for greatness, a career towards which this initial baptism of river crossings would prove a determining vector. The diary likewise gains meaning when joined to records kept by Washington of much more important western journeys to come, and it may be compared in that respect to the letters of Peter Fontaine, Junior, who was himself setting up at about the same time as a surveyor of the upriver Roanoke lands.

All we know about Peter is what we can learn from his few letters to England; all that interests us is the knowledge we can glean, not so much about Peter himself, as about the conditions of life in Virginia at mid-century. When his correspondence with his uncles in England ceased, so, in terms of history, did Peter Fontaine, Jr. By contrast, Washington's first diary is important chiefly, not as an account of an apprentice surveyor's experience in the unsettled regions above the falls of the Potomac, but as the record of how the man who was to be both father and savior of his country reacted to a signal, formative experience. Unlike Fontaine, Washington was actually a surveyor in practice, not through political appointment—a fact that provides the definitively pragmatic touch—yet the event had political implications of a symbolic sort. Where William Byrd with transit and line laid down the geopolitical division between Virginia and North Carolina, Washington's first surveying expedition parceled out what was then Virginia's westernmost territory; yet as Byrd's great Line was (as he himself saw it) a mechanism that would facilitate an orderly western movement of civilization, so Washington would thenceforth be associated with the territory beyond the Alleghenies. He would go on to a delineation of first colonial and then national empire, activities that made him, in effect, *the* Surveyor—or Surveyor-General—of an emerging nation.

"Nothing remarkable happen'd," as young Washington noted of his

Henry S. Sadd (active in America from 1840–1850). *Washington, Aged 18*. From *The Ladies' National Magazine,* Vol. VII (Philadelphia: 1845).

The nation's champion is shown as a young surveyor, transit set to one side as he strikes a ringletted pose, framed by a composition in which Romantic conventions struggle with notions of neoclassical rigor. Below lies a river, presumably the Shenandoah or Potomac. The picture accompanies a story in which the young Virginian leaps into the torrent to save a child, thereby exemplifying the "self-sacrifice" that the sentimental age would identify with Washington's famous devotion to duty. But it also certifies the riverine image, which the budding hero would be associated with from the start of his career to its end. The fictional plunge, like Washington's verifiable tumble into the Allegheny in 1753 (cf. following illustration), is a baptism of sorts, part of the deifying process by which the hero becomes a river god.

first day out, a hike over the Blue Ridge; yet we have a sense of something remarkable in the making, as the boy surveyor tastes the pleasures (few) and the inconveniences (many) of a backwoods journey (*Diaries*: I: 7). For the experience serves as a symbolic threshold to the much more important trip he was to take six years later, that "Journey to Lake Erie" celebrated by Nathaniel Ames, his embassy for Lieutenant-Governor Dinwiddie of Virginia to the commander of the French forces in the Ohio Valley. In the interval his brother Lawrence had died, leaving George the owner of Mount Vernon, and heir as well to the familiar round of a Tidewater planter's life—surveying, planting, hunting, and militia duty. Barely in his majority, and dignified for the occasion with the rank of major, Washington suddenly emerged from the particulars of provincial routines into the generalities of international plays of power. He was thrust, that is to say, into a much higher sphere than the world of Virginia gentry, entering the elevated realm of epic heroes and classical statuary at age twenty-one.

Governor Robert Dinwiddie had only recently arrived in Williamsburg, and though he never did learn to anticipate the jealousy with which colonial subjects of the King guarded their privileges, it was his virtue that he entertained large views, the largest of which took in the Ohio Valley. Dinwiddle was therefore one of the few men in power in the American colonies who understood the implication of French forces moving to take command of the major tributaries to the Ohio River. Having already encouraged (and invested in) the scheme of the Ohio Company (a plan similar to the one that had raised the hopes of the Reverend James Maury, Jr., and his father-in-law), Dinwiddie was acutely aware of the necessity for quick, concerted action if British claims to the lands west of the mountains were to have real rather than merely legalistic foundation. Quick action he got, thanks to George Washington, yet the eventual outcome of the young Virginian's two visits to the Ohio Valley was such that concerted action was needed much faster than even Dinwiddie had anticipated.

Whatever the circumstances, Washington's rise in profile was sudden and considerable. His embassy to the French prepared the way for his next errand, which, despite its ignoble outcome, guaranteed that the young Virginian, blessed by Dinwiddie's favor and the warrant of experience, would participate in the subsequent battles of the war to save the Ohio Valley for colony and country. This, despite his argument that the tributaries of the Ohio River should be secured in ways other than those that were proposed and carried out by men with greater rank but much less knowledge of the frontier than his own. And, it should be added, in spite of his continuing complaints about being shortchanged in pay and perquisites because of his provincial status. If his military experience laid the groundwork for subsequent and far greater exploits, his grievances turned the sharp edge of his heroic temper against British officers, perhaps even toward the king from whom their commissions came.

Still, in 1753 he was acting as the Royal Governor's messenger and was an embassy of British empire, carrying a message to the commander of French forces in the Ohio Valley protesting their presence on lands "notoriously known to be the property of the Crown of Great Britain" (*Journal*: 25). Governor Dinwiddie's letter to the French commander and the commander's polite but firm demurral were published in Williamsburg in 1754, along with Washington's journal of the trip. It was a slim book, little more than a pamphlet, but one that, like Peter Fontaine's map, sketched out an imperial plan. Though the currents of history seek untidy channels, we may chronicle the second great westward flow of American empire from that date of publication, for the forces of expansion would soon follow the path of George Washington along the rivers and creeks of Virginia into the mountains, and from there down along the creeks and rivers of the great valley beyond.

From the start, Washington's journey took on symbolic overtones, provided by his guide through the mountains into the western valley, Christopher Gist, a frontiersman who had served as surveyor for the Ohio Company. Moreover, once having crossed over into the Ohio Valley, Washington engaged the services of Indians, who were to act as guarantors of safe passage through the valley. Equally symbolic, however, were the several obstacles placed in the way of Washington's progress toward his assigned goal: because of French overtures, the loyalty of Indian tribes earlier pledged to the English was no longer certain, and Washington's journey was hindered along the way by the necessity of holding tedious negotiations with Tenarcharison, the "Half-King" of the Senecas, a powerful Iroquois chief whose control over neighboring tribes was considerable. He was delayed also by "excessive Rains, Snows, and bad Traveling, through many Mires and Swamps," yet with a planter's eye he noted the "good Land" through which they slogged, and—as per his instructions—he surveyed the territory for possible sites of forts (20). Washington paid particular attention to the confluence of the Allegheny and Monongahela Rivers, for the height of land there provided a location easy to defend, and access to the Ohio could be had by Virginians by means of the latter stream, which "runs up to our settlements and is extremely well suited for water carriage, as it is of a deep still nature" (4).

Although Washington took his errand very seriously (not, however, without some wry remarks concerning Tenarcharison's dilatoriness), there is a vein of unintended humor throughout the central section of his journal, commencing with the disparate and even disjunct character of a man who, in encountering all manner of natural adversity, manages to cast an evaluative, even acquisitive, eye over the landscape. This, however, is a great Virginian tradition that dates back to Captain John Smith, and much as Smith took advantage of his captivity amongst the Indians to study his enemies the better to deal with them, so Washington in his dealings with the French kept one eye on the main strategic chance. First at the fort at Venango, then at the fort at the head of French Creek, near Lake Erie, Washington met with vexing delays, yet he used his enforced leisure to study their fortifications, even while forestalling the attempts of the French to seduce the Half-King to their side. It was only with great difficulty that Washington managed to pry Tenarcharison loose from Fort LeBoeuf, and then he lost him when the party returned to Venango, reluctantly leaving the Half-King with the French commandant, warning the Indian to guard himself against flattery, "and let no fine Speeches influence him in their Favour" (20).

Having done so, Major Washington entered the most hazardous stage of his journey, and the comic element disappears, the triangular affair between French, Indians, and English giving way to an increasingly dangerous encounter with wilderness adversity. Abandoning affairs of state, the Virginian exchanged his diplomatic regimentals for "an Indian walking Dress," and set off on foot with Christopher Gist and a Dutch fur trader for home, their enfeebled horses being good only for pack animals. From that point on, the difficulties became less psychological and political and much more physical: "the Roads becoming much worse by a deep Snow, continually freezing" (20). The two Virginians were eventually forced to leave trader and horses behind, pushing on through the snow cross-country for Shanipan Town, an Indian village at the forks of the Allegheny and Monongahela. En route they were attacked from ambush by a party of Indians loyal to the French, one of whom they took hostage for a time as protection. Washington's situation had become quite different from what it had been at the start of his errand, he and Gist marching on in grim determination, virtually stripped of equipment.

During this stage of his errand, Washington takes on an outline seldom associated with his heroic image, his situation and Indian dress giving his ordeal a native American coloration. The two Anglo-Americans in buckskins, bent forward as they encounter increasingly cold weather and hostile Indian territory, lose their specific identities and merge with a pattern having no European coefficient, as Washington is absorbed into a metaphor of encounter that transcends his specific errand: at this early stage of his emerging public shape, Washington's classical outline merges with that of the American woodsman.

The wilderness adventure reached a climax of sorts when the two men arrived at the Allegheny River, which they discovered was not completely frozen over. Having spent two days and a night hiking, Washington and Gist rested until morning, then set to work building a raft "with but one poor Hatchet," a task that took up yet another day—the third since their departure from Venango—before they were able to launch their clumsy craft. Where delay before had a comic cast, during this stage it promoted crisis, and the two men had barely got themselves half over the river before the raft got jammed in ice floes, "in such a Manner that we expected every Moment our Raft to sink, and ourselves to perish" (26). As he leaned against his "setting Pole to try to stop the Raft, that the Ice might pass by," Washington was suddenly jerked off his feet by the force of the ice, dropping into ten feet of water: "But I fortunately saved myself by catching hold of one of the Raft Logs" (22).

Abandoning their useless craft, Washington and Gist waded to an island, where they spent the night, soaking wet and suffering from the bitter cold, but (such are the blessings of adversity) the river was frozen completely by morning, allowing them to cross to the side of the Allegheny where the Indians were friendly to the English. Their ordeal was not yet over, but the worst part, their dangerous baptism in wilderness waters, was behind them: they soon reached the English settlements, and after several more days of delay and travel, crossed through the mountains to Wills Creek on the Virginia side. Thus ended "as fatiguing a journey as it is possible to conceive, rendered so by excessive bad Weather" (22). To look at a modern map is to recognize the understatement here, for the two men had made a transit across the breadth of the Ohio Valley and back, an undeniably heroic exercise, at once a rite of personal passage for Washington and a tentative excursion demarcating empire: on the last leg of their return trip the two men encountered a party of Ohio Company settlers, carrying "Materials and Stores for a Fort at the Forks of the Ohio" (22).

Washington's wilderness crossing is recorded on a map he prepared for Governor Dinwiddie that accompanied his report, a continuation of Peter Fontaine's diagram of Virginia's westward reach. On it, the headwaters of the Potomac are joined by a short traverse of some seventy miles to the Monongahela and its tributaries, a wide waterway that winds its way due north, first as the Allegheny River, then as French Creek, coming within ten miles of Lake Erie. It is a sinuous diagonal that coincides with Washington's journey, and both the voyage and the line of waterways that determined it would delineate a major segment of imperial planning in Virginia for many years to come. For many reasons, including Leutze's heroic canvas, Americans are more than familiar with Washington's Delaware crossing, but in taking a larger-view, in stepping back to include the last and greatest of the French and

Daniel Huntington (1816–1906) and Denison Kimberly (1814–1863), after Huntington. *Washington Crossing the Allegany River. From Gem of the Season for 1849* (New York: 1849). (Courtesy of the American Antiquarian Society)

This scene was also engraved by Richard W. Dodson (1812–1867) for *The Gift* (Philadelphia: 1845) and other periodicals and texts. Huntington's depiction of a critical moment in Washington's great Ohio adventure was likewise copied by other artists, including William Sidney Mount, and was undoubtedly the most popular of the Washington "crossings" until displaced by Emanuel Leutze's monumental rendition of the wintry Delaware journey painted in 1851.

Indian Wars as integral to the geopolitical process making possible the Revolutionary War, then the much less famous crossing of the Allegheny River looms large as an archetypal experience. Less amenable to neoclassic canons of composition, stark, crude, elemental, it was essayed as a subject by the nineteenth-century artist David Huntington, with results less satisfactory than suggestive.

Like Leutze, Huntington used a pyramidal composition, necessarily less cluttered and dominated by the setting-pole that would in a moment turn Washington into something less than a heroic figure. As in the larger canvas, also, Washington dominates the composition: Gist stands to one side, a position confirming the dominance of his companion. As a tableau it clearly takes its composition from a historical perspective, viewing Washington from the far end of an intervening century. The young major, in presenting his report to Governor Dinwiddie, stated his hope that it would "satisfy your Honour with my proceedings for that was my aim in undertaking the Journey, and chief Study throughout the Prosecution of it" (23). We may justifiably conclude that he was pleased when Dinwiddie had the journal published; Washington like any young man in his position was anxious to advance his reputation (and rank). Yet at the far end of the process, with Huntington, we can now see the extent to which the book (and adventure) bound the advancing fortunes of George Washington to the imperial errand he had just run, linking forever the name of the soldier-surveyor with the rise of Anglo-American empire in the West.

The journal Washington kept on his second foray into the Ohio Valley is a much different document. The author, now promoted to colonel, was in command of a small army charged with taking possession of and defending the Ohio Company's fort at the forks of the Ohio. Forcing a military passage through mountainous terrain by felling virgin forest was a heroic-enough task, but it was never completed: the French had already captured the fort, the news of which halted Washington and his men partway. Hastily throwing up an earthworks in a mountain basin called the Great Meadows, Washington considered his next move, his deliberations fed by a series of reports and rumors, the last of which concerned a French scouting party spotted by the Half-King's men. Setting off through a dark and rainy night, Washington led a band of soldiers on a forced march to the Indian camp five miles away, and after a brief war council, the Iroquois and the Virginian militia joined forces and shortly after sunrise ambushed the French party, killing the officer in charge, a young lieutenant named Jumonville.

This encounter has been a source of confusion ever since it was erroneously reported by Peter Fontaine, Jr., and in the end it makes little difference whether Jumonville was indeed, as the French claimed, on an embassy to the English similar to that of Washington a year earlier, or whether he was leading an advance war party, as Washington maintained. The importance and definable fact is what happened as a consequence, for the French retaliated by attacking the Virginians in their little fort at the Great Meadows. Hopelessly outnumbered, the Americans were forced to surrender. Allowed to keep their arms, they lost all else, including Washington's journal of the event, which the French eventually published as "proof" of his warlike intent. The document is in all ways a counterpart to the journal that preceded it, being a record less of a single (if often frustrated) line of heroic endeavor than of uncertain advance, of confusing reports and tentative actions interrupted, and it breaks off suddenly, as Washington retreated to his hastily constructed earthworks and inevitable defeat. Authorities differ as to the wisdom of fortifying a meadow surrounded by tree-covered hills, but all are agreed that it was an expedient and an unfortunate step. Washington's "conduct" as a messenger to the French was exemplary; his actions as a young commander of militia were at best a mistaken display of impulsive heroics.

Still, his courage and quality of leadership were undeniable, and the consequences of his impetuousness when exercising his first command seem to have taught the future general an important lesson: Washington as a tactician would become a master of strategic withdrawals, a Fabius-like dealing from a position of weakness by constantly holding his forces in reserve, waiting for the right moment for a sudden advance, tactics that he himself called "partizan" and that are typified by his sudden surprise attack across the half-frozen waters of the Delaware. It is notable also, given the advantage taken by the French of his diary, that Washington would never again keep a journal while on the march, not at least until 1781, when within six months of victory at Yorktown. It was always Washington's talent to turn necessity into virtue—called by him "Duty"—and so Fort Necessity served its purpose as a salutary lesson. Moreover, what for many men might have been a place thrice cursed by ignominy—the ambush of Jumonville at Laurel Hill took place less than four miles from the spot, and Braddock was to be buried nearby—was for Washington primarily a piece of real estate, and when the opportunity presented itself in 1767, he bought the place.

iii

By 1753, young John Adams had also begun to keep a journal, a document dictated by the old Puritan necessity of keeping an eye not so much on his growing estate, like Washington, but on the state of his expanding mind. Like many of Adams' youthful undertakings, it was at first an irregular and often dilatory project, but about the time that George Washington was returning from his embassy to the Ohio Valley, Adams was recording a journey he had taken the previous summer. As Washington's trip evinced his Virginian heritage, which viewed the future as through a surveyor's transit pointed north, so the travels of young Adams demonstrated his own heredity, which (since the days of John Winthrop) had seen America from the top of Beacon Hill. Starting

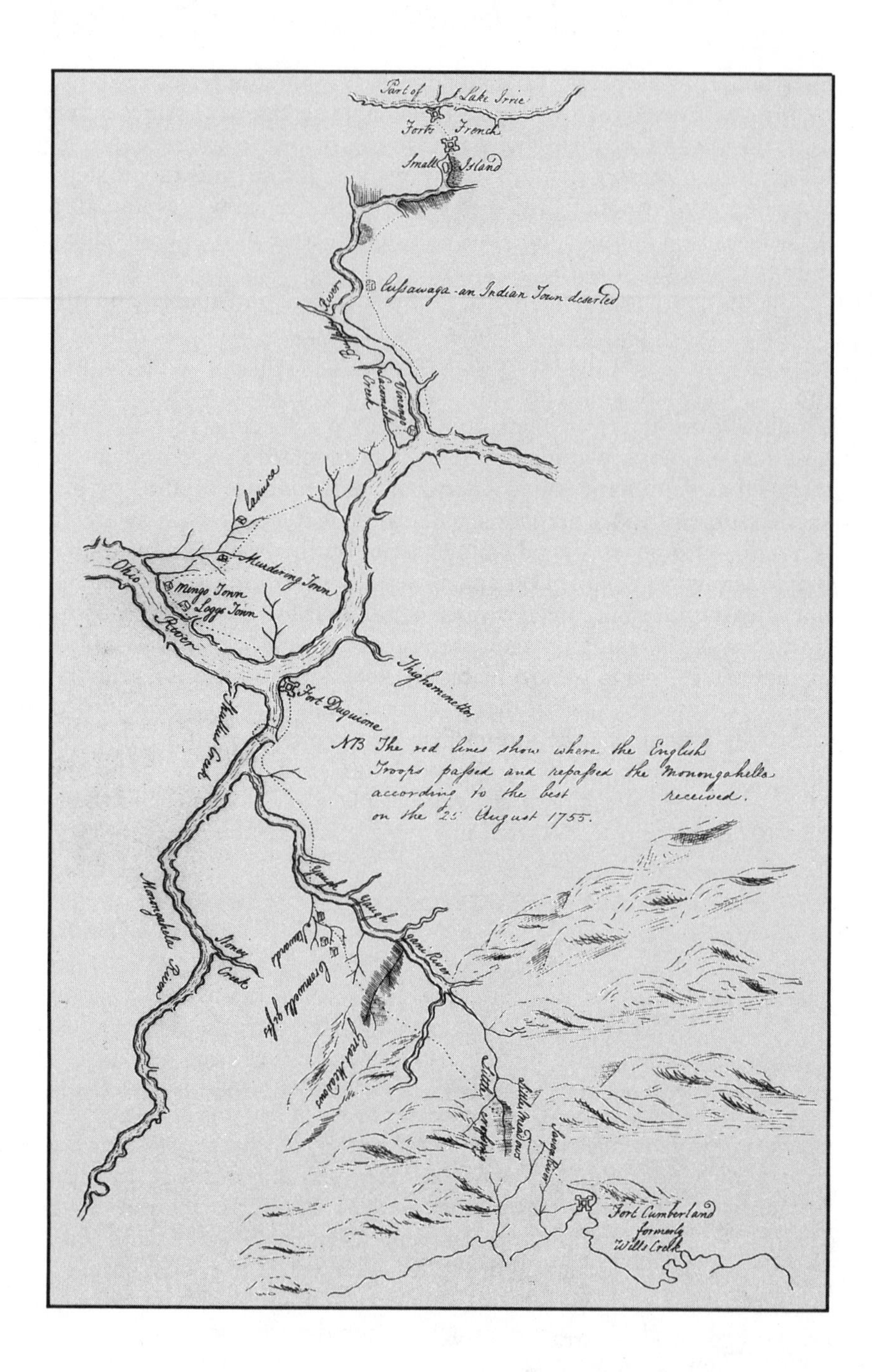
Part of Lake Irrie
Forts French
Small Island
Cussawaga - an Indian Town deserted
Murdering Town
Mingo Town
Loggs Town
Ohio River
Fort Duquione
NB The red lines show where the English Troops passed and repassed the Monongahella according to the best received. on the 25 August 1755.
Monongahela River
Stoney Creek
Great Meadows
Fort Cumberland formerly Wills Creek

out into the great wide world "by way of Litchfield" on the Merrimack, Adams traveled east to Newington on the Piscataqua, where he "tarried about a fortnight and visited Portsmouth" before returning home (*ED*: 49). "At the appointed time," Adams noted in his journal, he entered the gates of Harvard College as a freshman, "where I have been ever since."

Where Washington's journey was in the grand, heroic tradition of early Virginian exploration, carrying with it the future of England in the New World, Adams' was little more than a vacation trip, yet it was a journey that in its small compass asserted symbolic limits—in terms both of Massachusetts and of her chief representatives during the next half-century. Bound by the Charles and the Piscataqua, it was centered (in matters of mind if not fact) around the citadel of intellect at Cambridge. Where Washington's world is symbolized by the outreach of the Potomac, Adams' river-bound terrain is insular, self-contained, like the larger New England that contained it. The geopolitical contrast is dramatic, but so was the difference in the mental and bodily frame of the two men, the one becoming the politic General while the other became the general Politician of the Revolution. Moreover, what Adams was learning at Harvard would prove as important to the nascent nation as were the experiences of Washington in his wilderness journeys and campaigns, for as the squire-surveyor was mastering the harsh realities of frontier and military life, so the student-sophister was being transformed into a man not only of New England but of the Enlightenment.

The Enlightenment influence on Adams would be gradual, for the effect of intellectual irradiation on his inherited Puritan gloom was not unlike the single candle John Winthrop set up in Boston against the

(*Facing*) George Washington. *Map of the Ohio Country Traversed in 1753–54.* From Lloyd Arnold Brown, *Early Maps of the Ohio Valley* (Pittsburgh: 1959).

One of several versions drawn up or (as here) annotated by Washington, this manuscript map postdates the much more extensive one accompanying his published journal in 1754. As the notes indicate, it was made after the defeat of Braddock, and by showing the road and water routes into the Ohio Valley from Virginia, was an iconographic argument that the war against the French should be continued. A logical extension of the equally creative map drawn by Peter Fontaine, Jr., a few years earlier, it indicates a contingency, not only between the Potomac (via Wills Creek) and the Ohio Rivers, but between Virginia and the Great Lakes as well. Washington thus provides a diagram that would dominate his personal geopolitics for the next forty years, a vision shared by Thomas Jefferson.

heathen darkness. In 1756, the young Harvard student sounded a typical note, a reflection of his personal and regional bias: "The years of my youth are marked by divine Providence with various and with great events," he noted, including three military expeditions mounted by the British against the French in America, all three of which were "routed and destroyed without doing the least injury that we know of the enemy" (*D&A*: I: 35–36). But where Puritan clergy of an earlier period, like Cotton Mather, would have read the three signal defeats as signs of divine displeasure, evidence that Boston was not yet ready to receive the divine body of Christ, John Adams drew a slightly different conclusion: "Is it not then the highest frenzy and distraction to neglect these expostulations of Providence and continue a rebellion against that Potentate who alone has wisdom enough to perceive and power enough to procure for us the only means of happiness and goodness enough to prompt Him to both?" (36). Behind the piety, which was conventional enough, we can detect a characteristic retrograde movement, a shrinking from prideful extremes, which was the Adams brand of Enlightenment thinking. Mather would have concluded (as he had before) that the time was not yet right for the invasion of Carthage/Canada: more prayer and fasting were in order. Adams seems to feel that God would *never* favor the event, having blessed New England with territory enough.

The young Adams could occasionally rise to an enthusiastic view of the frontier as a symbol of civilized progress, but it was usually given a historical perspective, as in 1756, when he considered "the Changes produced in this Country within the Space of 200 years": "Then, the whole Continent was one continued dismall Wilderness, the haunt of Wolves and Bears and more savage men. Now the Forests are removed, the Land coverd with fields of Corn, orchards bending with fruit, and the magnificent Habitations of rational and civilized People. Then our Rivers flowed through gloomy deserts and offensive swamps. Now the same rivers glide smoothly on through rich Countries fraught with every delightful Object, and through Meadows painted with the most beautiful scenery of Nature, and of Art" (*D&A*: I: 34). This transformed landscape, in which the "narrow Huts" of Indians had been replaced by "fair and lofty Edifices, large and well compacted Cities," would soon enough be projected by native New Englanders into the Ohio Valley; but despite his reference to "the whole continent," "this country" was clearly for Adams the region bounded to the west by the Hudson. Celebrating the settlement of Massachusetts "with Reverence and Wonder as the Opening of a grand scene and Design in Providence," he seldom left his perch "upon Forefathers Rock," even when voyaging two decades later to Philadelphia to become a Forefather himself (I: 257, 258n.l).

In this as in other matters John Adams was somewhat out of phase with his age, yet in this also he was representative of regional tendencies, for during his lifetime New England grew increasingly insular as the western regions expanded, a centripetal ideal inherited likewise from the founding fathers of the Bay Colony. Still, Adams never entirely shared the sectional passions of hard-line Federalists, and his paranoia was limited mostly to personal, not geopolitical, matters—even where the latter related to the former. His Puritan heritage was witnessed mostly in his love of moral rigor and reasoned order, as well as by his growing inclination to doubt that the mass of men were capable of either. Moreover, even if the emerging Brahmin caste tended to stick close to Boston, its rise coincided with the emergence of something called a "Yankee," that almost mythic type much given to wandering far afield, whether pushing the frontier westward or peddling tinware and pedantry, and putting together machines that would transform the face of the continent.

iv

Such a one was the man who left Boston for New York in 1723, and who, finding no work there, went on to Philadelphia, a city scarcely twice as old as he. By 1753 the town was thriving, and so was Benjamin Franklin, who was at the point of attaining a certain eminence, not only in Philadelphia and the colonies but in England as well, for he would soon become the Enlightenment in America personified. Appointed that year Deputy Postmaster for British North America, Franklin received also a medal from the Royal Society for his discoveries in electricity, and honorary degrees from Harvard and Yale. New England was proud of her rising native son, regarded (in the words of the young John Adams) as having "a prodigious genius cultivated with prodigious industry" (*D&A*: I: 13).

If young Adams was at mid-century a student sophister, and young Washington a squire-surveyor and apprentice soldier, the middle-aged Franklin was his usual industrious, prodigious self, but he was busy also in extending his personal sway. Having made his reputation as first a printer, next a pundit, finally a projector, in 1753 Franklin was mostly a postmaster. The systematic organizer of lending libraries and fire departments was now turning his talents to a much larger sphere, surveying the postal routes throughout the colonies with a view toward improving the exchange of communications—in the seventeenth century mostly a matter of letters—which he identified with the lines of communication that would promote trans-colonial unity. With the outbreak of war, Franklin threw his energy and eye for order into military preparations, and having assisted General Braddock in mounting his disastrous invasion of the Ohio Valley, he was "prevailed upon" by Governor Morse of Pennsylvania to turn soldier, specifically "to take Charge of our Northwestern Frontier, which was infested by the Enemy, and provide for the Defense of the Inhabitants by raising Troops and building a line of Forts" (*Auto*: 743). Though mentioned in his *Autobiography,* this episode does not fit our popular notion of Ben Franklin, yet it is part and parcel of his passion for order, and gives a definitively martial edge to his notion of American geopolitics.

In the issue of the *Pennsylvania Gazette* for May 9, 1754, there appeared Franklin's famous cartoon of the severed snake of eight parts and the motto, "Join, or Die." The drawing was propaganda for unifying the colonies under one government in order to better direct "defense and other important purposes," more specifically the plan Franklin proposed to a congress gathered at Albany earlier that year to treat with the Six Nations in anticipation of war with France. What emerged was a "plan of union . . . that would best suit the circumstances of the colonies, be most agreeable to the people, and most effectually promote his Majesty's service and the general interest of the British empire" (*Papers*: V: 400). The syntactical order of priorities is significant, but the circumstances of the colonies in general were not yet sufficiently severe to earn the plan the approbation of the people—or at least of their representative. The war was therefore allowed to take its random and uncentralized way, with the colonies closest to the conflict carrying a disproportionate share of the cost. Like so many of Franklin's inventions, the Albany Plan was not an original creation but an improvement on already existing ideas, yet it is an essential expression of his belief in effective action through improved organization, a geopolitics of order that refers to a common center.

In 1751 Franklin wrote his influential expansionist essay, *The Increase of Mankind,* in which he pointed out the importance to Great Britain of the lands to the west of the existing colonies. As a tract, however, it was first published in 1755, when it was appended to William Clarke's sword-rattling *Observations on the Late and Present Conduct of the French, with Regard to their Encroachments upon the British Colonies in North America.* By that time Franklin was already framing a more specific proposal concerning territorial expansion in the New World, and like the Albany Plan, his scheme for "settling two western colonies," propounded in 1754, is both a logical extension of the New England rage for order and the expansionist Virginian (and Pennsylvanian) spirit, both combined in a typical expression of Enlightenment right reason: looking west to "the great country back of the Apalachian mountains, on both sides the Ohio," Franklin declared the region to be "one of the finest in North America," not only in terms of natural resources, but because of "the vast convenience of inland navigation or water-carriage by the lakes and great rivers, many hundred of leagues around" (*Papers*: 457). This territory promised "a great accession of power" to whatever country possessed it, and Franklin proposed that France be blocked by planting two armed colonies between the Ohio River and Lake Erie, a plan that would result in "a great security to the frontiers of our other colonies," and that in times of peace would facilitate the extension of trade through a vast territory, "by means of the lakes, the Ohio, and the Mississippi . . . greatly to the benefit of Britain" (458).

Franklin tied this project to his Albany Plan, for like the natural, interlinking system of waterways, a unified system of established colonies would be of great utility to the establishment of new ones, but his desire for expansion exceeded his love of intercolonial unity. Even if there were no such union, he noted, the two western colonies should be established, the one between the Ohio and Lake Erie settled by emigrants marching overland through Pennsylvania; the second, located farther down the Ohio, by settlers who would "assemble near the heads of the rivers in Virginia, and march over land to the navigable branches of the Kanhawa, where they might embark with all their baggage and provisions, and fall into the Ohio, not far above the mouth of Sciota. Or they might rendezvous at Will's Creek, and go down the Monongahela to the Ohio" (461). The propositions of projectors like Franklin are often at odds with the disposition of the actual landscape. At the very time Franklin was drafting his large-viewed scheme, Virginians were already gathering at Will's Creek, but the young colonel in charge could have told the Philadelphian geopolitician that obstacles to navigation on the Monongahela were considerable, including a fort now manned by the French on the Ohio.

Franklin's plan was both an expression of his particularism and in a general way prophetic: particular, in that it was designed to draw emigrant Connecticut Yankees out of the Susquehanna Valley where they did not (in Pennsylvania's eyes) belong and put them in the Ohio Valley as a human barrier against the French; prophetic in that Franklin not only appreciated the value of the Ohio Valley, but by demonstrating how old, established colonies could nurture new ones he anticipated the territorial plan by which the valley would be settled. He also foresaw two major routes along which that process would take place, *viz.* the great road cut by Braddock in 1755 and the road west from Philadelphia cut by Forbes in 1758. By then Franklin had left for England, and thenceforth his geopolitics would have a much larger scope. We are apt to remember most the Franklin of Philadelphia, London, and Paris, eminently an urbane and cosmopolitan man, the self-made American whose personal rise continued into the world of statecraft. But until 1774, when he returned to what was soon to be the United States of America (though perhaps never so united as he had hoped), Franklin continued to identify "the interest of Great Britain with regard to her colonies" with expansion into western regions, for in that direction lay the continental way to wealth.

Franklin's attempts to restore harmony between Great Britain and her rebellious colonies were famously futile, and when he returned to America at the start of the Revolution, the events of the past twenty years were in a certain sense canceled out. The orderly expansion into the Ohio Valley would be affected not by colonies of the Crown but by states of the Union. The details of the plan would be the work of others—most notably, Thomas Jefferson—but it would be along the lines proposed by Franklin, much as the Constitution resembled his Albany Plan. After 1755, Franklin did not much associate himself directly with western expansion, and like several of the Adamses he would increasingly be an American in Paris; but in casting his orderly, Enlightenment gaze westward during a critical moment in American history, he gave definitive shape to the unfolding American map. Where George Washington in the years 1753 to 1755 was an unwitting (though not necessarily unwilling) creature of geopolitical forces, his military career tied firmly to the welfare of Virginia, where John Adams saw the world as if through his dormitory window, Franklin stood (as William Penn is perched) on an eminence of his own making in Philadelphia, pointing west toward Ohio.

V

By 1755, the Reverend James Maury had joined his efforts to those of the Peter Fontaines, father and son, in brightening the chain of communication between America and England, and with the aid of a map of

Virginia drawn by his neighbors, Joshua Fry and Peter Jefferson, Maury helped his aged uncles locate their far-flung relations: "As for myself," he concluded, "I am planted . . . under the South West Mountains in Louisa [County], close by one of the head springs of the main northern branch of Pamunkey, which runs through my grounds—a very wholesome fertile, and pleasant situation" (Fontaine: 379). He also located on the map "the Great Meadows where our brave Washington was last year attacked by the French and Indians," and Fort Cumberland, "from which the brave but unfortunate, and I believe I may add, imprudent General Braddock marched this summer against [Fort] Duquesne" (385).

Like the news sent to England by his cousin Peter, the letters of James Maury reflect "the suffering and calamitous condition" of Virginia at the outset of the French and Indian War, but they also express a geopolitical view much larger than his cousin's, suggesting that the remoteness of his plantation placed it not in a backwater but in the forward-moving front of empire. Like Benjamin Franklin, he declared that the outcome of the war would affect the future not only of Virginia but of the other American colonies, in whose welfare the fate of Great Britain was invested: "The downfall of either must sooner or later be attended with that of the other" (380). Maury alluded to Franklin's Albany Plan, bemoaning that "there is no mutual dependence, no close connection between these several colonies; they are quite disunited by separate views and distinct interests" (382). Franklin's image was a severed snake, Maury's "a bold and rapid river, which, though resistless when included in one channel, is yet easily resistible when subdivided into several inferior streams and currents."

Early in 1756, James Maury sent his uncles news of another map, this one published "by Lewis Evans, Esq., of Philadelphia," who was the author also of "an instructive, curious, and useful pamphlet, explanatory not only of the map, but of many particulars, too, relative to the face and products, and natural advantages of the tract of territory which is the subject of it" (387). This was not only a more correct map than that of Fry and Jefferson, it was "much more extensive," giving "an attentive peruser a clear idea of the value of the now contested lands and waters" in Ohio. The map was graphic proof that whichever of the warring nations won possession of "the Ohio and the Lakes . . . must, in the course of a few years . . . become sole and absolute lord of North America." Here again (like Evans himself) Maury is echoing Franklin's wisdom, though he lacked the resources of Philadelphia for publishing his enthusiasms. There is something comic in his venting them on his ancient relatives in England, who could have had little effect on the course of the war.

Still, like his uncle and cousin before him, James Maury shares the Fontaine quality of consensus, and though he was himself in no position of power, his enthusiasm displays his agreement with the opinions of people who were. It may seem strange to find such ideas coming down from the foothills of Virginia in 1756, yet Maury's geopolitics share a mountaintop view, envisioning a day when "either Hudson's river at New York or Potomac river in Virginia" would become "the grand emporium of all East Indian commodities" (388). Maury was a latter-day dreamer of the old Virginian dream, and though he chided his uncles not to "marvel" at such news, once again we may amuse ourselves with their reaction to the tidings sent by this prodigy the Fontaines had sired in the New World, known to them only by his letters. "Perhaps, before I am done with you, you will believe it to be not entirely chimera," wrote Maury, and then hurried on with his proof, the which, like so many similar arguments engendered by earlier proponents of passage to India, deals not only in glittering but glorious generalities:

> *When it is considered how far the eastern branches of that immense river, Mississippi, extend eastward, and how near they come to the navigable, or rather* canoeable *parts of the rivers which empty themselves into the sea that washes our shores to the east, it seems highly probable that its western branches reach as far the other way, and made as near approaches to rivers which empty themselves into the ocean to the west of us, the Pacific Ocean, across which a short and easy communication, short in comparison with the present route thither, opens itself to the navigator from that shore of the continent unto the Eastern Indies.* (388)

The paragraph consists of a single unbroken sentence, by means of which Maury has carried his auditors from the Atlantic to the Pacific ocean, on the wings of such syntax as only the mountain muse could inspire.

Maury's argument here rests on the authority not of Lewis Evans, who was careful to gather only eyewitness reports, but of Daniel Coxe, who was not, and who published in 1722 a fabulous map and equally imaginative description of "Carolana," by which he meant Louisiana, conveying "intelligence from the natives" on the Missouri River "that its head springs interlocked in a neighboring mountain with the head springs of another river, to the westward of these same mountains, discharging itself into a large lake called Thoyago, which pours its waters through a large navigable river into a boundless sea" (Fontaine:

389–90). Coxe was an indefatigable promoter of the pyramidal concept of American geography, which placed the Mississippi and its branches at the center of a great basin at whose eastern and western extremes rose mountains of equal height—the height of the Appalachians. This theory of continental symmetry was kindly received by such as James Maury, who as a student and teacher of the classics was undoubtedly pleased to find nature arranged along orderly lines and whose plantation was located near to what was sure to become "the grand emporium of all East Indian commodities."

While admitting to his uncles that Coxe's book "hath in it something of the romantic air of the voyager," Maury was willing to rely on its general truth, most particularly since such an experienced surveyor and mapmaker as his neighbour Colonel Fry was also a true believer, to such an extent that "a grand scheme was formed here about three years ago" to explore the Missouri "in order to discover whether it had any such communication with the Pacific" (390–91). The outbreak of the war had put a stop to the expedition, but not to Maury's faith in an imperial if illusory syllogism:

> *As there is such short and easy communication by means of canoe navigation, and some short portages between stream and stream, from the Potomac, from Hudson's River in New York, and from the St. Lawrence to the Ohio, the two latter through the lakes, the former the best and shortest. As there also is good navigation, not only for canoes and batteaux, but for large flats, schooners, and sloops down the Ohio into the Mississippi,* should Cox's account be true of the communication of this last river with the South Sea, with only one portage, [*then*] *I leave you to judge of what vast importance such a discovery would be to Great Britain, as well as to her Plantations, which, in that case, as I observed above, must become the general mart of the European World, at least for the rich and costly products of the East.* (391–92, *emphasis added*)

From the shoulders of "should," as from the fabulous peak in Darien, the Reverend James Maury caught a glimpse of a glorious future:

> *What an exhaustless fund of wealth would here be opened, superior to the Potosi and all the other South American mines! What an extent of region! What a———! But no more. These are visionary excursions into futurity, with which I sometimes used to feast my imagination, ever dwelling with pleasure on the*

> *consideration of whatever bids fair for contributing to extend the empire and augment the strength of our mother island, as that would be diffusing liberty both civil and religious, and her daughter Felicity the wider, and at the same time be a means of aggrandizing and enriching this spot of the globe, to which every civil and social tie binds me, and for which I have the tenderest regard.* (392–93)

From our present perspective we can see a latent division here between the interests of the mother country and "this spot of the globe," given added meaning by Maury's little allegory, for in the Declaration of Independence, Liberty and Felicity are inseparable from Life itself as unalienable rights. Thomas Jefferson first wrote "pursuit of Property," then revised it to the less materialistic "happiness," but the intimate connection remains. If Felicity is the daughter of Liberty in America, then Property is the sire; and, in pursuing the father, Americans tended to head for the setting sun.

In 1756, Jefferson was a boy of thirteen, yet a year later, upon the death of his father, Peter, he came to man's estate. Still in need of schooling, he was tutored for a time by the Reverend James Maury, from whom he is supposed to have received his lifelong affection for the classics. From Peter Jefferson, Thomas inherited a deep love of Virginia, much as he continued to maintain a high regard for the map drawn by his father and Colonel Fry. But perhaps, as Bernard De Voto has suggested, during a very critical and formative period of his young manhood, he may have taken from James Maury a somewhat larger view of the Western world. For it would be Thomas Jefferson who, fifty years later, would actually mount the expedition that would explore the Missouri River to its source and beyond, making (in Maury's words) "reports of the country they passed through, the distances they travelled, what sort of navigation those rivers and lakes afforded, &c., &c." (391). Like John Adams, his senior by some few years, Jefferson was a child of the Enlightenment, and yet with signal differences, not the least of which was his expansive geopolitics.

In Jefferson there was combined Adams' holistic faith that mankind's best hope for happiness on this earth lay in rational instruments of balanced governance and Washington's imperial linearity. With Franklin, Jefferson shared a love of scientific inquiry for its own sake, the Enlightenment rage for order that, when imposed upon the map, put forth a distinctly rectilinear shape. But, perhaps most important, he took from his tutor James Maury an almost mystical enthusiasm for neoclassical ideals, in terms not only of literary and architectural but of geopolitical aesthetics, best symbolized by the domed structure he built for himself on a height of land in Virginia that commanded a westward view, a model in miniature for the capitol of Roman proportions that would rise on the bank of the Potomac as a symbol of national design.

The Fontaine family, once again, provides a relatively modest dimension to the movement of empires detectable at mid–eighteenth century, but in the linked connection between the two Peter Fontaines and between them and James Maury, we are provided with a suspension bridge of sorts. It is a family correspondence truly, joining the earlier movement westward initiated by the generation of William Byrd of Westover—undertaken to promote orderly expansion in the colonies of Virginia and North Carolina—to the westward thrust of Washington and his compatriots, by means of which the hostilities of the French and Indian War gave shape to an emerging republican polity, extending a plan of orderly empire into a further range. We may debate the extent to which the Fontaines were typical of all Americans during the years covered by their correspondence; we may doubt the degree to which Washington, Adams, Franklin, and Jefferson were representative of the regions to which they owed allegiance; but there is little question that in the period from 1753 to 1763, Anglo-Americans were on the move—which could be figured either as martial adventure, geopolitical outreach, or the expansion of the mind—and that westward-trending and eastward-bending rivers defined the direction of their march.

II

Fertile Ground

i

Cartography is often intimate with imperial necessity—witness Lear unfolding the diagram of his divided kingdom with a grand and generous gesture, or Hotspur arguing with Glendower over fine points of division on a map of territory they have yet to conquer. But whereas the earliest maps of the New World were often wishful projections of princely desires, by the middle of the eighteenth century, cartographers were chiefly occupied with filling in details of occupied territory and defining boundaries. Because of the influence of the Enlightenment, with the concomitant aid of improved technology, cartography had become a much more exact science. Lewis Evans himself traversed much of the territory he mapped or sought firsthand testimony from others. But the gathering of data in North America remained the responsibility of men who were by nature expansion-minded, and the maps that resulted were often explicitly designed to make the best possible case for increased domain. Moreover, as in Dr. Franklin's plan for planting colonies in the Ohio Valley—which Evans also proposed—the focus and thrust of Evans' map and essays reveal a certain particularism, reminding us of how many boundary disputes (the Glendower Syndrome) sprang up between the expanding colonies. If the nascent nationalism of Evans' argument looks forward to a republic that will inherit and greatly intensify Great Britain's imperial errand, then a careful look at Evans' map and essays (and the response to those essays) reveals a latent sectionalism that will threaten the unity of the nation yet to come.

It takes only a cursory examination of that map to understand a mighty geographical fact that history would have to accommodate. Running like a bend dexter from the upper right-hand corner to the bottom left is the vast cordillera of the Appalachians, which Evans aptly called (borrowing from the Algonquian) the "Endless Mountains." For

William Byrd of Westover, in 1728, the mountains were an effective "barrier against the French," but for the French in 1755, they were a natural fortification facing the English. And even after the great wall had been pierced and the French defeated, the mountains remained for a considerable period an impediment to easy communication between the eastern seaboard and the Ohio Valley.

The penetration of the mountains and the consequent establishment of trade routes east and west were problems the post-Revolutionary generation would find difficult to solve—the physical fact of the mountains exacerbating geopolitical concerns. Evans was very certain on one point in this regard, and his argument is as particular as his location on his map of the First Meridian, which bisects Philadelphia. Devoting his first "Essay," or "Analysis," to a description of the major rivers of both the eastern and western regions, with due consideration of their potential uses as waterways, he ends by concluding that though the "Inland Navigation" of the Potomac "is scarce begun . . . one may foresee that it will become in Time the most important in America, as it is likely to be the sole Passage from Ohio to the Ocean" (Evans: 167 [23]). It was this passage that elated James Maury, but not all Americans who read it were so pleased.

In the *New York Mercury* for January 5, 1756, there appeared a lengthy "Letter from an Gentlemen in New-York, to [his] Friend in Philadelphia," which turned Evans' map and "Analysis" to quite different ends from the ones he intended them to serve: "It shews in a very stricking [*sic*] Point of Light, that the *Virginia* Sollicitations for the Landing of the *European* Troops in that Colony, in order from thence to make a Descent upon *Fort Du Quesne,* were to the last Degree absurd, abstracted from their fitness to serve the Interest of a private Company, at the Expence of the Welfare of the Public. Had General *Braddock's* Army succeeded in that Enterprize, it would, in my humble Opinion, had been a useless and untenable Acquisition" (Evans: 179 [3]). The anonymous writer was a partisan of Governor William Shirley of Massachusetts, who as the commanding general of the northeastern forces had proposed that the most effective way of ending the French presence in the Ohio Valley was to seize control of the Canadian waterways that were the main avenues of supply—zones that, not incidentally, posed a major threat to the peaceful expansion of New England's trade. Reprinting this letter in his second "Essay," Evans devoted more than thirty closely printed (and reasoned) pages rebutting the New Yorker's main points, while not entirely refuting the larger argument.

In a postscript, furthermore, Evans took the opportunity to answer the charge of another hostile critic, the author of a pamphlet published in London in 1755, that the best way to Fort Duquesne was not the river route taken by Washington and Braddock, but one that lay overland from Philadelphia. General John Forbes would make the same decision, resulting in the first successful march on the French fort and proving the London writer's point, much as the course of the long war demonstrated Governor Shirley's perceptiveness concerning the importance of the Canadian contingency. Still, thanks to the existence of Evans' map and essays, the primacy of the Potomac to transmontane advance and the general superiority of waterways to roads in directing westward expansion would have an influence reaching well beyond the strategy determining the English victory over the French.

ii

The map carried by Braddock and praised by Thomas Maury was dedicated by Evans to the Honourable Thomas Pownall, Esq. Yet Pownall had only recently come over from England, sent in 1753 as secretary to the new Royal Governor of New York, Sir Danvers Osborne. If Sir Danvers does not figure largely in the great events that attended his advent in America, it is because he hanged himself a week after arriving, leaving his secretary likewise dangling but with no visible means of support. Yet Pownall made good use of the next two years: Ambitious to make himself an expert on colonial affairs, he traveled about the northeastern region, making friends of influential people in Philadelphia, including Lewis Evans, and especially Benjamin Franklin. He also became an enthusiast for the settlement of and investment in the Ohio lands. By May of 1755, Pownall had been appointed Lieutenant Governor of New Jersey, and he would in time displace Governor Shirley in Massachusetts, where he would play an important role in waging the war against the French in Canada. But he would become best known for his ardent defense of the colonies' right to direct their own destinies, a defense he mounted after his return to England in 1760, and that he would not abandon even after the outbreak of the War for Independence.

Lewis Evans, therefore, made a prophetic (perhaps inevitable) choice in dedicating his map to Thomas Pownall. Accompanying the engraved dedication, on the upper left-hand corner of the map, is Pownall's family crest, a rampant British lion over which a gloved hand holds a key. The heraldic implications of that blazon aside, it certainly has meaning where Evans' map is concerned, for the rising career of Thomas Pownall in America holds a key to the transfer of imperial designs from Great Britain to her colonies in the New World. That Pownall reissued a revised version of Evans' map and pertinent sections of his "Essays" in 1776, and that in 1784 he revised it yet again with the intention of publishing another edition, help to impress his personal seal more firmly into the forming shape of North America. Of all the circumstances promoting the primacy of the Potomac River route through the mountains, in 1755 and afterwards, perhaps none is more important to an understanding of how the British urge to empire in America was inherited with few modifications by the United States than is Pownall's ubiguitous connection with the publication and promulgation of Lewis Evans' map.

In revising Evans' text, Pownall added a great deal of original material based on his own experience and travels in America, moving away from strictly topographical considerations to descant on the natural (and potential) products of the countryside through which he traveled. His most distinctive contribution, however, is a number of anecdotal descriptions of American scenery, aesthetic marginalia that reflect Pownall's complex viewpoint. Typical, and exemplary, is his account of how it feels to travel on American rivers. Starting out with Evans' stress on the importance of waterways to settling and exploiting the lands, Pownall goes on to quite a different level of discourse:

> [*T*]*he General, and I had almost said, the only Way of travelling this Country in its natural State is by the Rivers and Lakes. . . . The general Face of the Country, when one travels it along the Rivers through Parts not yet settled, exhibits the most picturesque Landscapes that Imagination can conceive, in a Variety of the noblest, richest Groupes of Wood, Water, and Mountains. As the Eye is lead on from Reach to Reach, at each Turning of the Courses, the Imagination is in a perpetual Alternative of curious Suspense and new Delight, not knowing at any Point, and not being able to discover where the Way is to open next, until it does open and captivates like Enchantment.* (Pownall: 31)

There is a nascent Romanticism here, a delight in the landscape for its own sake, but it is only nascent, for there is also a Horatian slant to Pownall's view of natural beauty: "But while the Eye is thus catching new Pleasures from the Landscape, with what an overflowing Joy does the Heart melt, while one views the Banks where rising Farms, new Fields, or flowering Orchards begin to illuminate this Face of Nature"

(31). Natural beauty is fine, but the human heart really warms at the prospect of productivity, those useful and profitable works that catch the kind of light the Enlightenment preferred.

In the manner more of a tourist than a topographer, Pownall made two visits to the Cohoes Falls, where the Mohawk drops into the Hudson River. The second was necessary because the first trip was disappointing: "[T]here was but little Water in the River, and what came over the Fall ran in the Cliffs and Gullies of the Rocks" (35). Pownall had much better luck the second time, there being "a great Flood coming down the River," and the falls "were then a most tremendous Object":

> *The Torrent, which came over, filled the whole Space from Side to Side; before it reached the Edge of the Fall it had acquired a Velocity which the Eye could scarce follow; and although at the Fall the Stream tumbled in one great Cataract: yet it did not appear like a sheet of Water; it was a tumultous Conglomeration of Waves foaming, and at Intervals bursting into Clouds of Vapour, which fly off in rolling Eddies like the Smoak of great Guns. In that Part of the Fall where the large Rock shoots forward, the Torrent as it falls into the Angle formed by it seems to lose the Property of Water; if the Eye tries to pursue it in its Fall, the Head will turn giddy; the great and ponderous Mass with which it ingulfs itself makes the Weight of it (one may almost say) visible, however it makes itself felt by keeping the whole Body of the Earth on the Banks on each Side in a continued Tremulation; after having shot down as though it would pierce to the Center, it rebounds again with astonishing Recoil in large Jets and columns of Water to the very Height from which it fell. . . . This is not Poetry but Fact, and a natural Operation.* (35–36)

Burke's essay on the sublime and the beautiful appeared two years after Pownall visited the Cohoes Falls, and the governor's description contains little suggesting Burke's formulations or terminology. Instead of stressing the element of the awful obscure, Pownall attempts to be as specific as possible—"not Poetry but Fact" as he says—so that his description, once again, is more in the Enlightenment than the Romantic vein. What he seems to be after is less a purely aesthetic expression than one that emphasizes both the singularity and the sheer power of this American spectacle: "While we [were] contemplating this Object, there came on a most violent Thunder Storm: Any one who has been in

America knows how exceeding loud the Sound of these Explosions of the Thunder are: Yet so stunned were we with the incessant hoarse Roar of this Cataract that we were totally insensible to it" (36).

Pownall was a talented amateur draftsman, and a number of pictures that the governor drew of American scenes were engraved and published in London in 1768 as part of a series entitled *Scenographia Americana,* that appeared presumably in celebration of Great Britain's recent victory in holding on to the subject matter. Included was his view of Cohoes, but perhaps most relevant to Pownall's imperial (and topographical) élan are two pictures of the Hudson River, one taken at the entrance to the "Topan Sea" and the other near Poughkeepsie. They convey a curious bifurcation, one that is given an overall effect of unity intrinsic to the imperial perspective, and completely in harmony with the Horatian view.

Though Pownall's voyage along the Hudson is described in his prose account as "downriver," both pictorial views are from the south. The first depicts the entrance to the Tappan Zee, with the looming Palisades on the left and a contrasting scene on the right: "The western Banks are perpendicular rocky Cliffs of an immense Height, Covered with Woods at the Top, which from the great Height of the Cliff seem like Shrubs. The Eastern Coasts are formed by a gently rising Country, Hill behind a Hill, of fruitful Vegetation at the back of which lye the White-plains" (39). The second picture is of the Catskills and the fledgling town of Poughkeepsie, as seen from Esopus' Island, and by drawing into the foreground a rough tangle of rocks and trees, Pownall conveys a similar compositional contrast. Thus the western side of the river is uniformly an unsettled, even savage, border, while on the east lies evidence of settlement and domestication. But over all arches a very large sky, filled with dramatic displays of cosmic clouds, promoting an impression of openness and vastness: America, in Pownall's rendering, is a big and

(*Above, left*) Thomas Pownall. *A View in Hudson's River of the Entrance of What Is Called the Topan Sea.* Engraved by Peter-Paul Benazech (1730–1783). From *Scenographia Americana; or, A Collection of Views in North America and the West Indies* (1768). (Courtesy of the American Antiquarian Society)

(*Below, left*) Thomas Pownall. *A View in Hudson's River of Pakepsy & the Catts Kill Mountains.* Engraved by Paul Sandby (1725–1809). From *Scenographia Americana.* (Courtesy of the American Antiquarian Society)

Lending an imperialistic iconography to the Hudson River, these interior-pointing views posit a western route that rivals Washington's Potomac plan, and establish aesthetic signifiers that will become familiar through repetition over the next half-century.

vacant space, the upriver perspective acting to draw prospective settlers inland.

Given Pownall's preference for evidences of humanity's transformational presence, it is not surprising that the former Governor of Massachusetts saves his highest encomiums for the scenery of New England, most particularly that of the lower Connecticut Valley with its distinctly riverine culture. Standing on a high ridge near Wethersfield,

> *the Traveller has a view of the Great Vale of Connecticut up the River, a Landschape* [sic] *picturesque in every assemblage of beautifull Objects that gives a View of rich populous Inhabitancy of the Human Race enjoying in peace & Liberty every happiness that a* Heaven upon earth *can give. . . . Straight, as Milton says, the Eye catches new pleasure when it fixes in detail on the Multitude of towns, the innumerable farms & settlements, the Groups of Woods & rivulets, amidst cleared & cultured lands teeming with abundance. The River runs through the middle of it & Go which* [*way*] *you will on each side of it, It is as though you were still travelling along one continued town for 70 or 80 miles on end. . . . How other People may be made I don't know and what the Reader may think of I don't care, but I declare I could never view this scene of happiness nor do I now write the account of it without an overflowing of heart that putts a tear in My Eye, which Tear that was Sweet is now become a bitter one when I reflect that that Happiness has been destroyed; And most likely can never exist there again the same degree.* (59)

As the syntax and punctuation suggest, this is another rough-draft passage prepared for the 1784 edition of Pownall's *Topographical Description* that was never published, and his allusion to the disruptions of the Revolution is likewise a temporal note. Still, it gives further point through contrast (once again), and the allusion to Milton promotes the possibility that Pownall regards the remembered vision as a lost paradise. This divided prospect is distinctly Tory in sympathies, abandoning the theme of progress amidst abundance to evoke an abstract image of chaos and loss.

Although he never wrote his promised extended description and analysis of the process of colonial settlement, Pownall did execute a drawing that was apparently intended to accompany it: an ideal depiction of the beginning and completion of "an American Farm," it is a picture that is closer to an allegorical diagram than to an actual landscape. Once again a river is central to the composition: on one side

Thomas Pownall. A Design to Represent the Beginning and Completion of an American Settlement or Farm.* Engraved by James Peake (1726–1782). From *Scenographia Americana. (Courtesy of the American Antiquarian Society)

Reflecting the spirit of Crèvecoeur's *Letters from an American Farmer* (1782), which was in part written during the same period of proto-nationalistic optimism following the English victory in the French and Indian War, Pownall again establishes a set of symbols that will be essential to the American ideal of western progress. A river, notably, is central to the plan.

(again, the left) in the foreground is a crude waterworks, turning the sawmill essential to the initial stage of settlement, and lumber is stacked and drying nearby. Behind the mill is a log cabin, soon to become a universal symbol of frontier life in the New World; while across the river, on the right, we see the same farm at a later stage, as a much larger frame structure, presumably surrounded by a lawn, with numerous outbuildings, and a carriage approaching. What is suggested in the localized landscapes is here made general and abstract, but as in the Connecticut Valley, "the River runs through the middle of it," a geographical (and aesthetic) fact essential to the process and hence the definition of settlement. The river is therefore not only intrinsic to the frontier process, it *is* the frontier process, a fluid line that in every dimension connotes change and progress through contrast. Again, as in Pownall's description of travel on a river, where each turn brings up a

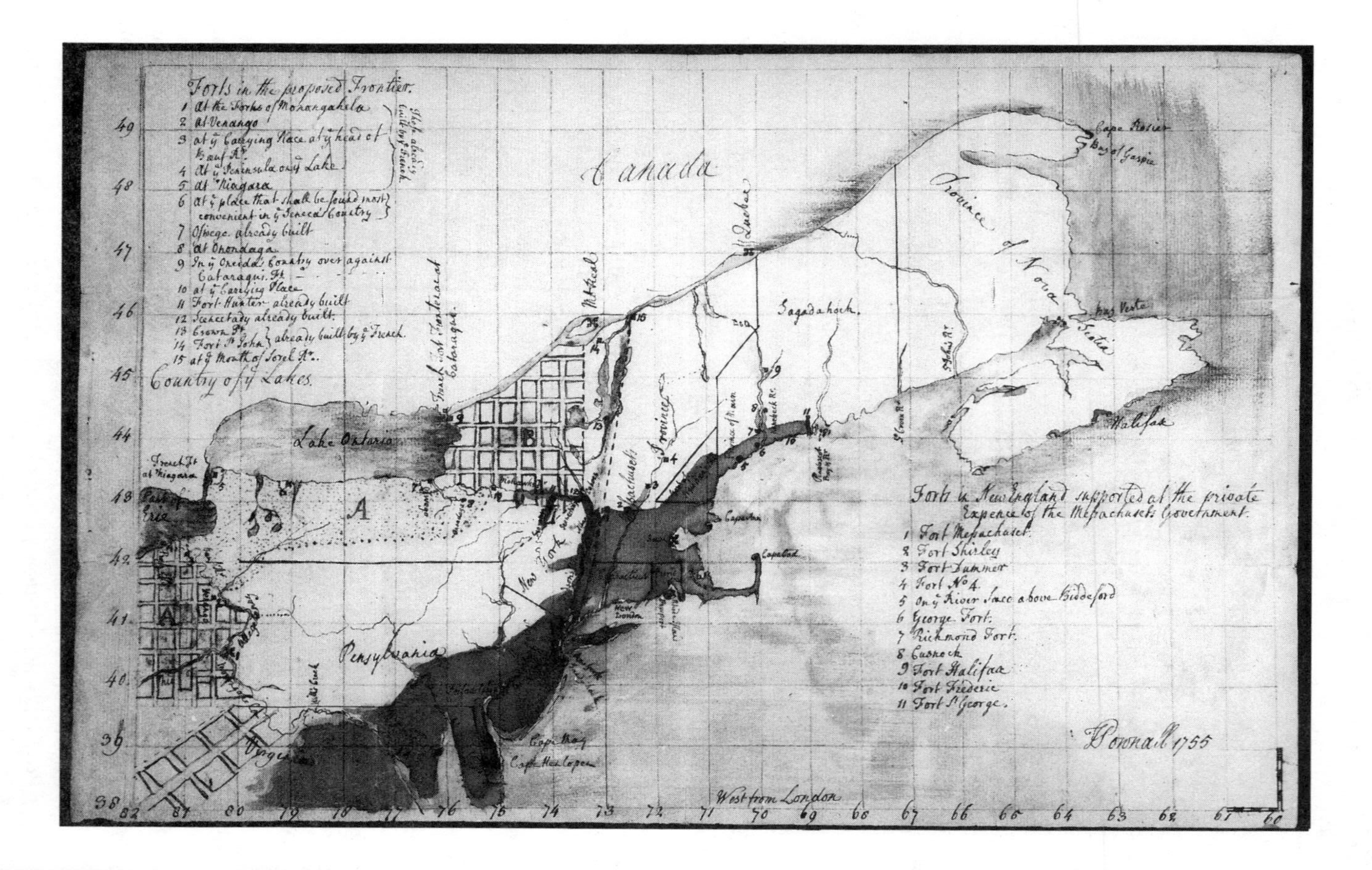
Forts in the proposed Frontier.
1 At the Forks of Monongahela
2 At Venango
3 at ye Carrying Place at ye head of
4 At ye Peninsula on ye Lake
5 At Niagara
6 at ye place that shall be found most convenient in ye Seneca Country
7 Oswego already built
8 At Onondaga
9 In ye Oneida Country over against Cataraqui Ft
10 at ye Carrying Place
11 Fort Hunter already built
12 Scenectady already built
13 Crown Pt
14 Fort St John
already built by ye French
15 at ye Mouth of Sorel Rr
Country of ye Lakes
Canada
Lake Ontario
Pensylvania
New York
Sagadahock
Province of Nova Scotia
Cape Rosier
Bay of Gaspie
Halifax
Cape May
Cape Henlopen
West from London
Forts in New England supported at the private Expence of the Massachusets Government.
1 Fort Massachuset
2 Fort Shirley
3 Fort Dummer
4 Fort No 4.
5 On ye River Saco above Biddeford
6 George Fort
7 Richmond Fort
8 Cusnock
9 Fort Halifax
10 Fort Frederic
11 Fort St George.
Pownall 1755

new and different vista, to journey through the American landscape is to capture a sense of the landscape moving, a movement carrying it from wilderness to town.

Moreover, though his interest in the Hudson River is largely aesthetic, Pownall's aesthetics are nothing if not imperial, for the strategic (Horatian) difference between the beauties of an untouched wilderness and one put to the uses of commerce contains a hierarchy of values giving added license of the frontier process. Pownall's imprint on the emerging shape of the American landscape is further illustrated by his map of the two colonies he, along with Ben Franklin and others, proposed planting in the Ohio Valley to supplant the Allegheny Mountains as an effective barrier against the French. Never published, this diagram cannot be said to have had much influence on subsequent ways of looking at "the Face of the Country," yet it serves as a remarkable demonstration of the way certain Americans, at the outset of the French and Indian War, envisioned the nature and function of transmontane settlement.

Pownall's map shows two intended colonies: one is placed between the headwaters of the Hudson and the eastern shore of Lake Ontario, the other along the confluence of the Allegheny and Monongahela rivers, extending all the way north to Lake Erie. By so doing, Pownall has fulfilled both the New England and the Virginia strategies, filling hotly contested zones with the Anglo-American imperial presence. Equally important, his map lays out both colonies in the manner of Philadelphia's streets, a neat reticulation of squares lending absolute order to the land, imposing a linear plan on the natural sites provided by rivers. As a nascent American, which his emotional paeans to the scenery of the New World reveal him to be, Thomas Pownall was also very much a man of the Enlightenment, projecting a mathematical (hence rational) order upon an as yet unsettled wilderness. If there was an American Horace before the emergence of Thomas Jefferson, then

(*Facing*) Thomas Pownall. *The Border Colonies Map* (1755). (Courtesy of the Huntington Library)

Drawn at the start of the great contest between England and France for North American empire, Pownall's map provides a virtual diagram of the opposing forces along the great northeastern imperial corridor, accommodating regional as well as international rivalries and tensions. His use of the grid plan, likewise, anticipates Jefferson's geopolitical geometry and the survey by Thomas Hutchins thirty years later of the Seven Ranges, which rectangularized the Northwest Territory. Though never published, Pownall's diagram certifies the linear (and neoclassical) basis of imperialistic plans.

Thomas Pownall was he, lining out the aesthetics for a subsequent imperial plan.

iii

Lewis Evans died in 1756 and did not live to celebrate the British victory his map had urged. In that same pivotal year his successor as mapmaker-general in North America enlisted to fight the French and Indians, and spent much of his time surveying for the British the rivers and streams of the Ohio Valley. This was Thomas Hutchins of New Jersey, who first distinguished himself by his sketches of the fortifications of Duquesne abandoned by the French. Hutchins, therefore, continues the process begun by Evans, and his subsequent career and life are associated with the maps he made of the territory west (and south) of the Appalachians, which would guide and promote the progress of settlement for which Lewis Evans and Pownall had argued.

If Pownall serves as a symbol of Anglo-American expansion, then Hutchins carries the transference of imperialism through to its completion: until 1778, the year in which he published his *New Map of the Western Parts of Virginia, Pennsylvania, Maryland, and North Carolina,* Hutchins enlisted his skills in the service of the king, but in 1779, while in England, he was arrested on a charge of treason. The charge proved false, but determining that his military career was irreparably damaged, Hutchins changed sides, and with help from Ben Franklin—who was always interested in encouraging champions of westward expansion and young men of ability—he volunteered his cartographer's talents to the American cause. In relatively short order he had become "Geographer to the United States." This turnabout, which took place as the fortunes of war likewise turned against the British, was both a practical and a symbolic act, for, like the captured forts and ships of Great Britain, Hutchins would continue to serve the old imperial cause while flying the stars and stripes.

Hutchins' first published work as "Geographer to the United States" was his influential *Historical Narrative and Topographical Description of Louisiana, and West-Florida* (1784). He ended his account with some reflections "upon the progress of empires" as it applied to the future of North America. Where Old World empires "were formed by conquest" and were a shaky union of disparate nations "by force jumbled into one heterogeneous power," the coming empire of the

United States would be an "immense continent . . . peopled by persons whose language and national character must be the same. . . . Further, the peopling of this vast tract from a nation renowned in trade, navigation, and naval power, has occasioned all the ideas of the original to be transplanted into the copy" (92–94). That is, the United States would be but a larger, stronger England, occupying an entire continent, not a tiny island, and though a large part of that continent "at present does not belong to our North America," Hutchins maintained, "If we want it, I warrant it will soon be ours" (93).

Further testimony to Hutchins' geopolitical reach is provided by the survey of the western territory he undertook, starting in 1785, implementing the Ordinance authored by Thomas Jefferson, which dictated the grid system of plotting as the most economical and efficient way of laying out and administering the settling of government lands. The result was a heroic exercise in imposing geometric order on the land: anticipated by Thomas Pownall's map of the two British colonies to be planted as a barrier against the French, Hutchins' Seven Ranges employed the rectilinear method by means of which Jefferson's fondness for architectonics derived from various versions of the perfect square would be extended across the entire western surface of the continental United States. Like Jefferson, also, Hutchins directed the flow of westward traffic through the Potomac River: *Topographical Description,* which Hutchins wrote to accompany his map of 1778, depends heavily on Lewis Evans' "Essays" for its account of the "branches of the Ohio and Allegany rivers" as well as for other matters, with the result that the primacy of the Potomac–Monongahela route is emphasized (Imlay: *TD*: 486). Hutchins may have had a more extended westward view than did Evans, but the needle's eye remained constant, and if a generation of geopoliticians heeded his advice, it was because, like Evans also, Hutchins was a careful cartographer whose maps were based on data collected by himself or by other reliable sources, including Evans himself.

Hutchins' counterpart in this regard is Jonathan Carver, whose *Three Years' Travels through the Interior Parts of North America* was published in London in 1778, the same year as Hutchins' map and *Topographical Description.* Carver's *Travels* was much read during the years following its first publication, and, like Hutchins' maps and descriptions of the western regions, his account of the headwaters of the Mississippi eventually served the cause not of British but of the United States' expansion. His book was not entirely reliable, and it was certainly adulterated by material borrowed from previously published books of questionable accuracy, but it was the only account available during the late eighteenth century of the upper Mississippi Valley. If

Hutchins put forth the hard lines of an early national imperialism, Carver graces expansion with an almost mystic aura, promoting not so much expansion as the notion of the upper Mississippi as a marvelous zone.

Still, like Hutchins and Evans, Carver actually explored the region he described, and the original journal he kept of his travels is a careful, factual account, for the most part concerned with promoting strategies that would help the British secure the fur trade of the lake regions, which, following the fall of Quebec, had been drawn south to New Orleans by the French. But Carver's London publishers apparently felt that this matter was not of sufficient interest to the general reader, and a Grub Street hack was hired to embellish the journals, bringing in extraneous material and expanding upon and touching up Carver's own anecdotes, the end effect of which was to promote the fabulous aspects of the Great Lakes region and to emphasize the possibility of finding the Northwest Passage. Carver is a colorful enough character in his own right, paddling into Indian country in a canoe that had a peacepipe as its figurehead and a Union Jack flying from its stern—as compact a symbol of empire as one could conceive—but he emerges in his revised *Travels* as something of a travel liar, and though this makes the narrative more interesting, the high coloration put off seriously scientific-minded men of the late eighteenth century.

Where Hutchins built upon the relatively accurate Evans, the publishers' hack resorted to Daniel Coxe's *Carolana,* by then more than a half-century old, and inserted in Carver's book the dubious information "obtained" (as always) from Indians "that the four most capital rivers on the Continent of North America, viz. the St. Lawrence, the Mississippi, the River Bourbon, and the Oregon or the River of the West . . . have their sources in the same neighbourhood. The waters of the three former are within thirty miles of each other; the latter, however, is rather farther west" (76). How "rather farther" it would be up to a later generation of explorers to learn, but the inclusion of Coxe's mythical geography in Carver's book served to sustain hopeful geopoliticians for another thirty years.

By 1782, therefore, the imperial diagram of the regions beyond the Alleghenies consisted of two parts, the relatively dependable if optimistic maps of Evans and Hutchins and the much more creative projection achieved by grafting Coxe's Missouri to Carver's Mississippi. By placing Coxe and Carver side by side with Evans and Hutchins, one could envision the great wealth of the Mississippi and Ohio Valleys as moving eastward along a series of confluences until it pushed through the Alleghenies by means of the Monongahela, Youghiogheny, and Potomac rivers. One could even peer over a farther range and see the wealth of India moving eastward along the very same route, but the dominant configuration was given shape and meaning by the confluence of the Mississippi and Ohio rivers, waterways giving unity to western regions, an imperial zone that, when linked to the states east of the Alleghenies, would become an empire on a truly Roman scale. The most Roman of all the forefathers, George Washington, was prominent among those who entertained large imperial views, and following his triumphant return to Mount Vernon at war's end, the father of his country staged an anniversary journey: in 1784, thirty years after his first trip to western waters, he headed once again up the Potomac toward the junction of the Allegheny and Monongahela rivers.

This journey, however, takes further point from a trip Washington had taken in 1783, scarcely more than a tourist jaunt, permitted by a period of enforced leisure as the American forces awaited the formal surrender of the British. Leaving his headquarters in Newburgh, New York, Washington first made a "tour through the Lakes George and Champlain as far as Crown Point—then returning to Schenectady, [and] proceed[ing] up the Mohawk river to Fort Schuyler." Crossing over Wood Creek, "which empties into the Oneida Lake, and affords the water communication with Lake Ontario," Washington next "traversed the country to the head of the eastern branch of the Susquehannah, and viewed the Lake Otsego, and the portage between that lake, and the Mohawk River at Canajoharie" (Chastellux: 391–92). Washington's route was in effect a survey of the waterways and portages connecting the Hudson with western waters, a counterpart to the Potomac–Monongahela contingency and one that opened (according to Lewis Evans) "Communications with the Inland Parts of the Continent," which Evans believed was "of the utmost importance to the [then] British Interest," but which in 1783 held out considerable promise of extending the United States' domain (Evans: 164 [20]).

Washington described his journey in a letter to the Marquis de Chastellux, in phrases so carefully chosen and composed as to suggest they were written with future publication in mind; and not only was the letter included in the first American edition of *Travels in North America,* published in 1827, but it had earlier appeared in New York periodicals and was frequently reprinted and alluded to by projectors and propagandists for expanded empire in the early years of the nineteenth century:

> *Prompted by these actual observations, I could not help taking a more contemplative and extensive view of the vast inland navigation of these United States, from maps, and the information of others, and could not but be struck with the immense diffusion*

> *and importance of it, and with the goodness of that Providence which has dealt her favours to us with so profuse a hand. Would to God we may have wisdom enough to make a good use of them. I shall not rest contented till I have explored the western part of this country, and traversed those lines (or a great part of them), which have given bounds to a new empire; but when it may, if ever it should happen, I dare not say, as my first attention must be given to the deranged situation of my private concerns, which are not a little injured by almost nine years* [sic] *absence, and total disregard of them.* (392)

Washington never did make his anticipated exploration of the western bounds of the United States empire, but his "first attention," as he called it, did result in a westward journey, partly undertaken with the intention of straightening out the "private concerns" alluded to in his letter to Chastellux—namely, the encroachment on lands he claimed were his own by settlers who had bought them from other claimants.

Where in his letter to the distinguished Frenchman Washington managed to maintain the disinterested patriotic pose he was increasingly careful to promote during his emergence as father of his country, his diary account of the journey taken in 1784 is much less sectionally neutral. As he moved into the Virginia mountains, moreover, Washington's perspective rapidly dininished, descending from the cosmic field of empire to the microcosmic arena of real estate; and as in his first military campaign thirty years since, the epic design becomes quickly tangled in the underbrush, as strategic lines disappear in the tortuous windings of creeks and rapids-torn rivers. His experience is a detailed demonstration that maps, however accurate, necessarily support larger views, which are often belied by actual encounter with the terrain.

iv

The opening pages of Washington's journal account of his westward journey in 1784 reveal a dual errand, the "indispensable necessity" that he visit his lands "west of the Apalachian Mountains," and the second "object"; namely, "to obtain information of the nearest and best communication between the Eastern & Western waters; & to facilitate as much as in me lay the Inland Navigation of the Potomack" (*Diaries*: IV: 1, 4). Where in his letter to Chastellux, Cincinnatus vies with

Caesar—the urge to return to his farm outweighing thoughts of empire—in his journal of 1784, the Caesarian impulse is in harmony with the Cincinnatian, given that the route in question would have brought the empire in question past his front doorstep. His fellow Virginian, Thomas Jefferson, had written him in March concerning "the opening the Navigation of the Ohio & Potowmac," pointing out that "this is the moment . . . for seizing it if ever we mean to have it. All the world is becoming commercial" (Bacon-Foster: 128). By "we," Jefferson clearly meant "we Virginians," as the rest of his letter shows.

James Maury's pupil went on to compare the northern to the southern route, and maintained that "Nature has declared in favor of the Potowmac [over the Hudson] and thro' that channel offers to pour into our lap the whole commerce of the Western world. But unfortunately [the route] by the Hudson is already open & known in practice; ours is still to be opened," Jefferson continued, declaring it was time "to open our doors" to the west. Washington, admitting he was interested in such matters for personal reasons, was grateful "to find that a man of discernment and liberality, who has no particular interest in the plan," was of the same mind as himself, "who have lands in the country, the value of which would be enhanced by the adoption of such a measure" (129). He agreed that no time should be lost in "this business, as I know the Yorkers will lose no time to remove every obstacle in the way of the other communication" (130). Scarcely a year after describing that "other communication" to Chastellux as a critical part of American empire, Washington indirectly confessed that he had surveyed the New York waterways by way of checking the enemy's advantage, a journey less nationalistic in purpose than regional. What he found along the upper reaches of the Potomac in all ways contrasted with the clean lines of the New York system of waterways and portages, yet so strong was Washington's local loyalty—reinforced as it was by personal considerations—that he never abandoned his faith in the Potomac route to the West, much as he had clung to that strategy thirty years earlier as the best way to defeat the French.

The first leg of his journey went smoothly, and the third day brought him to Berkeley County, where he intended to collect long-overdue rents from his tenants while conferring with a number of prominent men from the area about opening the Potomac to navigation. The gentlemen were enthusiastic about the project, and assured him of local support—including financial—but when Washington spent the next night at the home of Captain John Strode, "an intelligent Man, and one who had been several times in the Western Country," what he learned about the land routes through the mountains and the navigability of the rivers on the other side was not particularly encouraging (*Diaries*: IV: 6). Nor were Washington's tenants quick to pay what they owed in back rent, and those who did so used depreciated Continental notes. From these earliest encounters with the informed gentlemen and impoverished farmers, a pattern begins to emerge, in which expectations are not always matched by realities. Like the value of Continental currency, Washington's hope for a quickly opened door to the West tended to shrink in the face of insurmountable facts.

The next day brought him to the fledgling resort at Bath, where Washington obtained still more information about western waterways, but the most exciting news was conveyed by "the ingenious Mr. Rumsey," who demonstrated a model boat he had designed "for ascending the rapid currents by mechanism" (9). Rumsey's invention used the current to operate by means of a water-turned wheel a series of set-poles that pushed the boat upstream, and it worked well enough to earn him a written certificate of Washington's approval, a document that was of great help in obtaining the backing needed for a full-scale version. What worked in little, however, did not function when expanded, but Rumsey, still enjoying Washington's good regard, went on to experiment more successfully with steam-propelled navigation. He was also employed, again through Washington's agency, as the supervisor in charge of building the earliest Potomac system of canals, using the labor of slaves.

Rumsey was not the only petitioner for the general's patronage during his stay at Bath. A young lawyer from Pittsburgh, Hugh Henry Brackenridge, was in town, having composed for the occasion of Washington's visit a celebratory masque, in which the patriot hero was greeted by "The Genius of the Springs," who was joined by "the Potomac, the Delaware, and the Ohio," a trinity quite in keeping with the general's immediate errand but that was inspired chiefly by Washington's military adventures associated with those rivers (Newlin: 56–67). Along with Philip Freneau, Brackenridge had been the author of the Princeton College commencement poem in 1771, and prior to leaving Philadelphia a decade later for western regions, he had written a number of poems celebrating battles and heroes of the Revolution. He was hardly an unknown poetic quantity, but his bid for Washington's attention is not recorded in the general's journal, perhaps because of the great man's natural modesty, perhaps because arts domestic and industrial appear to have been the general's main concern at the time. Certain it is that Rumsey's boat gained his interest where Brackenridge's water poetry did not.

Moreover, though Brackenridge's Naiads of the Springs greeted

Washington with a song and dance, giving thanks to "that great chief . . . who made us free," the settlers of regions farther west did not share their grateful spirit, and insisted on the legality of their claims to lands to which he held title. Washington was only with great effort able to maintain the "equanimity" for which he was celebrated in Brackenridge's poem, for what was "a noble name" east of the mountains was to western farmers chiefly associated with a man who seemed intent on doing them out of their land and improvements. After discussing the matter amongst themselves, they declined his offer to sell or lease the land they had already bought and cleared, and declared "they meant to stand suit, & abide the Issue of the Law" (*Diaries: IV*:29). In the ensuing trial, the farmers were represented by Brackenridge, less perhaps as a result of authorial pique than from financial necessity, coupled with the young lawyer's tendency at this point in his career to take up the cause of the common people on his side of the mountains against the vested interest of absentee landlords in the East—a sectional division that would give order, but with quite a different emphasis, to his most ambitious literary effort, *Modern Chivalry* (1792–1815). Still, the wisest course is always to applaud the work of younger talents, especially if they are licensed to practice law and you are in their home country.

Washington's attempts to gain accurate or encouraging knowledge about the lay of the land were equally discouraging. His diary is full of data, but as the compilation of facts continues, the information becomes increasingly contradictory: "[I]t may be well to observe, that however knowing these people are, their accts. are to be received with great caution—compared with each other and these again with one's own observatns; as private views are as prevalent in this, as in any other Country; and are particularly exemplified in the article of Roads; which (where they have been marked [out]) seem calculated more to promote individual interest, than the public good" (47). Washington's tone here is in keeping with his quiet sense of humor, yet he seems not to have caught the larger joke, for his journey in the name of a "public good" was firmly anchored to "the individual interest" of Virginia and himself.

Despite the frustrations he encountered, Washington's faith in the Potomac route endured, and having returned to Mount Vernon from the West, he ended his diary account of the journey with "reflections" concerning the future of commerce on western waters, those "channels through which not only the produce of the New States, contemplated by Congress, but the trade of all the lakes . . . may be conducted according to my information, and judgment," both of which, predictably, declared the Monongahela–Potomac route to be "a shorter, easier, and less expensive communication than either of those . . . used with Canada, New York or New Orleans" (59–60). Relying in part upon Hutchins' table of distances and Evans' map—on which such matters leapt up with a clarity lacking when one was deep in the mountains—Washington easily convinced himself of the superiority of his favorite route over all others. Having done so, as in his letter to Chastellux describing the New York waterways, he allowed himself to meditate on the importance to the expanding nation of some link between East and West.

Where earlier propagandists like Evans and Franklin sought to extend British dominion into the Ohio Valley by planting colonies there, Washington now realized the importance of drawing the attention—and produce—of the western territories in an easterly direction: The opening "of a good communication with the Settlmnts. west of us . . . when [it] is considered in a political point of view, . . . appears of much greater importance" than ever (66). Given the fact that in 1784 the United States was still exposed on "flanks and rear" to "other powers"—Britain in Canada and Spain in Louisiana and Florida—it was necessary "to apply the cement of interest to bind all parts . . . together, by one indissolvable band—particularly the Middle States with the Country immediately back of them." The situation was all the more critical because "the Western Settlers—from my own observation—stand as it were on a pivet—the touch of a feather would almost incline them any way. They looked down the Mississippi until the Spaniards (very impoliticly I think for themselves) threw difficulties in the way, and for no other reason that I can conceive than because they glided gently down the stream, without considering perhaps the tedeousness of the voyage back, & the time necessary to perform it in; and because they have no other means of coming to us but by a long land transportation, & unimproved roads" (66–67).

Therefore, in Washington's reckoning, "a combination of circumstances make the present conjuncture more favorable than any other to fix the trade of the Western Country to our Markets" (67). These circumstances favorable to Virginia included the hostility of the Spanish to U.S. navigation on the Mississippi and the refusal of the British to abandon military posts along the interior waterways of New York, "which tho' done under the letter of the treaty is certainly an infraction of the Spirit of it, & injurious to the Union." But what hampered a more perfect Union in the northeast was to the benefit of Virginia to the south, and "may be improved to the greatest advantage by this State if she would open her Arms & embrace the means which are necessary to establish it." The alternatives seemed to be "a seperation" of the western states from the Union or "a War" to keep the Union together: the only way to avoid both, "happily for us, is easy, and dictated by our clearest interests. It is to open a wide door, and make a smooth way for the produce of that Country to pass to our Markets before the trade may get into another channel."

It is not necessary to see Washington as expressing here a narrow sectionalism: his concern for the sanctity and safety of "the Republic" is obviously sincere. But as Daniel Webster a half-century later would champion a Union whose integrity benefitted the mills of Massachusetts, so Washington and Jefferson saw Virginia as the chief beneficiary of western expansion and trade—if they could only work fast enough. Washington's interest in seeing the Potomac made more navigable warmed his interest in "Rumseys discovery of working Boats against [the] stream by mechanical powers principally," but he also was able to foresee how such "a fortunate invention" would profit "these States in general," being yet another of "those circumstances which have combined to render the present epocha favorable above all others for securing . . . a large portion of the produce of the Western Settlements, and of the Fur and Peltry of the Lakes, also—the importance of which alone, if there were no political considerations in the way, is immense" (67–68).

But however great was Washington's gift for geopolitical prophecy (the language of which is reminiscent of Jefferson's earlier letter), his high opinion of Rumsey's invention was misguided, both in the immediate outcome and its long-range effect. A year later he was visited at Mount Vernon by "a Mr. Jno. Fitch," who brought "a draft & Model of a Machine for promoting Navigation, by means of a Steam," and though Fitch hoped, like Rumsey, to get the General's endorsement for his invention, it was not forthcoming (*Diaries*: IV: 218). Washington's reluctance in this regard was primarily due to his sense of honor—having already given his blessing to Rumsey, he could not encourage a rival also—but other factors may have been involved. Fitch was a Yankee from Connecticut; Rumsey was from Baltimore and was currently living in West Virginia, and where the innkeeper was suave and handsome, Fitch, even in his own opinion, was neither. Whatever his reasons, Washington backed the young man, for Fitch's experiments ended in success, Rumsey's in repeated failures. Another visiting Yankee was likewise responsible, at least indirectly, for reversing Washington's predictions concerning the primacy of the Potomac system of canals, but this time it was the Virginian's hospitality, not his sense of honor, that turned the trick.

V

Several months after Washington returned from his western excursion, he was visited by an enterprising young businessman from Rhode Island named Elkanah Watson, and because of the wintry weather, his unexpected guest was invited to spend the night. The entry in Washington's diary for that day (January 19th) is brief, but from the event much subsequent matter would emerge, including Watson's own fulsome account of his visit: "To have communed with such a man, in the bosom of his family," he enthused a half-century later, "I shall always regard as one of the highest privileges and most cherished incidents of my life" (*Men*: 279). Watson was suffering from "a severe cold," and was both gratified and astonished when, after having retired to his room, he was visited by his host bearing "a bowl of hot tea" (280). But what most impressed the visitor from New England was the conversation earlier that evening, which "had reference to the interior country, and to the opening of the Potomac, by canals and locks." By Watson's account,

> *his mind appeared to be deeply absorbed in that object, then in earnest contemplation. He allowed me to take minutes from his former journals on this subject . . . worthy of the comprehensive mind of Washington. To demonstrate the practicability and the policy of diverting the trade of the immense interior world, yet unexplored, to the Atlantic cities, especially in view of the idea that the Mississippi would be opened by Spain, was his constant and familiar theme . . . Since my travels* [*abroad*] *in 1779, I had been deeply and constantly impressed with the importance of constructing canals, to connect the various waters of America . . . Hearing little else, for two days, from the persuasive tongue of this great man. I was, I confess, completely under the influence of the canal mania, and it enkindled all my enthusiasm.* (Men: 280–81)

According to Watson, "Washington pressed me earnestly, to settle on the banks of the Potomac," and at his host's suggestion, the guest rode his horse along the south bank of the Potomac, "to examine the proposed route of the canal" (281). But he declined Washington's invitation to remain in the area, and having in 1787 tried his hand at running a plantation on the Chowan River in North Carolina, Watson returned the next year to his native Rhode Island. From there, in 1787, he removed to Albany, where he set about effecting various improvements, including the paving of streets, chartering a bank, and—most important—promoting a system of inland navigation along the Mohawk River and beyond that would eventually be realized as the Erie Canal.

"There will . . . be," wrote Thomas Jefferson in his *Notes on Virginia,* composed in 1784 and published the next year, "a competition between the Hudson and Patowmac rivers for the . . . commerce of all the country westward of Lake Erie, on the waters of the lakes, of the Ohio, and upper-parts of the Mississippi" (15). As his mentor Maury may have instructed him, and as Jefferson earlier insisted in his letter to George Washington, the Potomac was the better route, not only because Alexandria was nearer than New York to the flow of the trade along the Mississippi and Ohio, but because the northern route froze over, "whereas the channel to the Chesapeake leads directly into a warmer climate" (16). Moreover, because of the hostile presence in Canada of "our neighbours the Anglo-Americans [and] the Indians, the route to New-York [is] a frontier through almost its whole length, and all commerce through it ceases from [the] moment" that war should break out between the two countries (16).

Still, observed Jefferson to his correspondent, "the channel to New-York is already known to practice; whereas the upper waters of the Ohio and the Patowmac, and the great falls of the latter, are yet to be cleared of their fixed constructions" (16). Precedent carries all after it, and the prior flow of traffic down the Hudson had the power of previous inclination: as Washington himself warned in his journal, should Virginia let the western settlers ally themselves commercially "with the Spaniards, Britons, or with any of the States of the Union we shall find it a difficult matter to dissolve them altho a better communication should, thereafter, be presented to them. Time only could effect it; such is the force of habit!" (67). Both Washington and Jefferson strove mightily to effect a counterforce to habit, and the Potomac route of canals and roads they envisioned in time became a reality, but too late to challenge the primacy—and supremacy—of the Erie Canal.

This in no way vitiates the force of their imperial vision, though subsequent events strategically changed the direction that force would take. There is no starker diagram of that vision than Jefferson's description of the rivers of North America that appears in his *Notes on the State of Virginia,* a description echoing James Maury's letter of 1756. Based on data compiled from the works of Evans, Pownall, and Hutchins, it is a diagram centered by the Mississippi, which Jefferson predicts "will be one of the principal channels of future commerce for the country westward of the Alleghany," especially since that great river is given a farther western reach by the Missouri (8). While admitting that since Virginia had ceded her rights to the western territory to Congress, the Missouri was "no longer within our limits," Jefferson's notes nevertheless include a lengthy description of all "the channels of extensive communication with the western and northwestern country" in his description of the rivers of Virginia.

Jefferson's account of the interior rivers is immediately followed by a chapter on Virginia's mountains—still an effective barrier to expansion west—and it is here that one of his most famous prose passages occurs, the description of the confluence of the Shenandoah and Potomac rivers at Harper's Ferry:

> *The passage of the Patowmac through the Blue ridge is perhaps one of the most stupendous scenes in nature. You stand on a very high point of land. On your right comes up the Shenandoah, having ranged along the foot of the mountain an hundred miles to seek a vent. On your left approaches the Patowmac, in quest of a passage also. In the moment of their junction they rush together against the mountain, rend it asunder, and pass off to the sea. The first glance of this scene hurries our senses into the opinion, that this earth has been created in time, that the mountains were formed first, that the rivers began to flow afterwards, that in this place particularly they have been dammed up by the ridge of mountains, and have formed an ocean which filled the whole valley; that continuing to rise they have at length broken over at this spot, and have torn the mountain down from its summit to its base. The piles of rock on each hand, but particularly on the Shenandoah, the evident marks of their disrupture and avulsion from their beds by the most powerful agents of nature, corroborate the impression. But the distant finishing which nature has given to the picture is of a very different character. It is a true contrast to the foreground. It is as placid and delightful, as that is wild and tremendous. For the mountain being cloven asunder, she presents to your eye, through the cleft, a small catch of blue horizon, at an infinite distance in the plain country, inviting you, as it were, from the riot and tumult roaring around, to pass through the breach and participate of the calm below. Here the eye ultimately composes itself; and that way too the road happens actually to lead.* (19)

Edward Savage (1761–1817) and David Edwin (1776–1841), after Savage. *The Washington Family* (1798). (Courtesy of the Metropolitan Museum of Art)

An early notice in the *Philadelphia Gazette* of this popular engraving called attention to the "perspective view of the river Potomac and of Mount Vernon, [which] forms an agreeable and appropriate embellishment in the picture." The notice also opined that the engraving was "executed in a style evincive of the rapid progress of an elegant art, which has hitherto been in a very crude state in this country," thereby associating Washington with the improvements in the young republic during a period when "art" could also refer to the work of civil engineers. See Wendy C. Wick, *George Washington: An American Icon* (Charlottesville, Va.: 1982), pp. 122–24.

Jefferson's aesthetic contrast between western wildness and eastern tranquillity is identical to Pownall's, and the cardinal directions associated with each are the same. Like Pownall's description of the Cohoes Falls, where the western Mohawk drops into the southward-flowing Hudson, the titanic marriage of the Shenandoah and Potomac connotes the terrific power rushing out of the wilderness interior, a natural force that will thenceforth be identified with American empire. As in Pownall's description of the entrance to the Tappan Zee also, the western vista is associated with "disruption and avulsion," while in the other, eastern, direction, the view is "placid and delightful." As the river flows, so runs the road, carrying the compositional eye—and commerce—toward the settled, civilized region beyond.

Jefferson's aesthetic eye was always attracted to the neat configurations of classical architecture—placidity made manifest in stone—but like Washington's his imperial eye was forever drawn up the Potomac, into the wilder zone, not for the sake of wildness itself, but toward those incipient channels of commerce that resolved themselves into yet another orderly design. It would be upstream that the course of empire moved, against the current, as up Pownall's pictured Hudson; yet the source of empire lay downstream, a placid and orderly zone toward which all waters flowed, carrying with them the tribute of the wilder regions in the West, an American version of the Horatian scheme. Because of the turns of chance and circumstance, Pownall's Hudson, not Jefferson's Potomac, would draw the flow of empire, and the consistency of the imperial aesthetic would as a result undergo strategic revision, as the Potomac gradually became a symbol of divided Union, less a gateway to western regions than a border between two ways of life.

vi

George Washington died in 1799, at century's end, putting an orderly finial to a monumental life. Shortly before his death, there was published in Philadelphia a family portrait of George and Martha, with the Custis children. Spread out before them is a map of the new city being laid out on the Potomac River, the neoclassical conception of Pierre L'Enfant, a French engineer, which would bear Washington's name. Above the map, between two columns suggesting a portico but hung with ornamental drapery, is obtained a view of the Potomac itself, which indeed ran past Washington's front door at Mt. Vernon, but in no such idealized way. As the artist has drawn it, the river in a symbolic sense runs out of and into the river on the map, certifying at the end of Washington's life his great hope for the Potomac as the entranceway to a nation he had fathered, which seemed at the end of the century about to be realized, even to locating the nation's capitol at the gateway to the opening empire beyond.

But we should also note the black man standing in the background, in compositional collusion with Mrs. Washington, his body merging with the column on the right. He, too, is part of the family, and his is a figure that appears in a number of Washington's portraits with riverine backdrops, as in John Trumbull's portrait of the victorious general standing above the Hudson with West Point below. If we can think of that map of Washington the city and the idea of Washington the man as casting regular shapes westward over the land, then we should think also of that ubiquitous black servant, surely a slave, who casts his shape westward as well, part of the orderly, columnar plan. Warranted on the bank of the Hudson by Washington's presence only, generic to the landscape as defined by the Potomac, the black man's future like the nation's would nonetheless be determined by the bifurcated diagram that the extensions of those rivers westward would make. That black man is Washington's shadow, the dark side of his dream of empire, living testimony to the human price paid for imperial schemes and of the deep resistance of mankind and nature to systems of absolute order imposed upon the land. In time he would become the central figure in another epic action, the first scene of which was set at the confluence of the Shenandoah and Potomac rivers, with an outcome disastrous to (though made inevitable by) the shadowy dimension of Washington's and Jefferson's plan.

III

Garden Bright with Sinuous Rills

i

Against the grand view of Washington and Jefferson, who saw the rivers of North America as a reticulated system of waterways giving order to the emerging United States, there must be placed a number of particulars that acted as impediments to such schemes. The larger perspective tended to obtain in views west of the Alleghenies, while, as both Jefferson and Washington realized, to the east, petty territorial imperatives prevailed. Washington's fear that Pennsylvania would challenge Virginia's use of the Youghiogheny, along with Jefferson's jealous fears regarding the Mohawk–Hudson contingency, are sentiments typical of the situation, in which the "common good" of opening channels of communication with western regions was often warped along sectional lines of interest. Moreover, because many of the major rivers east of the mountains flowed through two or more states and occasionally served as a boundary between two states—a divisive situation inherited from colonial days—cooperation through mutual interest dictated coordination of efforts, but, again, required the compromise of individual considerations.

The situation was analogous to the disruptiveness engendered by the Articles of Confederation, under which the new nation was governed for its first ten years, for after 1782 the cohesiveness (never perfect) necessitated by the exigencies of war disappeared, and the states fell to quarreling amongst themselves over their joint borders and rival interests. Where before the desire for independence had been a unifying force, now it tended toward separatism and disunion. That the use in common of waterways was intimately related to the need for a more centralized, better coordinated system of government is demonstrated by the example of the Potomac River, which served as the border between Virginia and Maryland: if plans for improving the river for navigation were to be realized, a common effort by the states was necessary. Because of his personal interest, Washington set in motion negotiations between the two states, which resulted in the Mount Vernon Convention of March 1785, which led directly to the Annapolis Convention of 1786, commonly regarded as the predecessor of the Constitutional Convention of 1787, over which Washington presided.

In arguing that Pennsylvania adopt the new Constitution, James Wilson acknowledged that as an instrument of government it was a creation of compromise, made possible by setting aside particular interests for the good of all, much as the plan that resulted dictated that individual states must surrender certain of their powers to the central government. The better to illustrate his point, Wilson employed an analogy, not particularly novel but certainly apt to the occasion: He compared the drafting of the Constitution to a newborn river, fed by "springs of opposition" that "poured forth their waters in courses so varying . . . [that] the stream formed by their conjunction was impelled in a direction somewhat different from that, which each of them would have taken separately" (*Essays:* 164). Vividly depicting the organic process of compromise, Wilson's metaphor implicitly put forth the nature of a confederated republic made possible by "the number, and greatness, and connexion of lakes and rivers with which the United States are intersected and almost surrounded," a geopolitical fact that dictated a final return to his fluvial image: "We shall find," he predicted, in ending his discussion of the separation of powers provided for in the Constitution, "the streams of power running in different directions, in different dimensions, and at differing heights, watering, adorning, and fertilizing the fields and meadows, through which their courses are led; but if we trace them, we shall discover, that they all originally flow from one abundant fountain. In this constitution, all authority is derived from THE PEOPLE" (174, 181). In his concluding figure, Wilson returns to the importance of American rivers that he had asserted at the start of his speech, when he pointed to "the Hudson, the Delaware, the Potowmack, and the numerous other rivers, that water and are intended to enrich the dominions of the United States" (164).

Out of diversity, unity: such was the wisdom from which the early republic took its shape and governing philosophy, less a static state than a desired end, generating growth that was, in effect, a future hope defined by a continuous extension of the process of compromise. The most far-reaching of these compromises first emerged from the Constitutional Convention of 1787, which permitted the system of slavery to continue, a bowing before regional interests in the name of a general good dictated by commercial as well as statist considerations. Like Washington's plan for canals along the Potomac—called the "Union System"—work on which was carried out by slave labor, the Union from which the system took its name and spirit was given shape by a notion of property that did not distinguish finally between human and other chattels. Establishing a mechanism of order that unified the existing states and guaranteed an intimate connection with the new states in the West, the Constitution contained a major flaw that would, in time, engender the very division the authors of *The Federalist Papers* (1788) uniformly deplored: in a way not foreseen by Washington and Jefferson, the Potomac River did finally feed into the Ohio, much as the Potomac itself would in time serve as the major battle line between North and South.

Counterpart to the constitution in lending order to the land was the Northwest Ordinance of 1787. Like the earlier Ordinance of 1784—the creation of Thomas Jefferson—it was intended to reduce the confusion resulting from antiquated methods of filing land claims in the regions west of the Alleghanies. It institutionalized the grid system of survey, exemplified by the Seven Ranges already imposed by Thomas Hutchins on the federal territory west of the Ohio River. But the Ordinance of 1787 went beyond its predecessor in setting forth the terms of land settlement, in effect lending a higher order to the process by guaranteeing civil liberties and requiring the establishment of public schools and universities.

By 1787, however, settlement had proceeded to such a degree south of the Ohio River that Kentucky was little affected by the rectilinear aspects of the ordinance, while the northern side (thanks to the efforts of Thomas Hutchins) quickly emerged as an ideal exemplification of republican right reason. Kentucky, having been in large part settled by emigrants from Virginia and Carolina, was a geopolitical extension for the slave-holding South, while Ohio took its spirit from an influx of New England–born settlers. As a result, the Ohio River became a complex geographical fact, a confluence of two rivers, the Allegheny and Monongahela, which flowed from opposite directions, connoting both union and division. As a great boundary stream extending to the Mississippi, the Ohio River became a border zone where South and North confronted each other across an ever widening and deepening channel.

From the beginning, because its lands were claimed under the antiquated regulations of Virginia, Kentucky had a disorderly image, while Ohio, with its Seven Ranges and their geometric sisters, became a paragon of orderliness. This was hardly a result of the power of a surveyor's transit, but the grid system was certainly symbolic of the situation. As a result, while the rivers of America served during the early years of the republic as rhetorical symbols and economic facts promoting the cause of national union, in time they would serve—as they often served during the War for Independence—as the battle lines of a subsequent continental war—a direct result of the compromises for the sake of the Union that James Wilson, in seeking to ratify the Constitution, had championed.

At the start of the process, however, a certain uniformity of attitude characterizes much of the literature associated with the settlement of both Kentucky and Ohio, a homogeneity that may in part be ascribed to the inherited tropes of earlier writings encouraging emigration into new regions. And yet, as we may make distinctions between the dominant stresses characterizing the initial propaganda for settlement associated with Virginia and New England, so the subsequent literature has its differences, traceable to the distinctions between what (for the purposes of simplicity if not strict accuracy) have been called the Cavalier and the Yankee impulses. Thus the settlement of Kentucky may be identified with the personal history of Daniel Boone, while Ohio is a much more corporate product, associated in the popular mind with no particular "legendary" figure, unless it be Johnny Appleseed, who was largely the invention a much later generation of writers. Still, the settlement of Ohio does have a representative man, the Reverend Manasseh Cutler, whose stay there was brief but who most definitely left (or imposed) his mark on the emerging state.

ii

In terms of his personal genesis and character, Daniel Boone would seem to be an unlikely example of the Cavalier type: he was not a Virginian but the Pennsylvania-born son of a restless farmer who followed the Quaker grapevine in a southwesterly direction until he came to rest on the Yadkin River in North Carolina. Yet Boone's Kentucky adventure is identified with the way West taken by a number of emigrant Virginians, and his buckskin-clad figure provides a martial outline for subsequent Virginia-born heroes of the western frontier, most notably the Clark brothers, George Rogers and William. And they, in turn, certify the subtle but definable changes that occurred when the amphibious Virginian type so well represented by the young George Washington—half-horseman, half-raftsman—crossed over the mountains to remain in the West. Thus the Virginia squire is father to the Kentucky settler; yet strategic differences remain. Perhaps because of his Quaker heritage, the young Daniel's role in the French and

Indian War was not as a soldier but a wagoneer for Braddock, a short-lived non-combatant experience. And Boone's strictly honorary title of "Colonel" was warranted by his experience fighting Indians during the later epoch of the Revolution, in this case largely a defensive exercise against tribes stirred up by the British. For a hero, moreover, Boone was remarkably unsuccessful, at least where real estate was concerned: his association with Kentucky ended in 1788 when the new Ordinance of 1787 invalidated his land claims, and he spent the next thirty years in regions farther west, living in relative poverty.

In truth, Daniel Boone is a tragic figure, and that we think of him in heroic terms (or, indeed, think of him at all) is because of a schoolteacher named John Filson, also from Pennsylvania, who crossed over to Kentucky in 1783 in search of land and fortune, with the result that Boone found fame. "Not a lie in it!" he swore, referring to Filson's as-told-to first-person account of Boone's adventure in the "Dark and Bloody Ground," which took up one-half of Filson's *Discovery and Settlement of Kentucke* (1784), a book which bore a subscription attesting to its truth, signed by Boone and two other early settlers. Despite their avowals, both parts of Filson's little work were shaped in the service of promoting real estate in Kentucky. Still, Filson yielded to an impulse other than mercenary as he set down a stylized, idealized version of Boone's adventures, and his readers were able to catch that impulse through the medium of his often stilted prose. If Washington carried the Enlightenment rage for order into the Ohio Valley, then, thanks to Filson's labors, he was followed there by an American Rousseau.

Where the Ohio Valley emerges from Washington's journals as a territory of adversity, for Filson's Boone it chiefly puts forth an accommodating landscape, a pastoral zone begging for settlers to occupy it: Kentucky is described as "the most extraordinary country that the sun enlightens with his celestial beams . . . situated on the fertile banks of the great Ohio" (Filson: 21, 50). And yet to reach this wonderful zone it is first necessary to pass through a forbidding region, the Cumberland Gap, past towering cliffs whose "aspect . . . is so wild and horrid, that it is impossible to behold them without terror (58). A testimony to Boone's courage, the mountain passage acts as an aesthetic contrast to what lies beyond, for having passed over the gloomy threshold, Boone beholds a vision of natural beauty that "expelled every gloomy and vexatious thought": "I had just gained the summit of a commanding ridge, and, looking round with astonishing delight, beheld the ample plains, the beauteous tracts below. On the other hand, I surveyed the famous river Ohio that rolled in silent dignity, marking the western boundary of Kentucke with inconceivable grandeur. At a vast distance I beheld the mountains lift their venerable brows, and penetrate the clouds. All things were still, I kindled a fire near a fountain of sweet water, and feasted on the loin of a buck, which a few hours before I had killed" (55).

This Pisgah view characterizes Filson's book throughout, and, as in Daniel Boone's prospect and on the map published to accompany the book, a river runs through it. By Filson's account, the entire region is defined by a series of "interlocking" waterways, a reticulation of streams that provides a convenient mechanism for commerce: "These rivers are navigable for boats almost to their sources, without rapids, for the greatest part of the year" (15). And the navigation of even rapid rivers will be made easy, according to Filson, by "the newly invented mechanical boats" of Mr. Rumsey, "it being their peculiar property to sail best in smart currents" (44). Filson refers to the water-propelled boat that so impressed Washington, and he also shared Washington's faith in other navigational improvements: Filson proposed facilitating traffic on the Mississippi by cutting through the narrow necks of its many meanders, after which, once Rumsey's invention was perfected, "a voyage from New Orleans to the Falls of Ohio will be attended with inconsiderable expence" (46).

Sharing Jefferson's aesthetic, Filson also echoed the Virginian's long view, regarding "the Mississippi and Ohio rivers to be the key to the northern parts of the western continent," a "great passage made by the Hand of Nature for a variety of valuable purposes," the chief of which was "to promote the happiness and benefit of mankind" (39). As for the happiness and benefit of Daniel Boone, they went into a perceptible decline thereafter, while a more rapid fate was accorded John Filson: in 1788, while engaged in surveying lands along the Little Miami—preparatory to laying out a town where Cincinnati now stands—he was killed by an Indian with prior rights to the place, and was thereby claimed by the darker, not the sunstruck, aspects of Boone's biography. His book, however, captured the early and ebullient mood of Kentucky settlement, and established a mode subsequent writers would inherit, blending Enlightenment optimism with the subsequent Romantic faith in the benevolent effects of nature (not including Indians) on men who maintained constant contact with the natural world.

In the year Filson died there arrived in the Ohio territory a man who was in all ways the antithesis of Daniel Boone, who was never elevated to the stature of folk hero, but who had a powerful effect on the emerging shape of the region north and west of the Ohio, the Beautiful River. In the person of the Reverend Manasseh Cutler from Connecticut, we have a bona fide son of the Puritans, who came trailing clouds of the kind of glory John Adams associated with Beacon Hill. And yet he was also of the Tribe of Ben, being very much invested with the Enlightenment spirit.

Cutler shared none of Boone's noble primitivism, and made the westward passage in a sulky, a vehicle with the apparent endurance of the Deacon's One-Horse Shay. A Yale-educated Congregational minister, he was licensed to practice law and had acquired sufficient medical knowledge to act the part of a physician when necessary. He was an amateur scientist who attained considerable reputation as a botanist, belonged to the foremost scientific societies of his day, and like Dr. Franklin carried on an extensive correspondence with his fellow savants. Like Dr. Franklin, also, Cutler became interested in the possibilities of western lands, geopolitically considered. Having served as an army chaplain during the Revolution, he formed an association with other veteran officers, who, under General Rufus Putnam's leadership, took the name in 1786 of the earlier Ohio Company of Virginia, which had relinquished the title along with its lands.

Under the scheme devised by Cutler and his associates, the federal territory along the north side of the Ohio River would be settled the New England way. Where Kentucky, like Virginia, had been taken up by individuals who moved into the area as impulse and circumstances permitted—"being collected from different parts of the continent," Filson observed, "they have a diversity of manners, customs, and religions" (29)—Ohio's early settlers were a homogenous and highly organized group for the most part, who brought with them from New England a strong sense of community. The Ohio Company was formed into a paramilitary unit, which marched as a body into the western valley, occupying a fortresslike stronghold at the juncture of the Muskingum and Ohio rivers called "Campus Martius." This pugnacious structure was designed to serve as the center of a community that would spring up around it, a town originally called "Adelphia" (taking the brotherliness from William Penn's city while leaving the love behind), but finally named Marietta after the wife of Louis XVI, the king who had been so influential in turning the tide of the War for Independence, in the hope of luring loyalist refugees from revolution-torn France.

The plan of the new settlement was distinctly neoclassical in conception, from the creek called Tiber that flowed through the settlers' garden plot, to the names assigned ancient Indian earthworks that were prominent evidence of former domain: Capitolium, Quadranau, and Cecilia. Built in the shape of "elevated squares," these remains were easily incorporated into the rectangular town plan, reinforcing the certainty of mathematics—if not eventual success. Cutler did not accompany this moving town, which arrived in 1787, but he did the settlers considerable service at a remove, proving himself an effective lobbyist in Congress by obtaining a much larger grant of land than the company had originally requested. But perhaps Cutler's most influential contribution was a pamphlet published in 1787, which described the company lands, a lengthy "explanation" designed to accompany a map of the region. Enjoying the great advantage of distance, Cutler was able to pass a transforming wand over the colorless (if optimistic) facts provided by Thomas Hutchins.

Although the dominant New England metaphor associated with the process of settlement during the colonial period was the Mosaic exodus, by the late eighteenth century, most particularly in Connecticut, the emphasis increasingly was on the Canaan prospect, not the wilderness crossing. "This country," wrote Cutler, quoting Captain Harry Gordon's journal from Pownall's *Topographical Description*, "may, from a proper knowledge, be affirmed to be the most healthy, the most pleasant, the most commodious and most fertile spot on earth, known to the European people" (Cutler: 29; Pownall: 159). Naturally productive of "vast quantities" of Escholian grapes (easily converted to "rich red wine"), the Ohio Valley also had abundant maple trees, which could by a familiar New England way be converted to "a sugar equal in flavour and whiteness to the best Muscavadao" (30). Welcome also to his New England audience was the news of potential mill sites, located on streams conveniently "interspersed, as if by art, that there be no deficiency in any of the conveniences of life" (30). Like Filson's Kentucky, then, Cutler's Ohio seemed fashioned by Providence for man's use, yet there was a subtle New England–bred difference; for although the western region was a fertile and accommodating zone, much work remained to be done: "The whole country above Miami," Cutler prophesied, *"will be brought to* that degree of cultivation, which will exhibit all its latent beauties, and justify those descriptions of travellers which have so often made it the garden of the world, the seat of wealth, and the centre of a great empire" (32, italics added).

Where Filson three years earlier had pointed out the potential value for navigation on the Mississippi and Ohio rivers of Rumsey's "mechanical" boat, Cutler made the even more remarkable prediction that "the communications between this country and the sea" would be greatly facilitated by "steam-boats," which "in all probability . . . will be found to do infinite service in all our extensive river navigation" (31). In 1787, Rumsey and Fitch were still experimenting with the first clumsy models using the power of steam, and during his visit that year to Philadelphia (where he had paid an obligatory visit to Dr. Franklin), Cutler seems to have heard news of these experiments. But so novel was the notion, that when his pamphlet was reprinted in Rhode Island, the printer, further confused by a broken slug of type that made the "t" in "steam" resemble an "r," converted the phrase into "stream boat."

Cutler's brief stay in Philadelphia resulted in much more than prescience regarding the steamboat on western waters, for while he was there, he was given a draft copy of the Ordinance of 1787, and revised it strategically, contributing (it is presumed) the provisions for the setting up of public schools and the abolition of slavery. Education was one of the traditional Puritan priorities that harmonized with the Enlightenment emphasis on improvement, and in his Ohio Company pamphlet Cutler expressed his hope that "the acquisition of useful knowledge [will be] placed upon a more respectable footing here, than in any other part of the world" (36). Cutler's faith in "science" and "useful knowledge" likewise blends Enlightenment emphases with Puritan traditions, as does his conception of the vital role of environment in the creation of a new and more perfect society: "There will be one advantage [to Ohio] which no other part of the earth can boast, and which probably will never again occur—that, in order to begin *right,* there will be no *wrong* habits to remove, before you can lay the foundation" (36).

From Cutler's pamphlet there emerges the image of an ideal community springing up beyond the Alleghenies, not only at the conjunction of the Muskingum and Ohio rivers, but throughout the entire western region. He "indulges the sublime contemplation" of the time when the great Ohio valley will be occupied "by an enlightened people, and continued under one extended government," a future moment that will warrant the location of the capitol of the United States—"the seat of empire for the whole dominion"—on the Ohio River, "and not far from this spot. . . . This is central to the whole; it will best accommodate every part; it is the most pleasant, and probably the most healthful" (37). Cutler was able to indulge in this geocentralism—part and parcel of his neoclassical aesthetic—because in 1787 the idea of establishing the nation's capitol on the Potomac was as nebulous as the future site was miasmic. His proposal seemed a logical extension of the republican necessity, moreover, geopolitical symmetry acting to assure westerners "that government will forever accommodate them as much as their brethren on the east," assurance that would dissuade them from "forming schemes of independence, seeking other connexions, and providing for their separate convenience" (37).

"Separatism" was ever a double-edged doctrine for Puritans, and like John Winthrop in 1630, Manasseh Cutler in 1787 was caught in a casuistical trap of his own design, urging the insular sanctity of the Ohio Valley while arguing the importance of making the region the active center of a unified whole: "It is the most exalted and benevolent object of legislation that ever was aimed at, to unite such an amazingly extensive people, and make them happy, under one jurisdiction" (37). He is here speaking of the Constitution, another Enlightenment product from Philadelphia, yet by conceiving of some spot near the Campus Martius as the future capitol of the new nation, he is in effect echoing John Winthrop, who conceived of his sanctified "Citie on a Hill" as a demonstration of the Puritan way to the entire world, as a "center" of a radically reformed notion of society, possibly even the site of the promised New Jerusalem. Both Cutler's Marietta and Winthrop's Boston were conceived as being at once separate from (thereby sanctified) and central to (thereby sanctifying) the rest of the world.

The Puritan continuity in Ohio is best illustrated by a sermon Cutler delivered to the settlers from New England, in Ohio, the import of which did not differ much from the exhortation preached by John Winthrop aboard the *Arbella*. However, because of the troublesome delay in the arrival of Winthrop's expected millennium, Cutler was forced to begin by addressing himself to the question rhetorically posed by Samuel Danforth in his famous "Jeremiad" of 1670, by observing that "advances to perfection are gradual and progressive" (442). Cutler then held out the hope that the long-awaited moment was soon to occur, indulging in the "pleasant contemplation that infinite wisdom and goodness, by a series of remarkable events, is preparing the way for the extension of that heavenborn, glorious, and benevolent religion, which consists in truth, righteousness, and peace—a religion most friendly to true freedom and happiness in the present world, and secures eternal felicity in a future and more refined state of existence" (443). As in his contributions to the Ordinance of 1787, Cutler's blend of Enlightenment optimism and Puritan millenarianism controls the rhetorical flow of his sermon, combining Lockean notions of "true freedom and happiness in the present world" and the old Puritan promise of "eternal felicity in a future and more refined state of existence."

But Cutler's "series of remarkable events" that hold out the promise of rapid improvements is purely Matherish in implication, being evidence of divine favor auguring well for the future—including the recent Constitutional Convention—and of similar origins is "a beautiful analogy" with which he concludes this part of his sermon:

> *The sun, the glorius luminary of the day, comes forth from his chambers of the East, and, rejoicing to run his course, carries light and heat and joy through the nations to the remotest parts of the West, and returns to the place from whence he came. In*

> *like manner divine truth, useful knowledge, and improvements appear to proceed in the same direction, until the bright day of science, virtue, pure religion, and free government, shall pervade this western hemisphere. . . . The Divine counsels, opened to us by the events of time, give us just ground to believe that one great end God had in view in the original discovery of this American Continent, and in baffling all the attempts which European princes have made to subject it to their dominion, and in giving us the quiet possession of it as our own land, was that a new Empire should be called into being—an Empire new, indeed, in point of existence, but more essentially so, as its government is founded on principles of equal liberty and justice.* (443–44)

Once again, Cutler instills Lockean notions in the old Puritan idea, identifying "principles of equal liberty and justice" with divine dispensation, and the net effect is to make Ohio a new-bottled Massachusetts—yet another Promised Land where "we see settlements forming in the American wilderness, deserts turning into fruitful fields, and the delightful habitations of civilized and christianized men. . . . We this day literally see the fulfillment of . . . prophecy . . . gradually advancing" (445).

Nothing like this appears in Filson's account of Kentucky, and where the New Jersey schoolteacher describes the fertility of the Ohio Valley with unqualified encomiums, Cutler regards "the liberality of the hand of nature in this part of the globe [which] seems to have distinguished it from almost all others" with a certain uneasiness (445). For "solid enjoyment and rational happiness" is not necessarily guaranteed by material plenty, but is derived from "the wise and judicious improvement we make of these natural advantages," the same New England emphasis on the necessity of hard work found in his pamphlet (446). Of similar origin is his concern for the younger settlers, who are most susceptible to the lure of "ease and plenty," whose boon companions are "dissipation, luxury, and vice" (446). If hard work protects the older people from the evil effects of abundance, the younger should be protected from harm by schools, which will lay "the foundations for a well-regulated society" by teaching "the rising generation" a proper regard for God and the importance of "virtue and righteousness" (449). To this conventional Puritan theme Cutler once again adds an Enlightenment touch, combining "the pure religion of the Gospels" with "civil liberty and the cultivation of the arts and sciences," producing a two-pronged instrument promoting "the civil and social happiness of a new settlement" (446).

In 1621, at the start of the Puritan experiment in New England, Robert Cushman delivered a similar sermon to the settlers of New Plymouth, ending with the admonition that the settlers set aside personal desires for gain in favor of the communal good, both for the sake of the new colony's survival and for their own salvation, so that "when that God of peace and unity shall come to visit you with death as he hath done many of your associates; you being found of him, not in murmurings, discontent, and jars, but in brotherly love and peace may be translated from this wandering wilderness into that joyful and heavenly Canaan" (44). Though Cutler's generation was more apt to identify Canaan with western than with heavenly regions, the Connecticut minister echoed Cushman in urging that "in our present circumstances we ought to consider ourselves as members of one family, united by the bounds of one common interest. . . . Then may we hope for the smiles of Heaven, the blessing and protection of a kind Providence," and the assurance that we shall "all be received to those regions of bliss in the heavenly world, where sorrow will not be permitted to enter, but uninterrupted happiness reign forever and forever. Amen" (449–50). The correspondences between these communal benedictions certify the continuity of the Puritan experience in America, and consistent also were the ensuing events: having delivered his sermon, Cushman took ship back to England, and Cutler himself soon boarded a boat for the first leg of his return journey to Connecticut, both men leaving behind high ideals that became increasingly impossible to uphold while struggling to survive in a land full of hostile heathen whose conversion to Christianity seemed increasingly unlikely.

Yet Cutler's sermon was an impressive performance, if only as a very agile intellectual exercise that testifies to the ease with which Congregational millenarianism could harmonize with the secular emphasis on earthly felicity promoted by Enlightenment thinkers, a mixture of Providence and progress that certified "a new Empire has sprung into existence, and there is a new thing under the sun" (444). Where Filson imposes a quasi-romantic aspect on both Boone and the Kentucky landscape, Cutler retains the traditional New England metaphor of a wilderness garden populated by a chosen tribe. The net effect is to oppose, at least in literary terms, two quite different modes of settlement, separated by the narrow boundary provided by the Ohio River, a pattern inherited from the first century of Anglo-American colonization in the New World, a projection beyond the Alleghenies of the Virginian and New England ideals.

iii

Manasseh Cutler's version of Ohio was given further circulation when, in 1789, Jedidiah Morse published his *American Geography,* which included generous exerpts from the pamphlet written by his friend and correspondent, including the good news about steamboats and Cutler's theory about the westward-moving center of empire. Like Cutler's sermon, moreover, Morse's book was a lengthy celebration of the certainties made possible by the new Constitution—the framing of which Morse described at length—a celebration filled with a nationalistic, millenialist fervor, reflected in language echoing Cutler's glorious analogy:

> [*I*]*t is well known that empire has been travelling from east to west. Probably her last and broadest seat will be America. Here the sciences and the arts of civilized life are to receive their highest improvement. Here civil and religious liberty are to flourish, unchecked by the cruel hand of civil or ecclesiastical tyranny. Here Genius, aided by all the improvement of former ages, is to be exerted in humanizing mankind—in expanding and inriching their minds with religious and philosophical knowledge, and in planning and executing a form of government, which shall involve all the excellencies of former governments, with a few of their defects as is consistent with the imperfection of human affairs, and which shall be calculated to protect and unite, in a manner consistent with the natural rights of mankind, the largest empire that ever existed. Elevated with these prospects, which are not merely the visions of fancy, we cannot but anticipate the period, as not far distant, when the AMERICAN EMPIRE will comprehend millions of souls, west of the Mississippi. Judging upon probable grounds, the Mississippi was never designed as the western boundary of the American empire. The God of nature never intended that some of the best part of his earth should be inhabited by the subjects of a monarch, 4000 miles from them.* (469)

Among Morse's purposes in publishing his book was the education of the rising generation, impressing "the minds of American Youth with an idea of the superior importance of their own country" (vii). Some notion of Morse's influence may be had by considering the ages in 1789 of the men who, a generation later, would be instrumental in guiding the shape of "the AMERICAN EMPIRE": Daniel Webster, John Calhoun, and Thomas Hart Benton were all seven years old, and we may assume—other evidence lacking—that they were all in some sense of the word students of Jedidiah Morse, who passed on to them a geopolitics similar to that learned by Jefferson from the Reverend James Maury.

Jefferson's *Notes on Virginia* was frequently quoted by Morse, as were the writings of Lewis Evans, Thomas Pownall, Thomas Hutchins, Jonathan Carver, and John Filson, from whose *History of Kentucke* Morse borrowed the vivid image of "Colonel Boone" viewing "from the top of an eminence, with joy and wonder, . . . the beautiful landscape of Kentucky" (407). All those who contributed descriptions of the territory beyond the Appalachians were propagandists for imperial expansion, and as a result, Morse's *Geography* is an *omnium gatherum,* a vest-pocket library of imperialist texts; and its publication occurred at a critical moment, not only influencing national development (and the lives of infant statesmen), but reflecting the cohesion of Enlightenment attitudes in the shape of national expectations.

The concept of the United States as a geopolitical process, like the Constitution itself conceived as a mechanism taking its meaning from a condition of constant change, permeates Morse's book, whether in statistics recording the facts of growth or in the identification of American empire with a westward-moving frontier. And, like Cutler, Morse identifies the frontier with the extension of a revived and modified Puritan errand: "New England may, with propriety, be called a nursery of men, whence are annually transplanted, into other parts of the United States, thousands of its natives. . . . They glory, and perhaps with justice, in possessing that spirit of freedom, which induced their ancestors to leave their native country, and to brave the dangers of the ocean and the hardships of settling a wilderness" (144–45). Where Daniel Boone, in Filson's account, viewed Kentucky from a Virginia mountaintop, Jedidiah Morse obtained his idea of the western Canaan from the vantage point of statistics gathered in New England—less, perhaps, a Pisgah view than a prospect from Mount Sinai, for the Yankee notion of the frontier process put a premium on mechanisms of law and order.

Another geographical compendium, published during the same critical period, while obviously indebted to Morse, introduces radical innovations to the idea of American empire, particularly the region west of the Alleghenies. Gilbert Imlay's *Topographical Description of the*

Western Territory of North America was first published in England in 1792; a second English edition, in 1793, added Filson's *Kentucke* as a second volume, and was reprinted in America the same year. The contingency was symbolic, for Imlay's topography provides the intellectual framework missing from Filson's impressionistic narrative, and puts forth what might be called a Southern strategy for American empire, albeit one that carries agrarianism to a physiocratic extreme. Like Morse's *Geography,* Imlay's *Topographical Description* promoted an anthology effect, giving ephemeral pamphlet publications additional circulation, writings that had (as his title suggests) a definite western bias: the third English edition (1797) contained in a series of appendixes some fifteen documents promoting settlement of the Ohio Valley—including Hutchins' two works entire—and by various inclusions and notes introduced excerpts from the works of Thomas Pownall and Lewis Evans.

Not much of a case can be made, however, for comparable influence. The third English edition was never published in America, nor was the only American edition ever reprinted. The regional particularism of Imlay's anthology may have contributed to this lack of circulation along with the waning fortunes of the Ohio and Scioto companies by 1797, a decline matched by the descent of Imlay's personal reputation. In the interval following the publication of the first edition of his book, Imlay had traveled to Paris at the height of the Revolution, where he became associated with the notorious Tom Paine, and had gained a further measure of dubious fame by consorting with, then breaking the heart of, Mary Wollstonecraft, whose *Letters to Imlay* were published posthumously in 1798. Imlay was also involved in Brissot de Warville's abortive attempt to launch a combined French and American invasion of Louisiana in 1793, resulting in the highly embarrassing Genêt Affair. Still, if Imlay's various adventures abroad brought him ill repute in America, they most certainly acted to strengthen the French connection, which is evident throughout his *Topographical Description.* Where Jedidiah Morse was never more than an enlightened Federalist, Gilbert Imlay was a republican with Jacobin tendencies, a radicalism that resulted in a view of New World possibilities with decided difference.

Imlay must be accounted one of the most fascinating people to surface during the expansive 1780s. He is, ironically, known best through Mary Wollstonecraft's letters, but as for matters of record, we know that he served as a lieutenent in the American army during the Revolution, that he may have been wounded, that he may have been promoted to captain—the title if not the rank he boasted—and that, with many other young officers, he headed west at war's end, into Kentucky, where he arrived in 1784. During the next two years Imlay was very busy, speculating in lands, serving as a deputy surveyor, and spending considerable time either involved in or avoiding court proceedings, legal troubles resulting from debts incurred while dealing in real estate. During this same period, moreover, Imlay was involved with concerns we know very little about, since they were purposely kept secret: after arriving in western regions, he made the acquaintance of General James Wilkinson, who was busy hatching plans regarding the destiny of Spain in the New World as it affected his own future, the first of several schemes by which the general would have led Kentucky away from the Union.

What we know about Wilkinson's subversive activities is the result of Aaron Burr's aborted conspiracy and the subsequent trial, by which time Imlay had long since left Kentucky, leaving behind a number of pending lawsuits and unpaid debts. Heading east, he was last seen in America late in 1785, in Richmond, Virginia, and next surfaced when his name appeared on the title page of his *Topographical Description*. By then he was one of a group of expatriate Americans and English who had gathered around the Girondist faction in Paris, and, after the death of Brissot, Imlay abandoned politics for shadowy commercial dealings and shady women of the theater, both of which strained his relationship with Mary Wollstonecraft. When their correspondence broke off, so did the record of Imlay's life, and his name next appeared on a gravestone on the Channel island of Jersey, along with a notation in a parochial registry that he was interred on the twenty-fourth of November, 1828. The long silence is all the more remarkable given Imlay's energy, talent, apparent personal charm, and his penchant for activities fostering notoriety.

It was Filson's kind fate to be remembered as the man who invented Daniel Boone, while Imlay is mostly thought of as the man who broke the heart of Mary Wollstonecraft. Yet in terms of American literature, during a very critical, because early and formative, period, Imlay should be known for much more. If he did not invent an equivalent to Daniel Boone, his writings, because of the influence of Brissot and Imlay's other radical associates in Europe, contribute the necessary philosophical backdrop to Filson's Rousseauvean version of wilderness heroics. For Imlay draws a sharp moral line between the Old and the New World, and between the eastern and the western parts of North America, and thereby provides a line of ideological continuity by means of which Filson's Boone becomes Cooper's Leatherstocking.

With Manasseh Cutler, Imlay regarded the option to start anew provided by the Ohio Valley as a singular opportunity, but his view of the western territory had little of the Old Testament overlay inherited from Puritan pulpit rhetoric. Instead, he imposed a scheme of values derived from the Physiocrats, a radical pastoralism quite in sympathy with Jefferson's agrarianism and perhaps in debt to Crèvecoeur's *Letters from an American Farmer* (1782). Like Crèvecoeur, Imlay, in his *Topographical Description*, used the epistolary mode and shaped his own identity to produce a persona, one calculated to convey the impression he was an ingenuous, honest reporter of things as they were in the Ohio Valley. He is introduced by the "editor" as "a man who had lived until he was more than five-and-twenty years old, in the back parts of America," and who was "accustomed to that simplicity of manners natural to a people in a state of innocence" (vi). This "Imlay" had since traveled to Europe, where he was "powerfully stricken with the very great difference between the simplicity of [American manners] and what is called *etiquette* and good breeding" in the Old World. This is the "contrast" that had recently been celebrated by Royall Tyler in his play of that name, and Imlay's is one of the earliest statements of what will become a dominant American theme in years to come, running a dividing line between the kind of natural nobility associated with types like Daniel Boone or George Washington and the corrupt products of European aristocratic society.

And yet, as in so much that Imlay associated himself with, his pose of the natural innocent was equivocal: he was born in eastern New Jersey, hardly the "back parts of America," and despite the claim of the editor, he was not writing letters from Kentucky in the early 1790s. Imlay did not develop this fictive element to any great degree, but used it chiefly to license a contrast between "the simple manners and rational life of the Americans, in these back settlements," and "the distorted and unnatural habits of the Europeans," which Imlay attributed to "bad laws" and "that pernicious system of blending religion with politics, which has been productive of universal depravity" (1). The picture of America he wished to promote was of a nation born, like Venus, on the first wave of the Enlightenment: "Happily for mankind, when the american empire was forming, philosophy pervaded the genius of Europe, and the radiance of her features moulded the minds of men into a more rational order: (1). Happily, also, the American continent was a fair field for agriculture, which for Imlay was the inevitable harbinger of "Arts and Sciences" (59).

As a physiocratic treatise, Imlay's book is a logical sequel to Brissot de Warville's *New Travels in America* (1791), another epistolary tract, whose penultimate letter states the author's wish that he had the time "to describe those new western territories that settlers enthusiastically call the Empire of the West" (413). Brissot's constant refrain that "Liberty can accomplish everything" might serve as the motto for Imlay's book: "O Liberty!" he exults, when considering the remarkable acceleration of the progress of invention associated with the discovery and settlement of the Western world, "how many blessings hast thou brought us!" (70). Imlay is also intoxicated with the potential for growth implicit in national newness: "In contemplating the vast field of the american empire, what a stupendous subject does it afford for speculation! Government, ethics, and commerce, acting upon principles different in many respects from those of the old world, and entirely in others!" (*TD*: 43). And he likewise delights in the intimate connection between American government and "the natural and imprescriptible rights of man," particularly the guarantee of "security of person and property, which is called freedom. Without such a preservation there can be no pure liberty" (215).

But the most pointed truths Imlay derives from Brissot have to do with the difference between America and Europe: "We have more of simplicity, and you more of art.—We have more of nature, and you more of the world. Nature formed our features and intellects very much alike; but while you have metamorphosed the one, and contaminated the other, we preserve the natural symbols of both. You have more hypocrisy—we are sincere" (179). Where Manasseh Cutler saw America as the last and best hope of Christianized mankind, the final stage in an ever-progressing series of civilized advance, Imlay established a dichotomy, insisting on discontinuity with the Old World, precisely the point of departure where separatism becomes utopian polity. Likewise, where Cutler regards the Ohio Valley as yet another Puritan asylum, a place in which Protestants may perfect their institutions, for Imlay the West is a pastoral paradise identified chiefly with Kentucky:

> *Everything here assumes a dignity and splendor I have never seen in any other part of the world. You ascend a considerable distance from the shore of the Ohio, and when you would suppose you had arrived at the summit of a mountain, you find yourself upon an extensive level. Here an eternal verdure reigns, and the brilliant sun of lat. 39°, piercing through the azure heavens, produces, in this prolific soil, an early maturity which is truly astonishing. Flowers full and perfect, as if they had been cultivated by the hand of a florist, with all their captivating odours, and with all the variegated charms that colour and nature can produce, here, in the lap of elegance and beauty, decorate the smiling groves. Soft zephyrs gently breathe on sweets, and the inhaled air gives a voluptuous glow of health and vigour, that seems to ravish the intoxicated senses. The sweet songsters of the forests appear to feel the influence of this genial*

> *clime, and, in more soft and modulated tones, warble their tender notes in unison with love and nature. Everything here gives delight; and, in that mild effulgence which beams around us, we feel a glow of gratitude for that elevation our all-bountiful Creator has bestowed upon us. Far from being disgusted with man for his turpitude or depravity, we feel that dignity nature bestowed upon us at the creation; but which has been contaminated by the base alloy of meanness, the concomitant of european education; and what is more lamentable is, that it is the consequence of your very laws and governments.* (28)

Such a place, perfect in all its aspects, differs strategically from the Ohio region described by Cutler—which still must benefit from the shaping hand of man—and thereby validates Filson's alternative view.

Still, Imlay has his plan, for this beautiful plenitude is valueless in itself. Like Brissot, he is mindful of commercial considerations, to which "Letter V" is devoted. This is the literal and figurative center of Imlay's book, for it is here that he sketches out an ambitious imperial scheme, remarkable for its scope: "this vast extent of empire is only to be equalled for its sublimity by the objects of its aggrandizement" (66). Commercial aggrandizement at the sublime level is associated with the many and conveniently placed waterways that characterize the western regions: "So friendly has nature been to this country," Imlay notes early in his book, "that, though it is without seas, the rivers run in such directions that there is scarce any place in all the back parts of America where art may not reduce the land carriage to a very small distance" (25). He then goes on to describe the nature of internal navigation along eastern streams, the difficulties of which are "merely imaginary" (68).

Starting with the "flat-bottomed boats" that carry produce downstream, Imlay maintains that "these boats must be worked up [stream] with steam and sails," exhibiting a certainty concerning the future of steam navigation even more sanguine than Cutler's, undeterred by the news that Rumsey's "invention of carrying a boat against the stream by the influence of steam" has thus far been prevented from being brought "into use" (69). Such setbacks are not discouraging, for "there can be no doubt of the success of his scheme," which has been guaranteed by "a certificate signed by general Washington" (69–70). And should that certainty somehow fall victim to other circumstances, "should we still be obliged to row our boats against the stream, it is not only practicable but easy" (70).

Imlay is here addressing the navigation of the Mississippi, a river that, by his account, is as accommodating as the green fields of Kentucky: the meanders Filson wished to bisect, Imlay sees as a natural mechanism facilitating the movement of boats powered by sails. According to Imlay's account, the meanders "produce in every bend eddy water; which, with the advantage the wind affords, that blowing the greater part of the year from the south-west, and directly up the windings of the river, by reason of the vacancy between the banks and rising forests on either side, afford a channel for the current of the air, [which] is sufficient with sails, keeping as much as possible in the eddy water, to carry a boat 50 miles a day up the stream" (70–71). Imlay's Mississippi is a vast, benevolent machine, powered by the reciprocal forces of eddying water and wind: "I have not the smallest doubt of the eligibility of the navigation of the Mississippi, which is proved from the experiments that are daily making" (71).

The "eligibility" of upriver navigation on the Mississippi was vital to Imlay's master plan, for the value of the western regions was integral to a vast system of interconnected waterways that link the valleys of the Mississippi and the Ohio to the rest of the United States: passing up the Illinois River from the Mississippi, the traveler gained access to the Great Lakes, a "plan of intercourse" that carried him "through Lake Ontario to Wood creek," and from there, by a portage and another creek, "to Fort Edward upon the Mohawk river a branch of Hudson's river. There are several carrying-places between that and its junction with Hudson; but very little labour would remove them, and which I have no doubt but that the state of New York will be judicious enough to set early about" (72). Disinterested where regional considerations were concerned (perhaps because of his European sojourn), Imlay smiled also upon "the portage between the Ohio and Potowmac," which he predicted would "be about 20 miles when the obstructions in the Monongahela and Cheat rivers are removed, which will form the first object of the gentlemen of Virginia when they have completed the canal on the Potowmac" (72).

Combining the alternative New York and Virginia strategies into one coherent plan, Imlay extended his scheme westward by means of the Missouri: Admitting to "little knowledge" about that river, he estimated it was "navigable for 12 or 1500 miles above its mouth, without obstruction" (74). More important, "it is not unlikely that in settling the country towards its source, we shall find it is not remote from the sources of the streams running into the Pacific ocean, and that a communication may be opened between them with as much ease as between the Ohio and Potowmac, and also between the settlements on the Mississippi and California." Imlay's accommodating geography was current wisdom in his day, and was derived from Carver and Jefferson, whose works he frequently quoted; while it added an expansionist note to this topography of the western regions, Imlay's view of easy westward communications chiefly important as further testimony to his overall meliorism.

Thus if "rivers are extremely favourable to communication by water," so "communication" has both a "social" and a "commercial" connotation, each inextricable from the other: "The intercourse of men has added no inconsiderable lustre to the polish of manners; and, perhaps, commerce has tended more to civilize and embellish the human mind, in two centuries, than war and chivalry would have done in five" (75). Imlay's benevolent view of commerce may be traced to Montesquieu and is an essential article also in Brissot's radical economics, but when applied to the promise held out by the great waterways of North America, it lent a transfiguring power to the topography, a mystical "creation bursting from a chaos of heterogeneous matter, and exhibiting the shining tissue with which it abounds" (44).

Cutler's optimism concerning the future of the Ohio Valley drew considerable strength from the certainties promoted by the new national Constitution, but Imlay went much farther than his Yankee predecessor. He put forth a complex geopolitical design, in which the river systems in the West combined with the Constitution to produce a vast reciprocating mechanism that would guarantee happiness through commercial exchange:

> *The federal government regulating every thing commercial, must be productive of the greatest harmony, so that while we are likely to live in the regions of perpetual peace, our felicity will receive a zest from the activity and variety of our trade. We shall pass through the Mississippi to the sea—up the Ohio, Monongahela and Cheat rivers, by a small portage, into the Potowmac, which will bring us to the federal city on the line of Virginia and Maryland—through the several rivers I have mentioned, and the lakes, to New York and Quebec—from the northern lakes to the head branches of the rivers which run into Hudson's bay into the arctic regions—and from the sources of the Missouri into the Great South sea. This in the centre of the earth, governing by the laws of reason and humanity, we seem calculated to become at once the emporium and protectors of the world.* (75–76)

By 1792, the District of Columbia had been established, but Imlay continued to promulgate Cutler's belief that the site of the "federal city" should not be permanently fixed until it could be removed westward, not to the Ohio but to the Mississippi River, "the most central, and consequently the proper place. By that means the efficiency of the federal government will act like the vital fluid which is propelled from the heart, and give motion and energy to every extremity of the empire" (76–77). Empire is the key word, as in Morse's *Geography*, requiring territorial expansion west of the Mississippi for that to become the central zone.

Imlay's espousal of an expansionist, neoclassical republicanism is clearly Virginian in spirit, albeit with several distinct qualifications. Though he was indebted to Jefferson's *Notes on Virginia,* Imlay's opinion of Jefferson himself is at best divided: he can characterize the Virginian as "enlightened and benevolent" but then dismisses his "political" judgment as "superficial . . . his mind attached to the theory of his own fabrication" (222, 81). And he devotes several pages to a sneering attack on Jefferson's weak defense of the abominable institution of slavery. Yet Imlay's physiocratic utopianism is closely allied to Jefferson's agrarianism, and contains elements likewise imported from the tidewater region. Where Cutler puts great stress on the importance of formal education to the process of settling Ohio, Imlay entrusts the civilizing process in Kentucky to a "genteel" class of men, mainly officer veterans of the Revolution with their families, who "chequered" the country with zones of influence "which operated both upon the minds and actions of the back woods people, who constituted the first emigrants" (168).

Despite his debt to Filson, Imlay has no high regard for the Daniel Boone type, but shares the dominant view, as expressed by other contemporary commentators on the frontier process like Crèvecoeur and Benjamin Rush, that the chief value of the hunting class of men is utilitarian: they lead the way in "the settling [of] a wild and infant country" (175). Having no particular virtue in themselves, they best accommodate civilization by moving on to the next wild zone, leaving the land ready for those American Cincinnati whose effect on the emerging frontier society is not unlike that of a finishing school:

> *A taste for the decorum and elegance of the table was soon cultivated; the pleasures of gardening were considered not only as useful but amusing. These improvements in the comforts of living and manners, awakened a sense of ambition to instruct their youth in useful and accomplished arts. Social pleasures were regarded as the most inestimable of human possessions—the genius of friendship appeared to foster the emanations of virtue; while the cordial regard, and sincere desire of pleasing, produced the most harmonizing effects. Sympathy was regarded as the essence of the human soul, participating of celestial*

> *matter, and as a spark engendered to warm our benevolence, and lead to the raptures of love and rational felicity.* (168)

Animated by the power of "sympathy," Imlay's scheme for civilizing the West owes as much to Laurence Sterne as to Adam Smith, sympathy being that mollifying emotion that is to "sentiment" closely allied.

By this natural, even organic, pattern of benign influence, Kentucky had become an Arcadian zone of pastoral bliss. Where Manasseh Cutler worried about the negative effects of abundant ease, Imlay maintained that the very power of "uncontaminated felicity" would guarantee the destruction of "sordid avarice and vicious habits" (169). Useful pastimes like gardening and fishing are "rational pleasures" that "meliorate the soul," and since "our young men are too gallant to permit the women to have separate amusements . . . we find that suavity and politeness of manners universal, which can only be affected by feminine polish" (169–70). A major mechanism of civilization is "evening visits," which "mostly end with dancing by the young people, while the more aged indulge their hilarity, or disseminate information in the disquisition of politics, or some useful art or science" (170). In passages like these, Imlay carried physiocratic idealism to a ridiculous extreme, yet as he parted with Jefferson on the matter of slavery, so he failed to share the Virginian's admiration of the American Indian. In his last letter, Imlay turned from describing the Kentucky paradise to recommending the proper way of dealing with the aboriginal inhabitants of the Ohio, and blamed the defeat of the forces under General St. Clair (governor of the territory settled by Cutler's Ohio Company) on a lack of experience in fighting Indians.

Like his mentor, Brissot, Imlay seemed to feel that the Indian's best hope was to "become civilized and be assimilated by the Americans," otherwise "a thousand causes will bring about their annihilation." But, come assimilation or annihilation, "there is . . . no need to fear that the danger of the Indians will check the drive of the Americans in their mass progress" (Warville: 421). In Imlay's words: "Certainly it is time that decided measures were taken; if possible, to civilize them; and if not, to confine them to particular districts; that is, by the vigour of our measures, to show them that we are not to be trifled with" (295). For whatever reason, the Indian cannot be accommodated to the symmetry of articulated benevolence, and falls beyond the pale of transforming "sympathy." Like his white counterpart, the frontier hunter, he will survive only so long "as there is left a wilderness in America" (175).

In summation, both Manasseh Cutler and Gilbert Imlay view the region west of the Alleghanies as an ideal territory in which to begin the world anew; but where Cutler sees the western valley as yet another sanctuary in which to prepare for the millennium, Imlay's plan is informed by the ideological forces that were released by the French Revolution. Both men reflect Enlightenment ideals, including a faith in material progress, and both view the institution of slavery as an unmitigated evil. But Imlay's picture of Kentucky is a physiocratic Utopia with an aristocratic bias, a plan that foreshadows the planter hegemony of a later period, while Cutler sees Ohio as a new and improved version of New England. And yet both writers argue that the nation's capital should be eventually located in the West, a centralism informed by the neoclassical aesthetic so essential to even the most radical of republican vistas in the closing years of the eighteenth century, a neoclassicism, moreover, with an expansionist élan.

Of the two, Imlay conveys a much more forceful sense of the West's uniqueness, conveyed in imagery suggesting an intensely pastoral zone. Cutler preaches the old millenialism in a new location; Imlay puts forth the promise of a new, distinct nation within a nation, one that imbibes virtue from the landscape itself. Cutler's views were easily assimilated to Jedidiah Morse's federalist plan, but Imlay's much more radical prospect was never incorporated into the New England version of American geography, and since that was the version that held, Imlay's book, like the author himself, simply disappeared. In the place of his intended agrarian Utopia there evolved a slave-supported planter society that was doomed from the start.

Imlay's projected vision for the western valley was never much more than a creation of rhetorical hyperbole, inspired by the heady environment created by the early days of the French Revolution, and it bore no real relationship to the emerging society of Kentucky. And yet, much as it approximates poetry, so Imlay's "plan" of accommodating rivers and fertile fields, a vast, reticulated, syncopated system made possible by the lay of the land, puts forward an ideal very much in harmony with the hope sustained by the Constitution, an ideal made real—if only in Imlay's imagination—by the orderly system of waterways that provided the "bright tissue" of a shining new land. In the end, however, the bright rills fed dark waters, for the twin gateways to the valley, the Hudson in the North and the Potomac in the East, became rivals for the rich commerce of the West, and the victory of the Hudson in that regard not only broke the delicate balance described by Imlay, but ensured that the Ohio River would be the watery boundary between two increasingly different and eventually antagonistic ways of life. A half-century of ideological conflict enforced the disjunction inherent in the confused notion of a separatist utopia joined to a grand continental scheme of union by a reticulated system of waterways. In time the idea of separatism hardened into secession south of the Ohio River, while union as an ideal was fostered chiefly in the North. Conceived of as a bright chain of unity, the Ohio would become a fiery line of division, as the expansionism inherent in imperial plans had unforeseen consequences. The center, finally, would not hold.

IV

Flashing Eyes and Floating Hair

i

As we move from Washington's embassy to the French in 1753, to his journey thirty years later in search of a practicable passage through the Allegheny Mountains; from Lewis Evans' map and essays of 1755, to Thomas Hutchins' of 1778; when we consider the contingencies of the documents gathered by Jedidiah Morse in 1789 and Gilbert Imlay in 1792—we become conscious of a consistency in geopolitical purpose, often promoted by multiple indebtedness, which considerably dwarfs the significance of the War for Independence. We can see that the womb of American imperialism was the French and Indian War, which generated terrific expansionist energies, and that the succeeding conflict, ironically fueled by Great Britain's attempt to defray the cost of the earlier struggle by means of taxation, was but a cutting of the umbilical cord.

We may be sure that thousands of Americans went about their daily concerns indifferent to, and even ignorant of, the cosmic events that engendered emerging epic figures like Washington. The historic record reveals that during the French and Indian War considerable numbers of merchants continued to conduct business with the enemy; and Benjamin Franklin attested to the difficulties he had rounding up horses and supplies to support the war effort. Even the anger aroused by the Townshend Acts suggests that many Americans were reluctant to support the cost of a war that had seldom touched their everyday lives. By contrast, the War for Independence both polarized and galvanized the nascent nationalism detectable in the propaganda engendered by the earlier conflict. It is a coincidence of large proportions that George Washington rapidly emerged as the commanding general of the Continental forces, his prominence warranted by the symbolic stature granted by his participation in the earlier conflict. We need not believe in destiny to collect the evidence of its manifestations.

Moreover, we have the correspondence of the Fontaine family to attest that among the general populace there were those who were quick to catch visionary fire from Evans' map and essays, that Washington's western expeditions were mirrored in a general movement in that direction; nor were such evidences of sympathy limited to Virginia. Dr. Nathanael Ames of Dedham, Massachusetts, the almanac maker who was one of the first Americans to celebrate Washington's exploits in verse, two years later seems to have had his imagination caught by Lewis Evans also. He wrote a prose rhapsody for his almanac in praise of the regions beyond the Alleghenies, which he identified with the "future State of North America," a nationalistic conceit coupled with the promise that it would become "The Garden of the World!" (284–86).

In 1765 the editorship of Dr. Ames' almanac was taken over by his son (also a doctor, also Nathanael), if anything a more ardent versifier than his father, and where Ames Senior had devoted his poetic energies in 1756 to a twelve-part sequence promoting the British cause at the outbreak of the French and Indian War, Ames Junior in 1769 wrote twelve monthly stanzas protesting the Townshend Acts. Looking to the Puritan past for sanction and strength, the "Forefathers" who had brought to New England "true religion, liberty and laws," and who had thrived on adversity and a meager diet of corn and clams, Ames Junior ended by looking into the future, standing on tiptoe to glimpse "Ontario's Shore" and "Art's fair empire rising" there. Where his father had envisioned cities springing up on "smooth OHIO's winding stream," Ames the Younger now populated them with a "future LOCKE" and a "second NEWTON," and likewise foresaw

> *Some MILTON plan his bold impassion'd theme,*
> *Stretch'd in the banks of Oxallana's stream,*
> *Another SHAKESPEAR shall Ohio claim,*
> *And boast its floods allied to Avon's fame.* (402)

Clearly, for the Ameses at least, if the emerging nation had a future, it lay beyond the mountains to the west.

The echoing themes in the writings of father and son attest yet again to the importance of the French and Indian War in producing what would become distinctly patriotic themes in the years immediately preceding the Revolution. It is a connection the younger Ames certified in his last poetic effort before the war necessitated his closing down the almanac in 1775, rallying the veterans of the earlier conflict to take up arms against the former protector nation, now regarded as an oppressor: "Stand forth the champions of your country's cause . . ." Ames cried, urging the seasoned soldiers to "Make ready then—and fierce begin the fray!" (457–58). Like the participants in the Boston Tea Party, Ames' veterans have a decidedly Indian look—"To scout and skulk to gain the scalp you've run"—that being a vital aspect of the emerging American identity, as subsequent, post-Revolutionary literature will show: but for our present purpose it is further proof of the transforming power of the Seven Years' War.

Moreover, since almanacs were the closest thing colonial Americans had to a popular literature, the contents of those produced by the Ameses over a half-century, from 1726 to 1775, suggest that the geopolitical schemes of Washington, Jefferson, and Franklin (himself an almanac writer of considerable impact) had a certain resonance amongst the general populace. Almanacs have been called "colonial weekday bibles" because of their popularity, and they were decidedly secular scriptures, whose prognostications of weather naturally inclined their compilers to prophecy, as did their function in the eighteenth century of passing on to readers the latest technological improvements, from formulas for home medicines to fertilizers.

Whether describing lightning rods or the workings of orreries, almanacs were pervaded with the influence of the Enlightenment, and such poetry as the Ameses added to the mix (Franklin preferred to borrow from the British muse) gave a literary dimension no less progressive. Tending toward visionary or hortatory modes (projecting a future, then helping Americans along toward it), the Ameses of Massachusetts provided a public complement to the private correspondence of the Fontaines in Virginia. Equally important, their mingling of agricultural technology, martial heroics, and geopolitical expansion—the whole irradiated with a sense of American progress firmly fixed to an idealized Puritan past—provided a forecast of much more ambitious poetry to come, which would likewise trace the future course of America's glory by means of shining rivers winding through the great valley in the West.

ii

Still more evidence of the visionary character of American life engendered by the French and Indian War may be found in John Adams' diary. Though Adams himself seems to have resisted the general afflatus, in 1763 he recorded the dream of his friend and neighbor, Deacon Benjamin Prat, who was preparing to leave Boston to assume a judgeship in the New York Superior Court. "He was seated on a Rock," wrote Adams, "in the Middle of the Sea, and reflecting on his Journey to N. York, leaving his family &c., when the Clouds began to rise from all Quarters of the Horison, and soon thickened and blackened over his Head. The Thunders began to roar And the Lightnings to flash. At last the Clouds opened and a glorious Luminary, in the shape of an Angel, made its Appearance and addressed Mr. Prat in these Lines: 'Why mourns the Bard? Apollo bids thee rise, / Renounce the Dust, and Claim [thy] native skies' "(*D&A*: I: 241). Prat died that same year, lending a degree of ambiguity to his dream, yet it would prove most prophetic where belles lettres in Revolutionary America were concerned. Adams may have preferred the view from his perch on Forefathers Rock, but many young poets of his generation shared the dreams of Deacon Prat, conjuring up sufficient glorious luminaries to bid a whole nation to rise and claim its own.

A mixture of Christian and pagan (neoclassical) imagery, Prat's remarkable dream dramatically demonstrates not only how obligatory heroic couplets were to poetic efforts of the day, but the extent to which those couplets were bent to visionary exercises. " 'Tis POPE, my Friend," wrote the Boston poet Mather Byles to an aspiring versifier,

> *that guilds our gloomy Night,*
> *And if I shine 'tis his reflected Light:*
> *So the pale Moon, bright with her borrow'd Beams,*
> *Thro' the dark Horrors shoots her silver Gleams.*
> (*Poems:* 54)

Written in 1744, Byles' advice demonstrates the colonial dependency of the muse in eighteenth-century New England, but the "glorious luminaries" heralded by American poets after 1763 tended increasingly to be associated not with light shed from the Old World but with the dazzling prospects irradiating from the New.

Alexander Pope continued to invite emulation, if only because the proto-nationalistic mood in America in the wake of the French and Indian War resembled the outburst of patriotism in England a half-century earlier, following the Treaty of Utrecht. The British anticipated a time of peace and prosperity, a euphoric mood that is reflected in Pope's *Windsor Forest,* a celebration of Queen Anne's reign that would serve as a model for a number of ambitious colonial efforts warranted by the Treaty of Paris in 1763. But if Pope provided the poetic mode and model, it was the Puritan sermon that helped sustain the visionary tradition in America, for by the middle of the eighteenth century, prophecy had become second nature to the natives of New England. Nor can we discount the sermon's influence on the secular exhortations of the Ameses. True, much Puritan gazing into the future tended to prefer the kind of "dark glass" produced by the smoke of Revelations, projecting the apocalyptic prognostications associated with the jeremiad. But the middle years of the eighteenth century saw a revival of the old millennial optimism that had characterized the early years of the Puritan experiment, in effect celebrating the centennial of Edgar Johnson's *Wonder-working Providence* by reviving in such as John Adams its evocation of a God-blessed "nation" built in a day.

This is not to imply that New England had a monopoly on prophecy, any more than Virginians alone appreciated the value of western lands. In 1771, for example, during the commencement exercises at Princeton and Yale, there was a coincidental eruption of aesthetic influenza, evidenced by two poems in the visionary mode written by graduating seniors from both the middle Atlantic and New England colonies. The most ambitious effusion was the one produced at Princeton, a consensus effort written by Philip Freneau from New Jersey and Hugh Henry Brackenridge of Pennsylvania, *The Rising Glory of America,* an exercise in spread-eagle flight given added wingspan by the recent repeal of the Townshend Acts. Though modeled on *Windsor Forest,* the Princeton poem is no mere imitation, the poets having thrown off the colonial chains of heroic couplets for the freer form of blank verse. They also took advantage of the public occasion and dual authorship to mount the poem as a dramatic dialogue.

Moreover, where Pope's is a topographical poem in the tradition of Denham's *Cooper's Hill,* the Princetonians took in a much larger view, the then-known bounds of Anglo-America as seen from New Jersey. In this particular, Freneau and Brackenridge many have been making a patriotic virtue out of a colonial necessity, for the traditional topographical poem celebrates the kind of well-known spot that is rich in historical associations and—as later Romantic writers would lament—such places were harder to find in North America than fairies in Secaucus. But the scope of the vista in *Rising Glory* most certainly expresses an expansionist élan, linked to anticipated, not historic, accomplishments; and where the most of Pope's poem bends under the burden of the past, with only a final and fleeting peep into the future, Freneau and Brackenridge devote most of their efforts to prophecy. They in effect begin where Pope ends, except that where for Pope the Thames, in entering the ocean, will bring Britain's "glory" to "Earth's distant Ends," for Freneau and Brackenridge "Glory" is preponderantly an American production.

At the very start, the poets dismiss the subject matter of Old World poetry ("No more of Memphis . . . no more of Rome . . . no more of Britain," etc.) and announce "A Theme more new . . . the rising glory of this western world . . . where freedom holds her sacred standard high, / And commerce rolls her golden tides profuse" (Freneau: 50–51). The American "difference" is immediately established by reviewing the events of a history that is (in contradistinction to Pope's bloody chronicle in *Windsor Forest*) a steady forward march of progress. Starting out with Columbus, Cabot, and Raleigh, the poets digress with a lengthy and unflattering portrayal of the American Indian—whose "listless slumber and inglorious ease" help point up the superiority of the white man—then move on to rehearse the "glorious cause" of the Forefathers. These, however, are not the Puritans of New Plymouth or Boston but the Quakers of Philadelphia, and other shifts of emphasis follow. Like the Ameses, the Princeton poets describe the French and Indian War, celebrating those who fought bravely and fell "By Monongahela and the Ohio's stream," not only Braddock but the lesser known "valiant Halkut" (who like Brackenridge was a Scot) and Sir William Johnson, the old Indian fighter from the Mohawk Valley (Freneau's family was from New York).

Still, the emphasis is not on the past but on the future; not on war but on the prospect of peace, and the poets' particularism has a generous perspective, as they look beyond New York and Philadelphia to "unnumbered towns and villages" yet to be founded, "embrio marts of trade . . . by commerce nurs'd . . . / On lake and bay and navigable stream, / From Cape Breton to Pensacola south" (69). Commerce as a nurse of embryonic towns is made possible by improvements in navigation, which in turn are indebted to science, another of the arts that have emigrated from the Old World to the New: "Ev'n now we boast / A Franklin skill'd in deep philosophy, / A genius piercing as th' electric fire" (71). The evocation of America's most famous contribution to the Enlightenment gives way to a twofold use of glorious imagery with a prophetic implication, the "noble light / Of golden revelation," and the even more intense gleam that illuminates "the mystic scenes of dark futurity" (71–72).

Snatching a "bright coal" from the altar of "seraphic fire" that inspired the biblical prophets, the Princeton poets envision "a thousand kingdoms raised" in America, and

> *. . . where the Mississippi stream*
> *By forests shaded now runs weeping on,*
> *Nations shall grow and states not less in fame*
> *Than Greece and Rome of old.* (74)

The rest of the poem, nearly a third of its total length, is devoted to an extended prophetic look beyond "the past and present glory of this empire wide," piercing "the mysteries of future days" (73).

As in the Ameses' almanacs, the landscape of the future is in large part defined by waterways and figured by the growth of cities: "Fair domes on each long bay, sea, shore or stream." In the North a new Petersburg rises, while in the West a new Palmyra, past which "the slow pac'd caravan return[s] . . . from the Pacific shore," and in the South "a Babylon, / As once by Tigris or Euphrates stream":

> *. . . And thou Patowmack, navigable stream,*
> *Shall vie with Thames, the Tiber or the Rhine,*
> *For on thy banks I see an hundred towns*
> *And the tall vessels wafted down thy tide.* (77)

Even "hoarse Niagara" will be diverted into "a better course," or else, her cataracts cleared away, "shall flow beneath / Unnumbered boats and merchandise and men" (77–78).

Along these streams "a glorious train" of patriots will arise, heroes whose deeds will be celebrated by an American Homer and Milton, and even

> *a second Pope . . . may yet*
> *Awake the muse by Schuylkill's silent stream,*
> *And bid new forests bloom along her tide.*
> *And Susquehanna's rocky stream unsung,*
> *In bright meanders winding round the hills . . .*
> *Shall yet remurmur to the magic sound*
> *Of song heroic. . . .* (79)

Another Denham will likewise be found to "celebrate" the Roanoke and the James in song, a rising chorus of "sweet music" that celebrates "the final destiny of things, / The great result of all our labours here, / The last day's glory, and the world renew'd" (79–80). This water music is a veritable *Messiah* heralding a millennial age, and the poets foresee a new "Canaan" far excelling the old, a landscape through which

> *Another Jordan's stream shall glide along*
> *And Shiloh's brook in circling eddies flow,*
> *Groves shall adorn their verdant banks, on which*

The happy people free from second death
Shall find secure repose.

Translated to the middle colonies, this is the old Puritan scheme, lending the rising empire its glory from a transcendent source and converting American rivers into sacred streams.

As an undergraduate exercise, *The Rising Glory* is quite an accomplishment: observing the deference due the British Crown (and with a respectful nod to Pope), the poets nevertheless go on to stress the potential of the American continent as a nursery of new "empires," and like such propagandists for expanding British empire as Franklin, Pownall, and Lewis Evans, they put forth an incipient nationalism. Much as the productions of the Ameses lead inexorably to a poetic declaration of independence, so Freneau's later revision of his own (and major) contribution to *The Rising Glory* demonstrates the ease with which the pre-Revolutionary poem could be converted to serve a subsequent republican cause. The same holds true for Yale's contribution to the spirit of 1771, Timothy Dwight's *America . . . A Poem on the Settlement of the British Colonies* (1772). Explicitly indebted to *Windsor Forest,* Dwight, like Freneau and Brackenridge, portrays the flow of empire and progress as a westward movement, and his poem ends with a "vision" of the future, which is conveyed by a glorious figure called "FREEDOM," who prophesies that America, "land of light and joy," shall grow in power, extending its "glory" throughout "earth's wide realms" (*Poems:* 11). But Dwight, true to his Yale education (and his intended ministry), tends to stick closely to the traditional Puritan vision of the future: where for the Princeton poets the millennium is rather general in outline and largely secular in implication, Dwight emphasizes the divine nature of American destiny, and sees Enlightenment improvements as only a foreshadowing of the glorious advent of Christ's return.

The conventional Congregational prospect of millennium is a major feature of Dwight's subsequent visionary exercises: a dogged loyalty that at times casts a chiliastic shadow over an unlikely landscape. In his most ambitious poem, the epical *Conquest of Canaan* (1785), Dwight inserted a lengthy episode into the biblical story of Joshua, comforted on the eve of his Jericho battle by an angelic visitor, who treats the hero to a sight not only of the Canaan he is soon to repossess on the east bank of the Jordan, but of a "Prospect of America" and a vision of "The Glory of the Western Millennium." This is typology with a twist, for not only is the Hebraic hero inspired to do battle by a prophetic glimpse of the messiah his people have yet to accept, but the glory of his forthcoming victory over the Canaanites is considerably dimmed by the western gleam, where "Empire's last, and brightest throne shall rise; / And Peace, and Right, and Freedom, greet the skies" (*Poems*: 275).

Dwight's anomalous angel provides a link of continuity with not only the Bible and *Paradise Lost* but Deacon Prat's dream, and would see considerable service in the works of Joel Barlow, the post-Revolutionary Yankee bard for whom the vision poem was almost his sole stock in trade. It was Barlow's fate (in several senses of that word) to mount, remount, and then mount again a developmental sequence of "visions" that blended geopolitical expansionism with the progressive thrust of the Enlightenment spirit but resulted in a poem monumental not only in scope but poundage. Thanks to the sheer bulk of Barlow's *Columbiad,* the vision poem emerged as a dominant genre of the post-Revolutionary period, but thanks to the sheer bulk of the *Columbiad,* it disappeared for a generation, to reemerge as Whitman's *Leaves of Grass,* which resembles the *Columbiad* in genesis, optimism, and size only.

But the *Columbiad,* like the continent that the poem's eponymous hero stumbled across, cannot be ignored, and the progressive process by which it grew is a veritable map of Enlightenment attitudes in America. In marked contrast to his academic tutor (and poetic mentor) Dwight, who increasingly espoused the parochial geopolitics that were typical of the Federalist mind during its last, most conservative phase, Joel Barlow moved steadily into ever more radical, even Jacobin postures; yet, like Dwight, he stuck to an increasingly outmoded and inadequate poetics. It would be difficult to find a better example of the perverse crosscurrents that characterized American literature during the early years of the republic than Barlow's poem, especially since his creative range seems so strangely narrow when compared with his intellectual versatility. If the emerging American muse can be detected in embryo in the Ameses' almanacs, then it is in the anthology-like *Columbiad* that the muse takes definitive shape, huge yet somehow infantile, like an exceptional child of Herculean proportions.

Born in 1754, on the eve of the French and Indian War, Barlow died in 1812, at the start of the second war between the United States and Great Britain, a conflict that ensured the territorial sanctity of the new nation. His life therefore was bracketed by wars that defined the course of the emerging American republic, and Barlow himself was a veritable personification of the ebullient, optimistic, idealistic mood that characterized the first twenty years of the United States. Houdon included Barlow's bust in his pantheon of republican heroes, and one need only place him in the company of Washington, Franklin, and Jefferson to detect his essential character: where Washington has a Caesarian profile, Jefferson a Voltairean quizzicality, and Franklin a tranquil air of cosmopolite wisdom, Barlow puts forth a quintessential American innocence supported by a sizeable jaw. An American Farmer truly, Barlow rose from plodding after the plow on the family acres in Connecticut to roving about the world as an ambassador with an ever-expanding portfolio (and widening wallet); and as he rose, his muse expanded her viewpoint while retaining her American visa. Barlow's poetry provides a chart of his political drift from enlightened Federalism to an expansive, even revolutionary republicanism, and his poems likewise reflect his abandonment of a conventional, Congregational chiliasm for a deistic celebration of the Enlightenment spirit of progress.

Awful as art, Barlow's *Columbiad* is a fascinating artifact: a mixture of neoclassical design and eclectic parts, like the expansionist geographies of Morse and Imlay it gathers an anthology of themes and tropes employed by earlier writers—in Barlow's case, his fellow poets of the republic. As a Big Book, it is a pivotal work, converting the millenarian chronicles of Edward Johnson and Cotton Mather to more secular business, instilling the experiment that was the United States with a manifest destiny, a transcendent, even a mystical, sense of unity that soon enough developed a necessity of its own: a mingling of territorial expansion by means of rivers and a technologically articulated union by means of canals that dictated, far beyond Barlow's intention, the geopolitics of later generations.

Joel Barlow's career as a professional visionary began, like Dwight's, at Yale, and his commencement poem—truly so-called—contains the seed of much that was to follow. *The Prospect of Peace,* composed and delivered in 1778, is in many ways a counterpart to *The Rising Glory,* for where Freneau and Brackenridge reflect a mood of nascent nationalism, Barlow shows the spirit of true independence. Drawing strength from the recent intervention on the side of the American forces by the French, the poet portrays a glorious future for the newborn republic:

> *At this calm period, see, in pleasing view,*
> *Art vies with Art, and Nature smiles anew:*
> *On the long, winding strand that meets the tide,*
> *Unnumber'd cities lift their spiry pride . . .*

There commerce swells from each remotest shore,
And wafts in plenty to the smiling store. (*Works*: II: 6)

Barlow's poem does not differ much from the visionary efforts that preceded it, and along with the Princeton poets and Dwight, he foresees that America will be chosen as the site for the millennium.

This derivative vein was continued in his next visionary exercise, composed in 1781 and bleakly entitled *A Poem, Spoken at Yale College*. Differing from his earlier effort only slightly, Barlow's second vision borrowed from Dwight the device of a seraphic visitor to serve as a visionary medium, here "Learning": prophesying peace, Learning consols gloomy Yalensians with a glorious prospect of America's future. As if from the top of West Rock in New Haven, Learning opens a wide geopolitical vista, for, turning his back on New England, he points to "Lands yet unknown and streams without a name," from the Gulf of Mexico, to Lake Ontario, and on to "Where Mississippi's waves their sources boast, / Where groves and floods and realms and climes are lost" (33).

This use of a prophetic intermediary, though imitative, permits a theatrical conceit that will endure in Barlow's subsequent poetry, for what the Genius of Learning unfolds for the instruction of Yale undergraduates is a grand entertainment:

On this broad theatre, unbounded spread,
In different scenes, what countless throngs must tread,
Soon as the new form'd empire, rising fair,
Calms her brave sons now breathing from the war. (33)

This panoramic trope becomes the controlling device in Barlow's next attempt at the vision poem, the much more ambitious *Vision of Columbus,* first published in 1787. Instead of the Yale student body it is now Columbus who is the audience, having been reduced to a passive spectator by his imprisonment at Valladolid. There he is visited by a heavenly spirit, who in the manner of Dwight's angel with Joshua, consoles the admiral with a vision of the consequences of his discoveries. Where Joshua stood on a mountain height, Columbus languished in jail, his "vision" brought to him as if by some divine predecessor of the television set. Thus both poems converge in a millennial prospect, signaled by the return of "the promised Prince" of Peace late in Book VIII of *The Vision of Columbus.*

Despite this Christian message, the penultimate Book IX of *The Vision* contains a detectable secular stress, demonstrating the extent to which the millenarian bias inherited by Barlow had already been taken over by the Enlightenment spirit. The poem throughout is animated by the ideal of progress, and though much of the "vision" revealed to Columbus was, by 1787, recorded history, the implication of the unfolding events is clearly (because literally) progressive. Thus the poem presents a panorama of the initial settlement of America, the successive wars between Indians and Europeans, the gradual spread of civilization, the pivotal French and Indian War, then the outbreak of the Revolution and the triumph of liberty and George Washington, followed by the postwar spread of science and the arts. The final book is doubly visionary—both for Columbus and the reader—and here, perforce, the angelic visitor becomes as gloriously vague as any campaigning politician. As in *The Rising Glory,* what is posited is a future when, through the agency of commercial exchange, the world will have attained a millennial state of universal peace. Barlow's chief addition is a putative League of Nations that elevates the republican plan to worldwide representation.

Much larger in scope than even the Princeton poem, Barlow's *Vision* permits considerable detail not found in his own or other earlier efforts. In Dwight's *America* the poet promised that the spread of "light and joy" would be facilitated by man's improvements: inland, "broad APPIAN ways shall wind, / And distant shores by long canals be joined" (*Poems*: 12). In Barlow's *Vision* the role of canals is expanded until it exceeds Gilbert Imlay's national system of articulated waterways:

> *He saw, as widely spreads the unchannel'd plain,*
> *Where inland realms for ages bloom'd in vain,*
> *Canals, long-winding, ope a watery flight,*
> *And distant streams and seas and lakes unite,*
> *From fair Albania* [Albany], *tow'rd the falling sun,*
> *Back thro' the midland, lengthening channels run,*
> *Meet the far lakes, their beauteous towns that lave,*
> *And Hudson join to broad Ohio's wave.*
> *From dim Superior, whose unfathom'd sea*
> *Drinks the mild splendors of the setting day,*
> *New paths, unfolding, lead their watery pride,*
> *And towns and empires rise along their side;*
> *To Mississippi's source the passes bend,*
> *And to the broad Pacific main extend.* (*Works*: II: 346)

Envisioning not only the Erie but the Panama and Suez canals, Barlow posits a grandly cosmopolitan geopolitics: if it is commerce that will bring about world peace, insuring that "mutual interest fix[es] the mutual friend," then it is technology that will make the way straight for universal democracy (244).

Nature in America is, if anything, more friendly in Barlow's *Vision* than in Imlay's *Topography:* Present at the creation, Columbus sees obliging "mountains ope their watery store" so that

> *Floods leave their caves and seek the distant shore,*
> *Down the long hills and through the subject plain,*
> *Roll the delightful currents to the main;*
> *Whose numerous channels cleave the lengthening strand,*
> *And heave their banks where future towns must stand.* (135)

The Chesapeake Bay is put forth as a veritable personification of accommodating geography, the capes of Virginia figured as arms embracing the "calm bosom" of the bay,

> *Where commerce since has wing'd her channel'd flight*
> *Each spreading stream . . . brightening to the light,*
> *York led his wave, imbank'd in mazy pride,*
> *And nobler James fell winding by his side;*
> *Back tow'rd the distant hills, through many a vale,*
> *Wild Rappahanock seem'd to lure the sail,*
> *While, far o'er all, in sea-like azure spread,*
> *The great Potowmac swept his lordly bed.* (140)

Barlow's fluvial catalogue brings in the major rivers of the eastern seaboard, and both the St. Lawrence and the Mississippi receive extended treatment as sublime spectacles. The Great Lakes appear as virtual avatars of the Enlightenment, displaying "their glittering glories to the beams of day, watery equivalents to Manco Capac, Sir Walter Raleigh, Ben Franklin, and—rivaling Louis XIV in this regard—George Washington, who, when he appears before the Continental Congress on the banks of "bright Del'ware's silver stream" is compared to the sun itself, rising "to gild our morning skies" and "pour the enlivening flame, / The charms of freedom and fire of fame," an absolute expression of the Enlightenment (263).

Barlow's activities subsequent to the publication of his *Vision* somewhat tarnished his own "effulgence": in 1788, having traveled about Connecticut with Manasseh Cutler's map and pamphlet as an agent for the Ohio Company, he departed for France in the service of the Scioto Associates. This was a speculative enterprise related to the Ohio

Company but rendered particularly dubious by its mastermind, William Duer, who seems to have had no real interest in anything other than quick profits, whereas Cutler and his partners were animated by a much more generous spirit. Barlow was apparently ignorant of Duer's shady dealings; indeed, his naïveté, along with his ebullient enthusiasm for hastening the settlement of emigré Frenchmen on the banks of the Ohio—both amply figured in a map he had published in France, showing towns where nothing but wilderness stood—helped hasten Duer's undoing. His innate innocence likewise enabled Barlow to step blithely clear of the wreckage of the Scioto scheme and next enlist himself and his pen in the service of political liberty, both in France and in England. These labors (largely in prose) earned him great popularity in Paris—like his friend Tom Paine, he was made a citizen of France—and brought him into the Girondist circle of Brissot de Warville, whose *New Travels in the United States* he translated into English.

Barlow's enthusiasm for the French Revolution nevertheless retained a Yankee cast, and in editing Brissot's book he elided anything unfavorable to the United States (not an extensive labor), thereby increasing its potential as propaganda for settling western lands. If his own heavy investment in government securities attests to his faith in the French Revolution, it also made him wealthy, and his association with Brissot involved him in the scheme to establish an independent Louisiana, from which he necessarily would have profited also. On the other hand, Barlow was no mere opportunist like Gilbert Imlay, but seems rather to have been the well-deserving recipient of a great deal of good luck. Thus, unlike Tom Paine, he escaped the consequences of the Girondists' fall, and thenceforth abandoned politics for business affairs, his advocacy of natural rights, as always, being intimately connected with commerce. Like Imlay, with whom he occasionally worked promoting obscure schemes, Barlow left France at the start of the Reign of Terror and relocated to Hamburg—yet another testimony to his ability to survive a general calamity. Where Imlay soon after disappeared from view, Barlow's good fortune and fame continued: as American consul in Algiers he was successful in his efforts to free American prisoners and end pirates' depredations of American shipping, and he had the good luck also to return to France just as Talleyrand imposed order and Napoleon emerged as the French champion. There he remained from 1797 until 1805, engaged in diplomatic matters and (with his friend Robert Fulton) scientific research, translating that Enlightenment classic of comparative religion, Volney's *Ruins,* into English, and, most significantly, expanding his earlier *Vision* into the massive bulk of *The Columbiad,* which was published soon after his return to the United States in 1807.

Barlow's amazing record of survival and success, both in America and abroad, suggests that he, like America itself, was surrounded during the early years of the republic with a magical aura of innocence that protected him, a singular blessedness not unlike the special grace that distinguished the Saints of New England. Imlay, in so many ways Barlow's shadow, lacked that aura, and suffered the ignominy that is the most painful death for seekers of fame; yet for a time the two operated almost as one, for Barlow, like Imlay was radicalized by his European experience. During his first stay in Paris, when the Girondists were in the ascendancy, he made a number of strategic revisions to *The Vision of Columbus,* the most important of which was almost totally expunging the Second Coming. These revisions left the millennarian element intact but entirely secularized, a change that necessarily emphasized the Enlightenment machinery. Published (in Paris) in 1793, the thoroughly revised *Vision* represented a major redirection of Barlow's energies, anticipating by more than ten years the poet's most ambitious visionary effort. In effect, there is little in *The Columbiad* that cannot be found, in terms of theme, in the 1793 *Vision;* the rest is mostly elaboration and expansion—sometimes unkindly called "bloat."

iv

Ironically, one result of the expansion process is the virtual disappearance of Columbus from entire episodes. Though the object of the poem remained, as Barlow announces in his preface, to "sooth and satisfy the desponding mind of Columbus," the author added a secondary intention: "to inculcate the love of rational liberty, and to discountenance the deleterious passion for violence and war; to show that on the basis of the republican principle all good morals, as well as good government and hopes of permanent peace must be founded; and to convince the student in political science that the theoretical question of the future advancement of human society, till states as well as individuals arrive at universal civilization, is held in dispute and still unsettled only because we have had too little experience of organizing liberty in the government of nations to have well considered its effects" (*Columbiad*: xi, xii; cf. *Vision*: xxi). The net effect is to have Columbus disappear beneath a mound of goose-down comforter, the kind of consolation equivalent to interment.

"I sing the Mariner who first unfurl'd / An eastern banner o'er the western world." So reads the opening couplet of *The Columbiad,* but the Aenean connection simply will not hold, and the titular hero, at best a passive spectator, is soon overwhelmed by the didactic burden. But the title suggests "Columbia" as well as "Columbus," and it is, finally, the continental presence that dominates Barlow's penultimate epic. His poem is a drama of metamorphosis that transforms a wilderness into an advanced civilization in eight books, then pushes it forward into a glorious future in the remaining two. Avatars of enlightened progress—like Raleigh, Washington, and Franklin—survive from the earlier version, but they, too, are dwarfed by the expanded panorama. Thus, in the *Vision,* Washington's decisive crossing of the Delaware (a patriotic commonplace apparently of Barlow's invention) is accorded two heroic couplets, whereas in *The Columbiad* another one hundred and fifty-four were added with the apparent intention of lending an epic monumentality to the episode—but the result is virtually to swamp Washington in a rising tide of bathos.

Like the Scamander in the *Iliad,* the Delaware becomes a personified stream—"assumes the offended god"—not because blood stains his waters but because Washington has broken "the holy truce of night" by making his crossing under the cover of darkness (*Works*: II: 613). Undaunted, the "fearless barges" move on across the storm-lashed waters, inspiring the "vanquisht Flood" to call upon his "ancient hoary foe, Almighty Frost," who "Shoots from his nostrils one wide withering sheet / Of treasured meteors on the struggling fleet" (an image that cannot stand too close an examination) so that the water freezes over ("conglaciates"). But Hesper, "guardian of Hesperia's right," speeds east "o'er California" and "Missouri's mountains" to do battle with Frost, "Hesper's heat" (brought apparently from California) melting the Delaware and freeing the flood (615–16). Barlow, deeming that the "base" of American empire stands on the bank of the Delaware, thanks to "the work of Washington's hands," proposes that a memorial be planted there, "a column bold its granite shaft shall rear"; but the heroism of the "great general" gets somehow lost in the supernatural warfare.

Moreover, to conceive of the Delaware as an unfriendly agent, though certainly warranted by history, goes against Barlow's usual stress on natural forces as accommodating the course of empire. More typical is another added episode, again with epic pretentions, that signals the advent of Lord Delaware on Chesapeake Bay. The arrival of Delaware's fleet will relieve the distresses of Jamestown, but it is the contributory rivers who provide much of the welcome: like Washington, Delaware arrives at night, but diplomatic courtesy apparently waives "the holy truce" thus broken, and river gods, instead of repelling the invader, "rise round the bark and blend their social urns" (140).

Chief of these is Potomac, who "towers and sways the swelling flood" as he rises to greet Delaware, delivering a lengthy peroration in which he foresees a future when "federal towers" will "my bank adorn / And hail with me the great millennial morn," for "On me thy sons their central seat shall raise" (555, 553).

It was the Potomac likewise that rose at Brackenridge's bidding at Bath to greet the returning Washington, sounding a roll call of the rivers who had witnessed the general's heroic deeds, a thoroughly Old World convention seemingly out of place in poetry designed to celebrate a brave New World. More in keeping with the notion of continental sanctity, perhaps, is Barlow's depiction of the Mississippi, which serves not as a welcoming committee but as a geopolitical center, at once "force" and "balance," "charged with all the fates / Of regions pregnant with a hundred states" (445). A regnant symbol of empire, the Mississippi accepts the "tributary stores" and the "copious treasure" of other rivers, the greatest of which is the Missouri, who comes out of "his world of woods," with a sovereign air: scorning "to mingle with the filial train," he "Takes every course to reach along the main" (447). But having (like Jefferson's Shenandoah) "Searched and sundered far the globes vast frame," the Missouri "Reluctant joins the sire and takes at last his name" (447).

The Missouri's bowing to the Mississippi's ascendency gives way to Hesper's prediction that the "future sons" of Columbus will be "a race predestined . . . to teach mankind to tame their fluvial floods." As in the *Vision*, internal improvements are important adjuncts to empire, speeding the movement of commerce; and as with other geopolitical features, their role is considerably expanded in *The Columbiad:*

> *Canals careering climb your sunbright hills,*
> *Vein the green slopes and strow their nurturing rills,*
> *Thro tunnel'd heights and sundering ridges glide,*
> *Rob the rich west of half Kenhawa's tide,*
> *Mix your wide climates, all their stores confound*
> *And plant new ports in every midland mound.*
> *Your lawless Mississippi, now who slimes*
> *and drowns and desolates his waste of climes,*
> *Ribb'd with your dikes, his torrent shall restrain*
> *And ask your leave to travel to the main;*
> *Won from his wave while rising cantons smile,*
> *Rear their glad nations and reward their toil.* (700)

Thus even the Mississippi (transformed for the occasion from an imperial figure of "balance" to a slimy "outlaw") is forced to submit to man's control, ushering in the millennial moment when all of Earth's streams have been rerouted along "new paths, some useful task to fill, / Each acre irrigate, reroad the earth / And serve at last the purpose of their birth" (762).

Barlow's final vision of the future promises a time when considerations of commerce will join "man to man" through "mutual succor," realizing a universal "progressive plan" (755). To illustrate this "enlarged compact," like James Wilson describing the Constitution-making process, Barlow employs the stock analogy of a river, which begins as "infant streams" that, when they first meet, create turbulence and floods, but with age "more tranquil grow," and move seaward with "steadier sway" (755–56). Thus "broader banks" permit the mingling of many waters, which converge from "different climes" to "join and spread" themselves for the general good of mankind. It is hardly an original image, and yet, as with Wilson's geopolitical rhetoric, it most certainly suits Barlow's poetic strategy, which has throughout emphasized the taming of rivers as an adjunct to the American millennium.

The Puritan errand, which at first involved converting Indians to Christianity but soon enough became a much more general program of conversion, is translated by Barlow into a technological mission—less evidence of the workings of grace (or Providence) than of Enlightenment reason. Thus, we are told, Hesper has created a new continent, a place set apart, not for the sake of purifying religion (it was at Jamestown, according to Barlow, that "the germ, the genius of a sapient race" was planted, not in Boston), but to provide a place where "social man a second birth shall find, / And a new range of reason lift his mind" (547). "Holy fire" is no longer identified with a Pauline candle of pure doctrine planted in a wilderness, but with sacred "liberty," that signal quality celebrated by the Enlightenment. More telling, the Connecticut-born cosmopolite relocates the place where the millennial Tower will be built, from Boston-on-the-Charles to Potomac-on-the-Chesapeake, aligning his neoclassical plan with the geopolitics of Jefferson and Washington.

This, finally, is the importance of *The Columbiad,* as a monumental assertion of consensus at a critical moment in American history. When viewed as a summation of so much that was written and said over the preceding half-century concerning the future of what would become the United States, Barlow's ponderous epic transcends its obvious faults. Long before Whig historians like Bancroft marshaled the past to sustain and extenuate the present, Barlow instilled the unfolding American panorama with a uniform, progressive pattern, guaranteeing a future in which an empire of pure reason would shape the landscape to man's notion of perfection. A monumental attempt at the "vision" poem, a genre that attracted such poetic genius as there was during the first quarter century of the new republic, Barlow's *Columbiad* is also of a piece with prognostications for America's future put forth by such disparate talents as Thomas Pownall, Franklin, Lewis Evans, Jefferson, Washington, Cutler, Imlay, and even Jedidiah Morse and his edifying machine.

Moreover, the intended function of Barlow's antique poetic mode may be better understood when we recognize that Pope's *Essay on Man* made the commonplaces of the early Enlightenment memorable by freezing them in heroic couplets, for Barlow likewise presumably intended his own message to be well remembered, locking into the echoing rhymes of his *Vision* the millennial moment when "every distant land the wealth might share, / Exchange their fruits and fill their treasures there; / Their speech assimilate, their empires blend, / And mutual interest fix the mutual friend" (*Works*: II: 344). Barlow's hope was articulated as Jefferson's foreign policy; both were informed by the Enlightenment faith in reason and a neoclassical love of harmony achieved by equipoise and balance. Similarly, Barlow's anachronistic poetics become less anomalous when seen as the literary equivalent to the architectonics favored by his fellow republican, who, in striving to express the purposes of the emerging nation, adhered to a formulaic Palladianism that expresses a neoclassical ideal in stone (and wood).

Like his grid plan for expansion westward, Jefferson's architectural idealism was a mechanism of order, borrowing authority and stability from ancient example, seeking cultural validation from the vestiges of a past of which the American continent north of Mexico was virtually innocent. So also with Barlow's use of the heroic couplet, borrowed from Pope, who borrowed it from Dryden, who borrowed it from Ben Jonson, who took it from the French and classical examples. Coupleted order, like white columns, gave stability to the American experiment, for, like the city designed by L'Enfant to serve as the new nation's capitol, the mathematical regularities of Palladian models seemed equivalent to the harmonies of the Constitution; but the Roman profile that gave shape to the "domes" rising above the Potomac promoted an imperial rather than a millennial errand. They were part and parcel of the mechanisms of orderly progress that soon began to realize Barlow's vision, the canals and aqueducts whose design was also based on Roman example, and likewise aided in the spread and consolidation of empire.

These forms most certainly (though not in Frank Lloyd Wright's sense) followed function, then conceived to be a thoroughly aesthetic manifestation, a combination of utility and beauty thought to illustrate the Horatian ideal, even while valorizing the American idea by means of

Roman tropes. Thus, like the geopolitical architectonics of the new republic, Barlow's poetics were imperialistic in design yet optimistic in implication, and both qualities would prove inadequate to the complexities of subsequent events, which inspired, and in literature and art yielded to, a gothic line. Jefferson's faith in geopolitical equipoise foundered in the years immediately preceding the War of 1812, and it was in 1812 also that Barlow's luck ran out: traveling to Poland that year on a diplomatic mission, he was caught in Napoleon's disastrous winter retreat from Russia, contracted pneumonia, and died.

Though no mention is made in *The Columbiad* of the French Revolution, whose excesses did not accord well with the Enlightenment faith in universal progress, Barlow—who as a citizen of France was actively if rhetorically involved in the Girondist phase of that revolution—was at last brought into direct confrontation with the most terrible fact it produced. Napoleon's Russian campaign was an aspect of expanding empire that had little to do with coupleted order, save in the savage satire with which Barlow, in his last poetic effort, attacked the ambitious general as a figure associated not with the glorious eagle but with the carrion-consuming raven. We may, however, accord Barlow the same comforts his visionary angel brought to Columbus languishing in prison, for although Jacobinism and all things French went out of favor in America following the Reign of Terror, yet in his *Vision* and *Columbiad* we can see the dim outlines of much that was to be championed, if not realized, by presidents and politicians in the years to come.

V

"Beware! Beware!"

i

While the college-trained poets of the young republic were attempting to direct the expanding energies of America through neoclassical verse, other writers, often lacking academic credentials but benefiting from equivalent cultural backgrounds, were employing a literary form that was more accommodating to the expansionist impulse: the loosely structured, circumstantial genre called the "travel narrative." Though not without its complexities, the travel narrative comes close to an impromptu, extemporaneous performance, at least to the extent that it takes its shape from the exigencies of event. Hardly a native American genre, the travel narrative did flourish in this country, particularly during the early years of the republic, for it satisfied (while expressing) the restless spirit that was part of the expansionist spirit, and at the same time answered a need, especially in Europe, for more information concerning the new nation. The travel narratives produced during this critical period often mingle elements of exploration with the kinds of observations associated with touring guides, resulting in hybrid forms rather than a uniform genre. What they do have in common, however, is a concern with particulars, often given shape and continuity by the rivers that served as main avenues of communication, especially in regions most recently settled.

Poetic visionaries and propagandists for expansion were alike in viewing rivers as imperial concourses along which cities would spring up and a great civilization flourish, while travel writers, unless in the service of land companies, tended to produce less hortatory and more balanced accounts of the opening American scene. But they do share with other writers of the period an optimism inspired by the opportunities for progress, whether on the individual or countrywide level, held out by a new and promising land. Moreover, the travel narrative often serves as a geopolitical conduit, conveying the imperial ambitions engendered by the French and Indian War forward into the earliest phase of nationalistic fervor that immediately preceded the Revolution, much in the manner of Lewis Evans' map and essays. This is especially true of the most significant example of travel writing produced during the critical period from 1770 to 1775, Crèvecoeur's *Letters of an American Farmer,* a work characterized by an ardent proto-nationalistic tenor, a quality all the more strange given the origins, history, and political bias of the author.

Hector St. John de Crèvecoeur, as I have already suggested, anticipates Gilbert Imlay as an articulate spokesman in America of the physiocratic ideology associated with Thomas Jefferson, but unlike both Imlay and Jefferson he was no radical advocate of revolutionary change. He was, in fact, a Tory, and because of this may be seen as a counterpart to Thomas Hutchins, whose career Crèvecoeur's early years in America closely parallel. Both men enlisted to fight in the French and Indian War at the start of hostilities, each on the opposite side, and while Hutchins was surveying for the British the rivers and streams of the Ohio Valley, Crèvecoeur was mapping the lakes of Canada. The young French officer gained royal recognition for his sketch of the captured British stronghold on Lake George, Fort William Henry, while Hutchins distinguished himself by his rendering of the fortifications at DuQuesne abandoned by the French. By war's end, however, Crèvecoeur had resigned his commission under a cloud and taken up residence in New York, and where Hutchins went on to establish himself as the delineator first of British then of United States empire, Crèvecoeur set up as a local farmer whose loyalty to the Crown made it necessary that he leave America in 1780, the year Hutchins came over to the patriot side. And yet Crèvecoeur, no less than his counterpart, attests in his writings to the ease with which British imperialism could be converted to nationalistic ends.

As we now know, much of the original manuscript of Crèvecoeur's *Letters* is made up of sketches decidedly hostile to the Revolution, not so much to the glorious cause of independence as to the inglorious liberties taken with private property by individuals using patriotism as an excuse to plunder their loyalist neighbors. But by the time the book was published in 1782, this material was viewed by a judicious editor and perhaps by Crèvecoeur himself as either irrelevant or harmful to the book's reception—most particularly in the United States—and it was excised. As a result, the *Letters* that finally saw print was so mildly and obliquely loyalist in sentiment as to read—as it often has been read—as a lengthy expansion of Jefferson's manifesto coupling liberty with the kind of happiness founded on the ownership of property. So sympathetic was Crèvecoeur's book to the spirit if not the absolute letter of

American independence that he was sent to the new nation at war's end as French counsel resident in New York. There he remained until 1790, a living symbol of the newfound love between France and America—made explicit in the name of his daughter, America-Frances—at whose wedding this country's best-known francophile, Thomas Jefferson, was present.

Those happy nuptials had a grim counterpart in Marietta, Ohio, the tragic experiment in transporting French loyalists to the American frontier, to which Crèvecoeur contributed his share, albeit indirectly. There is an extract in Manasseh Cutler's pamphlet of 1787 that was translated from the expanded French edition of *Letters*, a version that intensified the pastoral idealism of the original. Styling the Ohio River "the grand *Artery* of that part of America beyond the mountains . . . the centre where all the waters meet," Crèvecoeur in this extract describes a journey he (claims he) took down the river, a "sweet and tranquil navigation" that seemed an agreeable dream:

> *Every moment presented to me new perspectives, which were incessantly diversified by the appearance of the islands, points, and the windings of the river, without intermission—changed by this singular mixture of shores more or less woody; whence the eye escaped, from time to time, to observe the great natural meadows which presented themselves, incessantly embellished by promontories of different heights, which for a moment seemed to hide, and then gradually unfolded to the eyes of the navigator, the bays and inlets, more or less extensive, formed by the creeks and rivulets which fall into the Ohio.* (38–39)

Crèvecoeur's description is reminiscent of Pownall's account of river travel in the New World, a matching aestheticism that leaves an impression that the Ohio Valley is a lovely garden lacking only *cultivateurs* to tend it.

Inspired to "meditation" by the "majesty" of the rivers flowing into the Ohio, Crèvecoeur allowed his imagination to leap "into futurity"—that region favored by collegiate bards of the day—with a predictable alteration in the landscape:

> *I saw those beautiful shores ornamented with decent houses, covered with harvests and well cultivated fields; on the hills exposed to the north, I saw orchards regularly laid out in squares; on the others, vineyard plats, plantations of mulberry trees, locust, &c. I saw there, also, in the inferior lands, the cotton tree, and the sugar maple, the sap of which had become*

> *an object of commerce. I agree, however, that all those banks did not appear to me equally proper for culture; but, as they will probably remain covered by their native forests, it must add to the beauty, to the variety, of this future spectacle.* (39)

Crèvecoeur's Horatian "vision" of Ohio approximates the prospects of Imlay's Kentucky, and though in harmony with Cutler's emphasis on the necessity for useful transformation, signified by Jeffersonian "squares," Crèvecoeur's version of the future tends (like Imlay's) toward hyperbole:

> *—What an immense chain of plantations! What a long succession of activity, industry, culture, and commerce, is here offered to the Americans! —I consider, then, the settling of the lands, which are watered by this river, as one of the finest conquests that could ever be presented to man; it will be so much the more glorious, as it will be legally acquired of the ancient proprietors, and will not exact a single drop of blood—It is destined to become the source of force, riches, and the future glory of the United States.* (39)

Even more than Filson and Imlay, Crèvecoeur posits a highly accommodating Nature (as well as an extremely obliging Indian), for the Ohio landscape is already "useful to man," being furnished with "gentle risings," that have been "planted" by Nature with "the finest trees" (40). Ohio by Crèvecoeur's account is the golden mean of landscapes, having plains that are well-watered but "never overflowed," whose "fertility is most admirable," and recommends itself, indeed begs, for settlement:

> *If a poor man, who has nothing but his arms to support him, should ask of me, where shall I go to establish myself in order to live more at my ease, without the aid of oxen or horses? I should say to him, go upon the banks of some rivulet on the plains of Sioto; there you will obtain permission of the savages of the neighboring villages to* scratch *the surface of the earth, and deposit your rye, your corn, your potatoes, your cabbages, your tobacco, &c. leave the rest to nature; and, during her operations, amuse yourself with fishing and hunting.* (40)

This advice may be compared to the last episode in the *Letters* of 1782, wherein the rural epistler states his intention to resettle himself and his family "on the shores of ______" among the Indians of a farther West.

That decision was forced upon the "Frontier Man," and emerges from a situation of extreme adversity that dramatically belies the stability and ease of the agrarian life celebrated by Crèvecoeur in the opening chapters of *Letters*. Furthermore, when we restore to the book the sketches that were excised for the edition of 1782, the reduction in pastoral idealism is even more drastic, providing a sharp contrast between Crèvecoeur's original portrayal of the life of an American farmer and the revised (French) version he brought forth in 1784, whose intensely euphoric tone is epitomized by his description of life as it might be enjoyed along the Ohio River. Thus the English edition of 1782 may be seen as presenting a median version of the life of an American farmer, lacking the worst of the atrocities brought on by the Revolution and avoiding the euphoric hyperbole inspired by the author's turnabout at war's end.

In the edition of 1782, for example, Crèvecoeur mounts an interesting contrast by comparing the conditions of life enjoyed by the Quaker settlers of "New Garden" "in North Carolina, situated on the several spring heads of Deep River, which is the western branch of Cape Fear," with the life experienced by another group of Quakers who carry on the timber business along "the famous river Kennebeck" in Maine. In North Carolina (as in Ohio), the soil is extremely fertile and "rewards men . . . early for their labours and disbursements," whereas the land along the Kennebec requires "immense labour . . . to make room for the plough" (146–47). And yet it is the northern environment the Farmer prefers, for reasons completely in keeping with Manasseh Cutler's Yankee predeliction for the Protestant ethic:

> *If New Gardens exceeds this* [*Kennebeck*] *settlement by the softness of its climate, the fecundity of its soil, and a greater variety of produce from less labour; it does not breed men equally hardy, nor capable to encounter dangers and fatigues. It leads too much to idleness and effeminacy; for great is the luxuriance of that part of America, and the ease with which the earth is cultivated. Were I to begin life again, I would prefer the country of Kennebeck to the other, however bewitching; the navigation of the river for above 200 miles, the great abundance of fish it contains, the constant healthiness of the climate, the happy severities of the winters always sheltering the earth, with a voluminous coat of snow, the equally happy necessity of labour: all these reasons would greatly preponderate against the softer situations of Carolina; where mankind reap too much, do not toil enough, and are liable to enjoy too fast the benefits of life.* (147)

"Happy severity," like "fortunate Fall," manages to compress into an oxymoron the complex attitude toward adversity enjoyed by the original settlers of New England into a single pill of prevention. Urging those who disagree with him to "go and settle at the Ohio, the Monongahela, Red Stone Creek, &c.," the Farmer declares his preference for "the rougher shores of Kennebeck," where adversity guarantees a life "of health, labour, and strong activity . . . characteristics of society which I value more than greater opulence and voluptuous ease" (148). In similar fashion, the Farmer plays off the virtuous life of industry led by the natives of Nantucket against the highly civilized but decadent lives led by the slaveholding planters of Charleston, South Carolina.

Crèvecoeur's preference for the simple, hardworking life is part and parcel of the emphasis in the *Letters* of 1782 on the peculiar pleasures derived from travel in North America, an epitome of which is provided by the Farmer's clergyman in the opening chapter of the book. Drawing a distinction between the "reveries of the traveller" in the Old World and the New, the Minister observes that in America "everything would inspire the reflecting traveller with the most philanthropic ideas; his imagination, instead of submitting to the painful and useless retrospect of revolutions, desolations, and plagues, would, on the contrary, wisely spring forward to the anticipated fields of future cultivation and improvement, to the future extent of those generations which are to replenish and embellish this boundless continent" (43). This contrast is endemic to the chauvinist strand in American writing of the period, and it invariably links life in America with the rising glory of future accomplishment rather than the fading glories of a heroic past. "America is a large place," wrote Dr. Thomas Cooper, the emigrant physician, scientist, and radical agitator; therefore, it provides sufficient space for a man "to get up again" should he fall, whereas in crowded Europe he who stumbles is lost: "In fine, ours is a rising country . . . yours is a falling country" (Imlay: *TD*: 186).

For Crèvecoeur as for the academic poets of the revolutionary years, America was a Horatian vista with an aesthetic as well as an economic dimension. Thus the Farmer's Minister would rather admire "the ample barn of one of our opulent farmers, who himself felled the first tree in his plantation and was the first founder of his settlement, than study the dimensions of the temple of Ceres" in the Old World (43). For what could be more interesting to the Enlightened sensibility than "the progressive steps of this industrious farmer throughout all the stages of his labours and other operations"? Paradoxically, however, Farmer James devotes relatively few pages in his letters to such matters, and though the book is cast in the futuristic vein, we get very little of the visionary element that characterizes the poetry of the period or Crèvecoeur's subsequent description of the Ohio Valley. Despite his insistence on the importance of bustling industry, Crèvecoeur's picture of the life of the American farmer is one of static ease, for James himself benefits from the labors of his father, who originally cleared the land he now tills. Freed from the kind of hard work the Minister finds so interesting, Farmer James has the leisure to pursue his hobby of bee-hunting and to make the travels that provide the material for the letters.

The life led by the Farmer and his family resembles Franklin's "happy mediocrity" rather than the "happy severity" enjoyed along the Kennebeck. They inhabit the classically "middle landscape" located between the hardships of the frontier and the corruptions of the city, and though, ideally, "rapidity of growth" may be "more pleasing to behold than the ruins of old towers," the Farmer's chief pleasure comes from looking at his well-stocked property, fruitfulness ensured by fertility and the certain round of seasons (42). America may be associated with the promise of the future, but the Farmer associates that which is "modern" with "peacefulness and benignity," qualities derived from permanence, not continuing change. That is, despite the mild Jacobinism of the Farmer's pastor, the pastoralism of Crèvecoeur himself is profoundly conservative, and operates (as the ending of the book testifies) in the service of the status quo. For the "distresses" described in the last chapter are the result of loyalty to the Crown at the cost of loyalty to the patriot cause.

It is clear, moreover, that we are to sympathize with the "Frontier Man" who writes the penultimate letter, and who should not be confused with Farmer James. For he has cleared his own land, not inherited it, and has spent an early period of his life as a fur-trader. Moreover, he occupies a zone somewhat in advance of the "middle landscape," and is forced by circumstances to flee even further west. For the king can no longer protect him, and, indeed, is responsible for the attacks by savages who refuse to distinguish between loyalists and patriots. This kind of "severity" is clearly not "happy," nor are the Frontier Man and his family eager to participate in the kind of "future" that Farmer James and his Minister have found so interesting. He now has the opportunity "to begin life again," but instead of delight, the Frontier Man feels uneasiness. He joins the progressive process only because he has no choice. Where poets like Freneau and Barlow celebrated the American Revolution as a phenomenon guaranteeing that America would be divorced from the "falling" Old World, Crèvecoeur (in 1782) protested the war that disrupted his personal tranquillity, guaranteed in the past by the "easy reins" of benign colonial rule.

Letters from an American Farmer is a deeply disjunctive work, which suffers both from the author's divided loyalties and from the circumstances under which it was written (and published). To this we may add the problems resulting from the differences between the Farmer and Crèvecoeur, the one a second-generation American of English descent, the other a recent French immigrant. The Farmer occupies land in western Pennsylvania; Crèvecoeur settled in Orange County, New York, just north of Manhattan. Indeed, Crèvecoeur's latter fortunes in North America more closely parallel those of the Frontier Man than those of the Farmer, though his chief tormentors were not Indians but the kinds of "patriots" whose harassment of their loyalist neighbors is described in the sketches removed from his *Letters* and only restored in this century. Like the Frontier Man, he was forced to depart; not, however, in a westerly direction, where the future lay, but eastward, toward the past. The last chapter of *Letters,* reinforced by the excised sketches, transforms a celebration of the middle landscape as a sanctuary of permanent happiness into a tragic zone, conveying a sense of the world as a place in which no certainty is possible, and recalling Pownall's sad evocation of a "fallen" Connecticut valley. Change, which to the radical temper is usually identified with gain, is to the conservative associated with loss, and the ending of *Letters* therefore only reinforces the essentially reactionary tenor of the book as a whole, which identifies "progress" with the assertion of middling values and posits a future that is essentially more of the same. In 1784, Crèvecoeur was willing to rewrite his book to be more in keeping with the emerging optimism of post-revolutionary America, but such revisions were clearly expedient and gloss over the deeply conservative tenor of the original *Letters,* which increases with the addition of the excised material. Though *Letters from an American Farmer* seems to harmonize with the "Rising Glory" idea, it actually provides a counter-tradition, containing hints of the darkness that will be associated with Romantic writing of a later period in America. Crèvecoeur is often compared to Thoreau; Melville is closer to the mark.

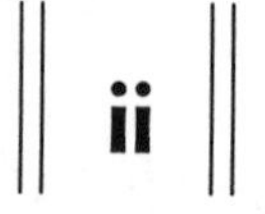

Letter XI (of XII) in the 1782 edition of Crèvecoeur's book is written not by the "Farmer" but by "Mr. Iw—n Al—z, a Russian Gentleman," and describes "the visit he paid at my request to Mr. John Bertram [*sic*], the celebrated Pennsylvanian botanist." The shift in point of view seems to have dictated by Crèvecoeur's desire that Bartram and his farm be described from a European perspective, for the simple Quaker botanist emerges as a physiocratic ideal, whose way of life contrasts sharply with the Europe from which the Russian has come. A central feature in the American landscape inhabited by Bartram is the Schuylkill River, which "winds through delightful meadows" and is kept in its place by a manmade bank or levee, which not only confines the stream but frees more land for cultivation. The humble, industrious Bartram is first encountered as he works on his bank, and Ivan, once he realizes he is speaking, not to a field hand, but to the proprietor of the farm (a "European" mistake), inquires about the purpose of Bartram's labor. The answer is a summary of Crèvecoeur's Enlightenment vision of the marriage of utility and beauty in America, limning the proper relationship between man and nature:

> *Friend Iwan, no branch of any industry was ever more profitable to any country, as well as to the proprietors; the Schuylkill in its many windings once covered a great extent of ground, though its waters were but shallow even in our highest tides: and though some parts were always dry, yet the whole of this great track* [*tract*] *presented to the eye nothing but a putrid swampy soil, useless either for the plough or for the scythe. . . . It is owing to this happy contrivance* [*of banking*] *that so many thousand acres of meadows have been rescued from the Schuylkill, which now enricheth and embellisheth so much of the neighborhood of our city.* (190)

In sum, if the meadows through which the river winds are "delightful," it is because they are no longer "putrid," having been "rescued" from the river by the "art" and "industry" of banking. Like the land itself, the Schuylkill is under man's control, part of the rational order and regularity imposed by the good Quaker, whose farm is a benign machine, a Horatian paragon of useful design and a paradigm of productivity. Counterpart to Bartram's farm is the Quaker community of Nantucket, where reason and order also prevail—and a factory-like organization dominates the community—an urban equivalent of the agrarian ideal. Throughout his *Letters,* Crèvecoeur celebrates a stable, orderly, often insular mean, an ideal that derives its values chiefly from stability and emphasizes domestic and collective, not individual, values.

Quite a different emphasis emerges from one of the sketches removed from the 1782 edition of *Letters,* an account of the settlements along the sister river of the Schuylkill, the Susquehanna. A number of considerations may have dictated the removal of the Susquehanna section, including political ones (the episode ends with an account of the Wyoming Massacre, in which the British forces played a scandalous part); but what is of interest here is that the narrative takes the form of an actual journey, travel defined by the course of the river, whereas all of the "Travel" letters gathered in the 1782 volume involve little or no physical movement. Equally important, the river along which the Farmer travels is in all respects the antithesis of Bartram's Schuylkill, and is associated with aspects of the pastoral life in America that are not easily accommodated to Enlightenment optimism. Though explicitly intended to provide an actual example of the "pleasure" Farmer James receives from "following a branch of humanity shooting up all round and replenishing in the course of a few years those beautiful shores hitherto savage and wild and entirely uncultivated," the "progress" motif is qualified throughout (354). Moreover, the agrarian metaphor of the transplanted shoot, which works so well as a consistently physiocratic figure in the published *Letters,* seems here less propitious, given the eventual fate of the Susquehanna settlers.

The "branch of humanity" in question was Yankees who abandoned the exhausted soil of New England, crossing over into Pennsylvania lands enriched by abundant waters, equivalent to "the sources and springheads of those immense rivers everywhere traversing the great continent." It is on these "innumerable streams" that "Ceres and Pomona have fixed their pleasing abode," part of a rational, providential scheme: "Bountiful Nature seems purposely to have given this soil a deree of fertility proportioned to its distance from navigable rivers, in order, no doubt, that men tho' so far removed from markets might afford in their extreme plenty the means and expense of an unavoidable transportation." The richer the land, the farther the distance, a reciprocal arrangement testifying to the benignity of the mechanism Crèvecoeur calls "Nature."

To benefit from that mechanism, however, the settlers from New England have had to undertake an arduous journey and shoulder "Herculean labors" once they arrive. In order to convey an accurate sense of the settlement process, Farmer James elects to follow the same "path which these people had made," traversing the country defined by the course of the Delaware River, which passes through "the midst of . . . desolate ridges" (357). Like other rivers in America, the Delaware has been "indulged by Nature," for obstacles to its passage "obsequiously open" and make way for the stream (357–58). But travelers afoot find the going less easy, especially because of the maze of wild laurels that block the way, "which by their low size and the extreme ramifications of their crooked limbs are the greatest and most unsurmountable impediment a traveler can meet with" (358). The traveler, especially one who is not used to the deep woods of America, is seized by a "secret uneasiness and involuntary apprehension," and since "hearing," not sight, has "the preeminency" in the woods, "I was alarmed at every distant sound" (359). The Farmer's apprehension is augmented by ample evidence provided of "the fury of the waters" along which he moves: "Immense trees lodged sometimes across its stream, at other times deep ponds it had dug by carrying away all the earth" (360).

And yet, in the midst of this frightful terrain, man has carved out an occasional spot of "open'd grounds" on which the sun shines, "clear'd fields" that are a "feast for an unexperienced traveller" to behold (359). One such tiny farm is made possible by a freak of nature: "a fortunate bend in this river and . . . a few buttonwood trees which Nature had planted on these banks" acted to preserve a bit of low land from being torn away "by the impetuosity of this torrent" (360). Called "Blooming Grove," this unlikely and perilous bit of pastoral terrain is the home for a family of "happy hermits," whose "round of labour and perpetual industry" not only "fills the measure of their time" but supplies all their wants (361). Beyond this brief sanctuary of domestic happiness lies an unpleasant region of swamps and sterile soil, chiefly valuable for its timber, and a maze of ridges and rivers, with the usual impediments of roots and fallen trees, which the Farmer must cross before finally arriving at "the great swamp of which I had so often heard," where both the Delaware in the east and the Susquehanna to the west take their rise.

Pausing in this watery center, the source of two great rivers, the Farmer delivers what would appear to be an unlikely soliloquy, given the terrain he has just passed through: "When the age, the wealth, the population of this country will arrived to such a pitch as to be able to clear this immense tract; what a sumptuous, what a magnificent sight it will afford! . . . Here imagination may easily foresee the immense agricole richesses which this great country and this spot in particular contain. I never travel anywhere without feeding in this manner on those contemplative images" (363). Here, as elsewhere in the sketches left unpublished during his lifetime, Crèvecoeur's style resonates with a heavy French accent, no more so than in the exercise of physiocratic optimism capable of converting a huge swamp into arable land, sustaining a vision of the future that seems belied by all the difficulties he has up to that point endured. But, in some senses, the "agricole richesses" are warranted by what lies ahead, for, once again, the swamp is also the source of the Susquehanna, a vast watery womb near which he finds "the embryo of a settlement," and farther on down the westward-flowing river, the much larger settlement of "Wiomen" itself. For, in marked contrast to the upper regions of the Delaware, the Susquehanna near its source is an accommodating stream, which "bends itself in an amazing number of creeks and rivulets" and thereby imparts to "mankind a greater degree of benefits. Few rivers in this part of the world exhibit so great a display of the richest and fertilest land the most sanguine wish of man can possibly covet and desire" (364–65).

This is the "New Canaan" sought by the emigrant New Englanders, who in four years' time have made good use of the rich and fertile land, characteristic Yankee enterprise combining with an obliging landscape to produce a Horatian paradigm of aesthetic and moral dimensions:

> *This fine river is . . . interspersed at every little distance with pleasing islands, points of low lands, some of which seem to be detached from the main. That pleasing variegated mixture of high, low, and still lower grounds, that alternative vicissitude* [sic] *of extensive plains and high promontories view'd at every angle as you either ascend* [*or descend*] *the river, the prodigious number of houses rearing up, fields cultivating, that great extent of industry open'd to a bold indefatigable enterprising people afforded me a spectacle which I cannot well describe.* (366–67)

The industrious settlers had already set up a number of those mechanisms Thomas Pownall had regarded as essential to a well-ordered community, sawmills supplying boards for houses yet to be "reared up" and lumber to be floated downriver in rafts.

Thus the Susquehanna is not only the source of the necessities of life, but it provides power for mills with which to improve natural advantages and a route by means of which commerce may be carried out; or, in Crèvecoeur's Gallic terms,

> *a debouche by water to exchange their exuberancy for what they want. What a pity that this and other branches and ramifications of this immense river, all possessing on their shores low lands proportioned to the size of their streams cannot be permanently settled and be made to unite the advantages of peace, political tranquility with every other which nature offers them with the most liberal hand. . . . It is here that human nature undebased by servile tenures, horrid dependence, a multiplicity of unrelieved wants as it is in Europa reacquires its former and ancient dignity,—now lost all over the world except with us. May future revolutions never destroy so noble, so useful a prerogative.* (368)

Confronting the infant community (the future Wilkes-Barre) on the banks of the Susquehanna, the Farmer delivers a familiar encomium on the signal difference between life in the New World and the Old, keyed to the pleasing spectacle of a people creating civilized order in the midst of a wilderness, an order moreover that evokes a universal golden age. This is precisely the sort of thing the Farmer's Minister has praised, and when James revisits Wilkes-Barre two years later, the situation has been improved still further: "A better conducted plan of industry prevailed throughout," log huts have become "more substantial habitations," and the valley exhibited everywhere "the strong marks of growing wealth and population. . . . Nothing could be more pleasing than to see the embryo of future hospitality, politeness, and wealth disseminated in a prodigious manner of shapes and situations all along these banks" (378–79).

This paragon of orderly progress has its problems, however: the Farmer has by this point in his narrative described the battles over proprietorship of the Wyoming Valley—a result of conflicting claims by Connecticut and Pennsylvania—which had already erupted in violence. "The names of Yankees and Pennamites were invented and became two words of reproach not only among the two rival provinces but even among themselves" (374). Where other pioneer settlements were threatened by Indians, here the source of conflict was sectional, traceable to the claims of Connecticut based on "the charter words of Charles the 2d," which the Farmer dismisses as "credulous." In sum, these Yankee exemplars of order and industry have set up their "New Canaan" in a territory to which they had no rightful claim, so that along with "the embryo of 16 townships" had been planted "the rudiments of . . . misfortunes and unhappiness" (374).

Where Crèvecoeur's account of Bartram's farm on the banks of the Schuylkill is a paradigm of physiocratic right reason, the Wyoming Valley on the Susquehanna is a paradox of unreason, for in the midst of a region where "nature in her most indulgent hours" has supplied man the materials with which to satisfy all his needs, civilization has brought with it disruptive discontents. Crèvecoeur's narrative history of the Wyoming Valley ends with the outbreak of the American Revolution, which further exacerbated the situation, resulting in a terrible massacre, led by loyalists who had been driven out to live among the Indians by patriot agitators. "Thus perished one fatal day most of the buildings, improvements, mills, bridges, etc., which had been erected there with so much cost and industry. Thus were dissolved the foundations of a settlement begun at such a distance from the metropolis, disputed by a potent province; the beginning of which had been stained with blood shed in their primitive altercations" (386).

Still, like the Happy Valley created by John Bartram, the Fatal Valley settled by interloping Yankees can be reconciled with the Enlightenment spirit that irradiates the Farmer's letters: the massacre is directly traceable to *irrational* claims (based on a "credulous" interpretation of an ancient charter), claims that in turn fostered "political iniquity," traceable to "the leaders of this unfortunate settlement" (387). As a result, many innocent people died, and the survivors were forced to return through the harsh wilderness to the place from which they had come, now bereft of even the property they had brought with them: "They had nothing to carry . . . but the dreadful recollection of having lost their all . . . afflicted with the most pungent sorrow with which the hand of heaven could chastize them" (388–9). Intentional or not, this final sentence carries a certain irony, the notion of a chastizing divinity being an important part of the New England tradition. The published book also ends with an exodus, but the Frontier Man and his family are guilty of no sin. They are victims of forces beyond their control, caught between contending parties and divided allegiances, innocents whose sufferings illustrate the price to be paid whenever established order is disrupted.

But both the forced exile and the lesson of the Wyoming Massacre would seem to call into question the basic premise of *Letters*, which is that the American landscape is a regenerating force, producing a New Man, who imbibes a sense of freedom from the ownership of property: "Could I have ever thought," wonders the Farmer in another of his sketches, "that a people of cultivators, who knew nothing but their ploughs and the management of their rural economies, should be found to possess, like the more ancient nations of Europe, the embryos of these propensities which now stain our society?" (342). The history of settlement along the Susquehanna provides a dark alternative to the idea that in America, nature and humanity work in perfect harmony to bring forth a new kind of nation. To restore the Susquehanna material to *Letters* is to provide a disjunction equivalent to the divided landscape along the Ohio River, i.e., yet another alternative to the notion of transsectional unity, and a discordant concourse that would, in time to come, increasingly prevail in the literature through which American rivers run.

iii

In positing an ideal figure for his physiocratic landscape, Crèvecoeur made a wise choice in John Bartram. Because of his agrarian emphasis, it is Bartram the "worthy citizen who united all the simplicity of rustic manner to the most useful learning" who is emphasized, at the expense of the "enlightened botanist." Bartram was certainly Ben Franklin's rural counterpart, devoting to his fields the same enlightened energies Franklin expended on Philadelphia; but much as Franklin's talents as a scientist transcended his printerly skills, so Bartram the botanist is of far more consequence than Bartram the simple farmer. That he traveled through the wilderness along the Susquehanna and Delaware rivers in 1743, more than thirty years before Crèvecoeur made his trip, gives the Quaker seedsman a much more heroic profile, and that his companion on that journey was Lewis Evans connects his botanizing activities with the imperial energies then expanding westward of the Alleghenies.

But perhaps the most symbolic journey taken by John Bartram occurred in September 1753, for this was the first of the botanist's extended travels in which he was accompanied by his son, "Billy," then fourteen, who would become a much more famous traveler than his father, identified as much with the emerging literature of North America as with scientific knowledge. This trip through the Catskills of New York ended at the Hudson Valley home of Cadwallader Colden, Surveyor-General of New York and himself a botanist of note, where the Bartrams were introduced to yet another botanist, Alexander Garden of Charleston, South Carolina. The coming together of the three botanizing representatives from the Northern, Middle, and Southern colonies may be said to epitomize the scientific Enlightenment in colonial North America, an intellectual voluntarism expressing the cooperative idealism of Dr. Franklin, who was a patron of Bartram, a friend to Colden, and would soon (and predictably) make the acquaintance of Dr. Garden.

Franklin and his friends made up a very large committee of correspondents, extending lines of communication across the disparate North American colonies and across the Atlantic to England and Europe as well. By means of letters, packets, and bundles, they disseminated information about the New World, contributing greatly to the intellectual life of the Enlightenment. But surely the most important person attending the impromptu conference at Coldenhamia in 1743 was young Billy Bartram, who would carry curiosity about the natural world to a transcendent level. Franklin attempted several times to raise the necessary funds to permit John Bartram to compile an American botanical dictionary, but without success. It was William Bartram who would eventually write the most complete botanical guide of its day: not a dictionary, but a book of travels, an account recording his journeys through the Carolinas, Georgia, Florida, and even to the bank of the Mississippi, a book that, when it was published in 1791, was seized on by geographers like Jedidiah Morse and poets like Coleridge and Wordsworth with equal avidity, for the useful information it provided was set forth in a manner unlike the usual narratives of the day, and contrasts sharply with the few published travel accounts by Bartram Senior.

In the relationship (and differences) between John and William Bartram we have a compact demonstration of the strategic changes in that complex, evolving spirit of rational inquiry and invention we for convenience call the Enlightenment. John Bartram's love for order seldom exceeded a utilitarian view of botany as a practical science and a moneymaking trade. His son, by contrast, was a poor hand at practical affairs but had considerable artistic talent, and would in time become a traveling seedsman with an errand far greater than his nominal purpose for journeying about the South. Father and son are virtual personifications of the early and late Enlightenment in America, at one end a persistent, even obsessive, gathering of information, at the other an elevation of natural phenomena to symbols of quasi-Romantic intensity. For John Bartram, nature was a storehouse of commodities created by a beneficent deity for man's use, but for William the natural world shone with a mystical aura, a divine light that gave a religious meaning to Linnean order. The father was Quakerism in its practical, no-nonsense guise, which saw no disparity between piety and profits; the son expressed the qualities associated with his fellow journal-keeper John Woolman, for whom the material world was irradiated by the inner light of simple faith. As a rustic Franklin, John Bartram in his writing spoke directly and helpfully to his contemporaries, but William's most appreciative audience was a newer generation even than his own. At first criticized for containing little science and much poetry, within a decade his book was praised for the same ratio.

Based upon journeys taken even as the opening battles of the Revolution were being waged, Bartram's writings evince a creative counterpart to that martial epic, an unselfconscious expression of the inquiring spirit that matches Washington's herioc exploits. For that reason, Billy's journey to the Catskills in 1753 may be seen as a symbolic counterpart to Washington's much more hazardous trip into the Ohio Valley the same year, as a rite of passage into a world that held both his own and America's future. That trip took the Bartrams along the Delaware to the Hudson, rivers Washington would also be associated with, but the river invariably linked to William Bartram is the St. John's in Florida, a peculiar stream in its perverse, north-running current, and an exotic river in all respects. Where Washington's riverine associations involve heroic crossings, Bartram traveled up the St. John's to its source, a journey identified not with the epical shape of an emerging republic but with a mystical, highly subjective quest. As a literary trope, Bartram's journey was not derived from neoclassical models, but provided instead a prototype for many Romantic narrative quests to come. Yet the same disjunctive dimension that characterizes so many accounts of American rivers during the late eighteenth century may be found also in Bartram's account of the St. John's. As in Crèvecoeur's contemporaneous account of the Susquehanna, we find a divided consciousness, putting forth contrasting accounts describing not so much the process of settlement as the world of nature itself.

William's first trip to Florida, in 1765, was with his father, and it is the senior Bartram's account of that trip that survives, but it is clear from the son's later journal that a generation gap of sizeable dimensions separated the two. For the prosaic father, Florida was just another, albeit exceptionally interesting, place from which to harvest plant samples and information, but it held William in powerful thrall, drawing on something deep within his psyche, a hunger for the unknown and the mysterious that transcended an urge to gather and classify "curiosities" of nature. So strong was that thrall that William refused to return home with his father in 1765 and stayed on for a year—much to John's vexation—setting up a quixotic venture in indigo planting that soon failed. This brought him back to Pennsylvania, but when the opportunity presented itself to return to Florida, in 1773, he did so, and the result is the first of the journeys that make up the title page of his *Travels*.

William's specific mission in Florida was to gather plant specimens for his patron, Dr. John Fothergill of England, the same sort of errand his father so often made. Indeed, from 1773 to 1774 he took the same route through Florida he had earlier taken with his father, and freely borrowed from his father's journal of that voyage. But we have only to lay one journal next to the other in order to measure the generational difference. In terms of errand, John Bartram's journal reflected the imperial spirit that had engendered the French and Indian War, for the purpose of his journey to Florida was to assess the importance of the region to Great Britain, which had recently acquired it from Spain as part of the massive geopolitical realignment of North American territories occasioned by the Treaty of Paris. In the words of Dr. William Storck, who wrote the introduction to Bartram's journal, "a country unknown, must, if a paradise, still continue a desert" (xi). In sum, matters of quality were contingent upon quantification. By contrast, William Bartram was in the hire of a single individual, on a particular errand, serving as the vicarious agent of another man's curiosity, his purpose tied to neither colonial nor commercial considerations. For him, a desert was paradise enough.

John Bartram's Florida journal is nominally a record of his "search for the head of the river St. John's," but the chief thrust of his inquiry is revealed by the long catalogue of those aspects of the landscape along the river that "may be improved by culture" (1–2). If Bartram Senior's journal differs from the usual imperial survey, it is because of the author's fascination with the abundance and sheer size of tropical flora and fauna, aspects of natural plenitude that first stagger and then baffle man's capacity for mensuration, like "the monstrous grapevines, 8 inches in diameter, running up the oaks 6 foot in diameter, swamp-magnolias 70 foot high strait, and a foot diameter, the great magnolia very large, liquid amber, white swamp and live oaks, chinquapines and cluster-cherry, all of an uncommon size, mixed with orange-trees, either full of fruit or scattered on the ground, where the sun can hardly shine for the green leaves at Christmas [this entry was for the twenty-seventh of December], and in a mass of white or yellow soil 16 foot more or less above the surface of the river" (5). John Bartram's emphasis on sheer size extends to his account of the great sinks of Florida, those "fountains" that his son would make famous among the poets of England, but for Farmer John the springs were reckoned chiefly in Horatian terms of usefulness, being "big enough to turn a mill" (12).

Bartram Senior was not without his capacity for wonder, but here again it turns on the fact of dimension: "What a surprising fountain it must be, to furnish such a stream, and what a great space of ground must be taken up . . . to support and maintain so constant a fountain, continually boiling right up from under the deep rocks, which undoubtedly continue under most part of the country at uncertain depths" (12). Equally characteristic was his observation that the water from this great fountain had "a loathsome taste," a quality never ascribed by his son to the Helicons of Florida (19). Instead, William wrote an account of the Alachua sink holes that lifts his *Travels* into the company of Crèvecoeur's *Letters* as an example of the hyperpastoral mode engendered by the American experience. Both books have a lyricism that operated on the imagination of young Romantic poets in Great Britain, yet, like Crèvecoeur, Bartram is every inch a child of the Enlightenment. Much as the Farmer's letters reveal a love of rational order and benign stability, so Billy Bartram never strays too far from the family farm that provided Crèvecoeur a perfect model of physiocratic life in America. If we look closely at his *Travels* we find the marks of neoclassical order all along the lush banks of the St. John's and a quality of wonder that persistently yields to an impulse to explain.

That order and impulse is essential to the form and purpose of Bartram's book, which was—its effect on Coleridge and Wordsworth notwithstanding—scientific, as witnessed by the voluminous catalogues and classifications that make up a large percentage of its total bulk. This material, which often disrupts the flow of narrative and counteracts the lyric intensity of the most-often-quoted passages, has been left out of a number of modern editions, yet it is an important and even essential element of the Enlightenment aspect of Bartram's book, with a meaning that transcends its nominal function. As his periodic apostrophes to the "supreme author" suggest, Bartram regarded nature's plenty as divine in origin, having been dispersed over the earth for the good of God's favorite creature, Man. Thus it is Man's sacred duty to determine the proper use of his bounty, for to neglect divine intention would be to sin by omission. Moreover, by piling up knowledge we get closer and closer to the source of ultimate truth, for as the outlines of the pattern of plenty become clearer, so does the providential plan, and as the beauty and fitness of the design become more apparent, God becomes all the more worthy of our worship.

William Bartram was especially fond of a line from Pope's *Essay on Man*—"Hills peep o'er Hills, and Alps on Alps arise"—whose meaning changes when it is removed from the original context. Pope meant to suggest the futility of learning, which serves chiefly to reveal how much more there is left to know, but Bartram was exhilarated rather than discouraged by the prospect, and the discovery that more mountains lay beyond the first range of hills only served to drive him on: "Continually impelled by a restless spirit of curiosity," is how he describes his quest, which was a constant "pursuit of new productions of nature" (*Travels*: 48). His notion of "happiness" had nothing to do with the possession of property, but rather with the acquisition of knowledge, with the twofold desire of "tracing and admiring the infinite power, majesty and perfection of the Almighty Creator, and . . . the contemplation, that through divine aid and permission, I might be instrumental in discovering, and introducing into my native country, some original productions of nature, which might become useful to society" (48). Though apparently foreign to the poetic impulse, Bartram's assiduous accumulation of data is an expression of his obsession with the unknown, and is intimately connected with his frequent apostrophes to natural beauty and his hymns to divine creation. His *Travels,* therefore, is at once religious and scientific in spirit as well as occasionally poetic in effect.

Most authorities regard the second part of Bartram's book, his account of his voyages on the St. John's River, as the most interesting in terms of science, and it has been that part also that has been regarded most highly by poets. There is, moreover, a natural order involved in any narrative of a river voyage, an upward and downward movement essential to those actions we call plots. And the unique character of the northward-flowing St. John's serves to increase the narrative interest, an exotic dimension further enhanced by exciting events and sudden discoveries. How much of Bartram's narrative is a genuine record and how much manipulated and exaggerated fact we will never know, but there are numerous indications that he compressed, expanded, and rearranged his material at will. Whether consciously imposed or purely coincidental, there is an order in Bartram's account of his voyage on the St. John's that encompasses and transcends his catalogues of Linnaean classifications.

First of all, like Crèvecoeur's Susquehanna and the Ohio of Cutler and Imlay, the St. John's River plays a geopolitical role, for it separated the lands claimed by the British from Indian country, and this function is expanded in Bartram's account to something very like the division promoted by Pownall's Hudson, a symbolic demarcation between aesthetic categories with powerful historical associations. Where the wild, western side of the Hudson as described by Pownall is merely a descriptive circumstance, the western side of Bartram's St. John's is literally *the* West, for the river served as a frontier line separating savagery and civilization. Indian territory lies to the west, the white settlements to the east, and throughout the narrative this distinction is enforced by a series of contrasting vistas much like Pownall's: "The Eastern coast of the river now opens, and presents to view ample plains, consisting of grassy marshes and green meadows, and affords a prospect almost unlimited and extremely pleasing. The opposite shore presents to view a sublime contrast, a high bluff bearing magnificent forests of grand Magnolii, glorious Palms, fruitful Orange groves, Live Oaks, Bays, and others" (88–89). The contrast here is classically Burkean, and is elaborated upon in subsequent passages: at other times the west side of the river is "swamps and marshes" or one "endless wild desert," while the east side opens to "the beautiful appearance of green meadows," an "Elysium" that bids the observer pay a visit (58, 103–4, 93).

It is, moreover, on the eastern side of the river that Bartram discovers the wonderful fountain containing an earthly "paradise of fish," "all in their separate bands and communities, with free and unsuspicious intercourse performing their evolutions: there are no signs of enmity, no attempt to devour each other; the different bands seem peaceably and complacently to move a little aside, as it were to make room for others to pass by" (105). For Bartram the religious man and poet, the "amazing and delightful scene" seems at first to be a marvelous work of art, "a just representation of the peaceable and happy state of nature which existed before the fall," but Bartram the scientist is committed to breaking the illusion: "In reality it is a mere representation" only, not a miracle, for the peace is maintained by the perfect clarity of the water, which places all the fish "on an equality with regard to their ability to injure or escape from one another," so that "the trout freely passes by the very nose of the alligator and laughs in his face" (106). The laws of nature have not been suspended; quite the contrary, for it is the absolute clarity of the water (a natural phenomenon) that enforces the marvelous peace, resulting in a microcosm in which the discrete species of fish make up a republic of restraint. Thus miracles yield their secrets to the power of rational inquiry, while reinforcing the decorum of harmony that holds on the eastern side of the river.

The wildness of the western side is correspondingly symbolized by another marvelous fountain located there, in which are found "incredible numbers of crocodiles [*sic*], some of . . . enormous size," all gorging themselves on the multitudes of fish that crowd into the deadly sink "from the rivers and creeks, draining the savanna" (130). Bartram calls this a "fatal fountain," and like the "paradise of fish" it is an "extraordinary place and very wonderful work of nature"—and explainable (131). The common denominator of both fountains is the alligator, "voracious" but motionless as a log in the first instance, and in the second animated by an insatiable appetite, for the giant American reptile plays a defining role in Bartram's *Travels,* serving the region watered by the St. John's as a presiding Genius of the Place: "Behold him rushing forth from the flags and reeds. His enormous body swells. His plaited tail brandished high, floats upon the lake. The waters like a cataract descend from his opening jaws. Clouds of smoke issue from his dilated nostrils. The earth trembles with his thunder" (75). Something other than detached scientific observation, Bartram's description evokes nature in its most horrific aspects, for if the scenery along the St. John's is characterized by abundance, so also is it a terrain of giganticism: John Bartram's excitement over relative sizes is elevated by his son into proto-Romantic awe in confronting the sublime. Like Pownall's Cohoes Falls, the St. John's is a great field of force in which fountains suddenly erupt with the sound of "a mighty hurricane," tearing the earth asunder; a river as apparently inimical to civilized tenure as is the flood-ravaged basin of Crèvecoeur's Delaware (150–51).

The alligator is a fit, even symbolic, denizen of such a titanic arena, but even the great saurian and the explosive fountains are overmatched by the hurricane itself, which is the most powerful tropical creature of all. Bartram encountered one such at the furthest reach of his upriver voyage, a horrifying experience first announced by the ominous sounds of approaching thunder and by the answering response of the alligators, "sure presage of a storm!" (89). Terrified, Bartram attempted in vain to flee the storm, as the skies above him "appeared streaked with blood or purple flame," while "the flaming lightning stream[ed] and dart[ed] about in every direction around," and the thunder shook the earth (89–90). Though he took shelter under his boat, he was soon drenched by rain, which fell "with such rapidity and . . . in such quantities, that every object was totally obscured, excepting the continual streams or rivers of lightning, pouring from the clouds; all seemed a frightful chaos. When the wind and rain abated, I was overjoyed to see the face of nature again appear" (90).

This last figure of speech establishes a disjunctive dualism, which calls into question Bartram's otherwise neatly articulated trinity of Religion, Science, and Poetry. The storm, having hidden the face of nature from sight, is therefore not to be considered part of nature, but distinct. By "nature," in this instance, Bartram means the benign aspects of the natural world, yet it is a telling limitation: the hurricane, like the alligator, is not only part of the natural world but it is to the source of the St. John's what the alligator is to the great fountains of Alachua, at once the dominant denizen and the ruling spirit. And yet where evidence of natural order and beauty is apt to inspire a rhapsodic hymn to the supreme wisdom of God, Bartram remains silent on divine purposes when describing the alligator or witnessing a hurricane, aspects of the natural world that are difficult to reconcile with a benign interpretation of universal creation.

In his longest hymn or "anthem" to the "great Creator," Bartram prays that "the universal sovereign" will "grant that universal peace and love, may prevail in the earth, even that divine harmony, which fills the heavens, thy glorious habitation" (65). This prayer is intrinsic not only to Bartram's Quaker religion but to the Enlightenment faith (keyed by terms like "great Creator") in the harmony of ultimate order. Bartram declares himself "an advocate or vindicator of the benevolent and peaceable disposition of animal creation in general," and, in specific, is willing to ascribe natural virtue to a rattlesnake and a renegade Indian alike (168). Throughout his travels, Bartram entertains a primitivistic vision of nature and natural man, depicting "sublime enchanting scenes" that permit "visions of terrestrial happiness." Florida is a natural "Elisium" in which the "wandering Siminole, the naked red warrior, roams at large, and after the vigorous chase retires from the scorching heat of the meridian sun. Here he reclines, and reposes under the odiferous shades of Zanthoxilon, his verdant couch guarded by the Deity, Liberty, and the Muses, inspiring him with wisdom and valour, whilst the balmy zephyrs fan him to sleep" (69). Yet the clumsy mixture of native American and neoclassical imagery betrays the unlikeliness of the scene, and Bartram is also capable of describing the Seminole in much less Arcadian terms.

Thus, having depicted the Alachua savanna as a natural paradise, in which deer, wild horses, turkeys, and "civilized communities of the sonorous, watchful crane, mix together, appearing happy and contented in the enjoyment of peace," he next conjures up a disruptive image of "warrior man": "Behold, yonder, coming upon them through the darkened groves, sneakingly and unawares, the naked red warrior, invading the Elysian fields and green plains of Alachua. At the terrible appearance of the painted, fearless, uncontrolled and free Siminole, the peaceful, innocent nations are at once thrown into disorder and dismay" (120). The effect is similar to combining the waters of the two marvelous fountains, setting loose the "fatal" qualities of the western sink in the "paradise" of the eastern.

Keepers of slaves, the Seminoles (those Sons of Liberty) also "wage eternal war against deer and bear" and make "war against, kill and destroy their own species" (135). They are however no more "savage" than other men, no different from "all other nations of mankind," and with the hurricane and the alligator must be distinguished from the paradisiac terrain they inhabit, with which they harmonize only when at rest "after the vigorous chase" is over. As for that paradise of wildness on the western side of the St. John's, Bartram is quite willing to subject it to the "united ingenuity and labour of men" other than red ones, for although civilized man could never "imitate" such a "natural" paradise, he most definitely could improve it: "Under the culture of industrious planters and mechanicks," the western side of the river "would in a little time exhibit other scenes than it does at present, delightful as it is; for by the arts of agriculture and commerce, almost every desireable thing in life might be produced and made plentiful here, and thereby establish a rich, populous, and delightful region" (148). In fairness to due proportion, such observations occupy a very small place in Bartram's book, yet they are consistent with the whole, and are not mere token gestures. They belong instead to that Horatian urge to make any natural landscape more beautiful by introducing the element of utility that William Bartram prized in common with Pownall, Crèvecoeur and other travel writers of his day. He was in this regard his father's son.

Bartram's orderly sensibility tranforms the landscape of the Floridian savannas in other, subtler ways, for with Euclidean eyes he sees cones and pyramids everywhere, from the spire-shaped magnolias to the funnel-form nests of alligators, rats, and crayfish. Despite the Romantic-seeming excesses of style, Bartram's book evinces an orderly, neoclassical aesthetic by abstracting a triangular precision from the abundance of teeming nature. This disposition to derive order from abundance is intrinsic to the Enlightenment imagination, as witnessed by Crèvecoeur's notion of the Schuylkill as a benign machine or the descriptions by Imlay and Cutler of the Ohio Valley landscape. Like those other apparent paradigms of orderly dispensation, moreover, Bartram's Florida has its disjunctive aspect, symbolized by the presence of disruptive forces, whether figured as hurricanes or Seminole Indians.

Perhaps the most distinctive feature of the Florida landscape is the symmetrical sinkhole, which Bartram describes as "being mostly circular or elliptical," and which, being frequently "surrounded with expansive green meadows" or "a picturesque dark grove of Live Oak, Magnolia, Gordonia, and the fragrant Orange," generally brings to the civilized mind a pastoral image (110). As at Pope's orderly garden at Twickenham, the sight of "a rocky shaded grotto of transparent water" suggests the poetic notion that it is "the scared abode or temporary residence of the guardian spirit, but [it] is actually the possession of a thundering absolute crocodile." Fountains of light, the sources of the St. John's are also basins of darkness, evoking the revelations of the wilderness saint for whom the river was named, and containing likewise an apocalyptic beast, which lends the illusory pastoral garden its one absolute but not very utilitarian fact.

VI

The Fountain and the Cave

i

In the travel writings of Crèvecoeur and Bartram, as we have seen, a latent design lurks in the observed landscape, an implicit symbolism incidental (even antithetical) to the factual burden of the narrative. By contrast, in the fiction produced by American writers during the same period, the abstract quality of the landscape dominates, virtually obliterating any vestiges of realistic depiction. This is especially true of works in which the landscape plays a minor role, like the sentimental novels of the period, in which the use of scenery resembles that of contemporary portraits—a distant glimpse is had of a parklike patch of trees or a meandering stream, seen over a shoulder or through a window. The proportion is not surprising, since the sentimental novel is a thoroughly domestic (if often violent) genre, dictating that the action take place mostly withindoors. But even in the picaresque novel the landscape plays a minimal role: it is an assumed rather than an accentuated presence, flat and featureless as a painted stage set, being a stock and generalized scene against which the plot is enacted. Yet in both instances the landscape is highly charged with symbolic meaning.

There are, however, transatlantic differences: where in the English picaresque tradition the tripartite division of town, road, and country dominated, Hugh Henry Brackenridge's *Modern Chivalry* (1792–1815) yields to Pennsylvanian necessity, and the action is divided between the eastern and western sides of the Alleghenies. Still, though the division is essential to Brackenridge's argument, it is largely a geopolitical or topographical one, and does not result in any detailed account of the scenery along the route taken by the aristocratic (and quixotic) Captain Farrago and his ad hoc squire, the Sancho Panza–like Irish servant, Teague O'Regan. When Teague first becomes ambitious for political office, the captain chides him for his lack of qualifications: "Do you understand anything of geography? If a question should be, to make a law to dig a canal in some part of the state, can you describe the bearing of the mountains, and the course of the rivers?" (16) The same vacuity may be charged to the book as a whole, which moves in general terms back and forth across the mountains between the new settlements on the western side and Philadelphia to the east, but with no acknowledgment of the peculiarities of the landscape.

As a young Princeton poet, Brackenridge celebrated the great rivers of North America as adjuncts to national growth, but in his extended fiction, the publication of which took place over almost twenty-five years (from 1792–1815), it is not rivers but the Allegheny Mountains that play the most important role, as just the sort of barrier that geopoliticians like Jefferson and Washington hoped to surmount. On one side lies the civilized East, on the other a much more dubious region, in which the democratic spirit takes such extreme forms that, toward the conclusion of *Modern Chivalry*, brute animals are elevated to full citizenship in the United States. The West as a literary territory will often be regarded with uneasiness, if not fear and loathing, by those eastern writers who subsequently chose it as a fictional territory—a disjunctive tradition first established by Crèvecoeur—but as a realized landscape the region beyond the Alleghenies would have to wait for a writer who had benefited from the Romantic movement, particularly as it was manifested in Scott's Waverley novels. As we have already seen, writers educated in Enlightenment ideals had difficulty incorporating wilderness scenery into their celebrations of rational progress, nor did personal contact with the new inhabitants of western regions do much to foster Enlightenment optimism.

Brackenridge is occasionally a celebrant of the western landscape, praising "the country in its virgin state," where the rivers "run clearer than in the old," and where the people "appear to have more intellectual vigour, and in fact more understanding in the same grade of education, than [do] the inhabitants of an old settlement, and especially of towns and cities. The mind enlarges with the horizon" (555–56). But in general the book belies this, since in Brackenridge's increasingly conservative assessment, the western settlers are particularly susceptible to demagoguery, and we see little evidence of men's growing to match their widened vistas. What does grow is the ambition of the stupid and vainglorious O'Regan, whose unsuitability for office is no obstacle to political popularity.

The promise of the new country is mostly vested in representatives of the eastern aristocracy, like Captain Farrago, who builds his western home on a symbolic site, orienting it "east and west upon a ridge of ground like a whale's back, with a stream on each side running a direction contrary from each other, but falling into two sister rivers on the east and west, which joined their silver currents at a small distance, and in prospect of the building" (789). Farrago has been elected governor of a new western state, and his barrackslike log house is designed for the democratic "accommodation of individual families for a night, in their emigration to a new settlement" (789). Yet its location on a height of land suggests a certain aristocratic removal from the people he governs (nor is the "whale's back" an image of stability), while the symmetry of design is a purely intellectual construct, like the layout of the capital city below it a geometric projection westward of the Philadelphian ideal: "The new town, as it was yet called, stood in sight [of the governor's house], and had begun to shew two streets of houses at the confluence of the two rivers, and parallel with each, with the public buildings at equi-distance from the banks; and towards the base of the right angle which the two streets formed" (789).

A far different version of the Ohio Valley emerges from Gilbert Imlay's *The Emigrants* (1793), which employs the same geopolitical division as *Modern Chivalry* but with a decidedly Jacobin bias. Informed by the utopian view imposed on Kentucky real estate in his *Topographical Description,* Imlay's novel posits a symbolic geography in which virtue inhabits "the region of innocence" west of the Alleghenies, while vice thrives in Bristol, Pennsylvania, and points east. The Jacobinism that informs his earlier work is brought to a finer and more pointed instrumentality in Imlay's novel, which may have been directly influenced by the author's intimate association with Mary Wollstonecraft and her circle. An epistolary novel of considerable complexity, *The Emigrants* is set, variously, in London, in Philadelphia, and along the Ohio River, from Pittsburg to Louisville, with a concluding and climactic excursion to the Illinois River. The tone of the book is that of the sentimental novel, with a concomitant emphasis on the more painful aspects of young love, and though there is considerable description of natural scenery, it is highly idealized and in service of Imlay's chief thesis regarding the salutary influence on character of the western climate and landscape. What Brackenridge declares but then proceeds to disprove, Imlay demonstrates at length. Though sharing the epistolary mode and sentimental effusiveness of contemporary erotic fiction, Imlay's book was written in the service of a radicalism that was antithetical to the conventional morality preached by Samuel Richardson and his American imitators. Moreover, as an early work of native fiction, it resembles Royall Tyler's play, *The Contrast,* (1787), drawing invidious distinctions between American and European manners and morality. At once singular and prophetic of much fiction to come, *The Emigrants* is in all respects a fascinating exercise, in none more so than in its presentation of the western landscape and its rivers.

ii

Where the action in *Modern Chivalry* is defined by a reciprocal movement back and forth across the mountains, in *The Emigrants* (as the title suggests) the predominant direction is from east to west, the course of the book carrying the characters from Philadelphia to Pittsburg and thence down the Ohio to Louisville. And yet, while this movement is aligned along the course of western emigration, Imlay chose as his main vehicle, not the narrative of adventure, but the sentimental novel, producing a curious hybrid. We may assume that Imlay was motivated by contemporary literary fashions, but the result is an interesting instance of subversive design, in which the substance of the novel is at war with the traditional implications of the form. Mixing adventure with the materials of the novel of manners, lending the genteel tradition of the sentimental romance an erotic edge, Imlay produced a mixture of genres that perfectly expresses his radical view of the Western world.

Sentimental fiction in the late eighteenth century was to sensationalism closely allied, with an emphasis on seduction and betrayal, yet the genre was dedicated to promulgating middle-class norms, specifically the sanctity and desirability of marriage. The moral thrust was given a certain spiciness by parading bad examples before the reader, but the desired effect was to impress the audience—generally conceived as female—with the dangers of yielding to temptation. William Hill Brown's *The Power of Sympathy* (1789), generally recognized as the first American novel, is a case in point, a short, epistolary fiction that is crowded with disastrous examples of the consequences of seduction—including suicide, incest, and bad dreams. Against this panorama of fatal folly is placed the shining example of a good marriage, based on sensible precepts rather than an excess of sensibility. Notably, the assorted victims of lust act out their fates in the purlieus of Boston and Rhode Island, while the spokespersons of virtue send out letters from "Belleview," a pastoral asylum whose inhabitants, when not writing moral epistles, read aloud to one another from *The Vision of Columbus* and *The Conquest of Canaan*—less a guide to reading habits in the early republic than an exercise in chauvinistic zeal, with a clearly moralistic bias.

In Imlay's novel, the pastoral zone is translated much farther west, yet the division between urban vice and rural virtue remains constant. Still, vice for Brown is a relatively simple (if disastrous) yielding to the power of sympathy, while for Imlay it is a Godwinian mixture of bad environment (including monarchical governments), faulty education, and flawed character. While his plot centers on a pair of lovers who are eventually united in lawful wedlock, equal emphasis is given to a number of staggeringly bad unions that only endure because divorce is not legally possible. In sum, Imlay's book, while masquerading as a celebration of marriage, is actually a tract encouraging legislation permitting ease of divorce. Along the way, Imlay denounces the commercial spirit, debates the "perfectability" of society, and demonstrates the beneficial effects of the Western environment on a wayward and effeminate (because Europeanized) young man.

His is, in short, a thoroughly Jacobinized fiction, informed by Rousseau and sustained by Sterne, in which the cast of virtuous characters—including the Wise Old Man, the Reformed Youth, the Admirable Young Couple, and several subsets of Loyal Friends and Loving Relations—end the novel by setting up a utopian community on the banks of the Ohio near Louisville. Where other "moral" American novels of the period, like *The Power of Sympathy, The Coquette,* and *Charlotte Temple,* end tragically, with a terminal emphasis on the sufferings of the victims of sensibility, *The Emigrants,* having disposed of sundry bad examples, thereby purifying society, ends on a positive note, like a Shakespearean comedy celebrating the triumph of right reason via a symbolic pair of marriages. Thus, where *The Power of Sympathy* posits a pastoral center in opposition to urban vice, Imlay's novel ends by removing all his virtuous characters to a radicalized Edenic zone where bad marriages are no longer possible, giving new life (and a much different meaning) to the old Puritan tradition of separatism.

Where, in *Modern Chivalry,* folly may be found on both sides of the mountains, geography in Imlay's novel, as in his topography, has a distinctly moral watershed, establishing zones of influence that will become familiar terms in much American fiction—and many utopian tracts—to come. What is unique about *The Emigrants,* in terms of contemporary writing as well as of much that is to follow, is its erotic emphasis, a celebration of love with an associative geographical backdrop. Thus the heroine, Caroline T———n, is the daughter of an impoverished English emigrant, while the hero, Captain James Arl——ton, is a native-born veteran of the Revolution, and their courtship an international affair that has its beginnings when the lovers first meet in crossing the Alleghenies into the Ohio Valley. From the start, Caroline is depicted in terms evoking the landscape through which she passes: "When I first caught sight of Caroline," Arl—ton writes to his wise and benevolent friend, Mr. Il—ray, "she was resting upon a large stone on the road side . . . leaning with one hand upon her cheek, and [holding] a handkerchief, which I thought had been applied to her eyes, that were glimmering like the rays of the sun through the mist of an April shower" (14). Comparing a fair maiden to the delightful aspects of nature is a common analogy, but by repetition Caroline becomes a virtual personification of the "innocent realm" toward which she is heading.

Through a series of misunderstandings, Caroline is separated from her lover, Arl—ton, and while she remains in Pittsburg to pine exquisitely, he travels down the Ohio to Louisville in a fit of expansionist pique. (In Imlay's world view, the Wertherean pangs of disappointed love are best buried in real estate speculation.) Though Caroline rejoins Arl—ton in Louisville, further misunderstanding results in the hero's hurling himself into an even wilder zone: "Let me look," he writes Il—ray, "in the wilds of this extensive region for that peace of mind, which Caroline, I fear, has for ever destroyed. Come my friend, let us together explore the country until we find the sources of the Mississippi, and the limits of the more impetuous and extensive Missouri; for I will live in this uncultivated and uncivilized waste, until my person shall become as wild as my senses" (234). In sum, Arl—ton elects to follow Crèvecoeur's Farmer into a wilder zone, and what the Farmer fears (the uncivilizing influence of the wilderness), Arl—ton seeks.

Hoping to assuage her own sorrow, Caroline makes her own—albeit unintentional—incursion into the wilderness: crossing the Ohio to the other (Indian) side with the intention of viewing the falls at Louisville, she is taken captive by a band of marauding savages and is carried westward through the wilderness. Arl—ton, arriving at the Illinois River in a cross-country trek to St. Anthony's Falls, comes coincidentally upon Caroline and her captors. He rescues her and finds she is literally all the better for wear (or lack of it):

> *Ah!* [*he exults in a letter to Il—ray*] *how did my swelling heart beat with joy, which was instantly succeeded by sorrow, when I first caught a glance from the brilliant eyes of the most divine woman upon earth, torn into shatters by the bushes and briars, with scarcely covering left to hide the transcendency of her beauty . . . She is at this moment decorating the gardens of this place while I am writing to you, and seems to give enchantment to the whispering breezes that are wafted to my window, and which in their direction as they pass her, collect from her sweets*

the fragrance of ambrosia, and the exhilerating charms of love itself. (251)

Stripped by the thorns of adversity, the nearly naked Caroline becomes type and symbol of the virgin land she will eventually inhabit; yet she is also a representative of civilizing order, for her presence converts "the horrors of a wilderness . . . into elysium." She becomes (as in her gardening activities) a central and organizing feature of the landscape, "her smiles like the genial hours of May when nature blooms in all its eradiated charms . . . [her] bosom more transparent than the effulgence of Aurora, when robed in all her charms" (253). In no other example of early (or late) American fiction will the landscape take on such a distinctly erotic coloration, for the Illinois country clearly is transformed by the addition of Caroline to the scenery:

Every thing seemed to be enchantment as we passed the extensive plains of the Illinois country. The zephyrs which had gathered on their way the fragrance of the flowery riches which bespangle the earth, poured such a torrent of voluptuous sweets upon the enraptured senses, that my animation was almost overpowered with their delicious and aromatic odours. The fertile and boundless Savannas were covered with flocks of buffalo, elk, and deer, which appeared to wanton in the exuberance of their luxurious pastures, and which were sporting in the cool breezes of the evening. . . . But when the scene was embellished by an image so fair and beauteous as that of Caroline's, we seemed to have regained Paradise, while all the golden fruits of autumn hung pending from their shrubs, and seemed to invite the taste, as though they were jealous of each other's delicious sweets. (258–59)

"Flocks of buffalo" keys the pastoral note, but the diction carries a sensual burden also, which Caroline echoes in a complementary description of the same territory sent to her sister, Eliza, in England: "The wild regions of the country of the Illinois, where the sweetened breezes attune the soul to love, and nature exuberant, in her extensive lap, folds the joyous meads which enraptured smiled around, and every shrub seemed pregnant with her charms" (274). It was here, she tells her sister, that she was rescued by Arl——ton, an act that converted the "wilderness" into a "heaven," for while "the soft gales of nature, which then appeared to unbosom and display her every charm" were blowing, the "senses" of Caroline were "ravished with the modulated symphony of the feathered choir" (274). It becomes difficult here to determine who is being stripped and ravished by whom: "Beauty smiled in all its mild effulgence, and when Arl——ton snatched kisses from my lips, all my soul hung lambent to the ambrosial touch. —Heaven! to what an ecstacy are our feelings brought when they have been wantonly sported with, and thus unexpectedly to taste the rewards of their virtuous struggles?" (274). This, plainly, is not the sort of reward virtue customarily receives in the moral fiction of Imlay's American contemporaries, nor does his notion of the uses of adversity comport with that promoted by the Puritans' severer muses.

Imlay's unsubtle use of sexual imagery when describing the western landscape extends to other, less obviously erotic, descriptions, as in Caroline's account to her sister of the two rivers that join at Pittsburg to form the Ohio:

> *The Allegany is not so broad as the Monongahela, but its current is much more impetuous, and from the fierceness of its aspect, and the wildness which lowers over its banks, it appears to be what it really is, the line between civilization and barbarism. . . . On one side of us lie the wild regions of the Indian country; on the other side our prospect is obstructed by the high banks of the Monongahela, beyond which lies a beautiful country that is well peopled and cultivated—behind us a considerable plain that is laid out in orchard and gardens, and which yields a profusion of delicious fruits,—and in our front the Ohio displays the most captivating beauty, and after shooting forward for about a mile, it abruptly turns round a high and projecting point, as if conscious of its charms, and as if done with an intent to elude the enraptured sight.* (61–62)

Here again we have the theme of two rivers, which from Pownall to Bartram defined the American landscape in disjunctive terms, the savage Allegheny and the civilized Monongahela joining to become the Ohio. Yet in Caroline's description the union becomes another erotic metaphor, put forward in the standard pastoral language of seductive flight. Imlay later reinforces this sexual metaphor, this time through the agency of Mr. Il—ray, who writes to Arl——ton that "in contemplating the fierce Allegeny . . . I could not help reflecting upon the impetuosity of your passion, which it appeared to me must have been influenced by the current of that rapid river which seems to be hurrying, to intermingle its waters with the more gentle Monongahela" (206). After Arl——ton has left for Louisville, Caroline finds herself "constantly mortified with the projecting headland, that obstructs my view down the river," the same feature of the landscape that before was "used" by the Ohio "to elude the enraptured sight" now a symbol of interrupted desire (87, 62).

All features of the Ohio conspire to promote a sensual and fruitful union of civilization and the wilderness, as at "Bellefont," the community that Caroline and her Captain establish on the banks of the river near Louisville, a "delightful spot" that

> *has a combination of charms, that renders it altogether enchanting. . . . This tract is bounded by the Ohio to the west; and here the expansive river displays in varied pride the transparent sheet, that gushing, shoots impetuous over its rocky bed; which, as if in a rebellious hour had risen to oppose its genial current, presents a huge, but divided barrier; and while nature seems to scorn its feeble power, the repercussive thunder proclaims her triumph, and the ethereal hills on the adjacent shore give lustre to the rising moisture, which creeping through the vistas of the groves, the country round, high illumined, in blushing charms its sweets diffuse, and nature shines effulgent to the joyous sight.* (311)

The scene evoked is one of those symbolic junctures characteristic of landscape description in late eighteenth-century America, and, as in other instances, it seems essentially confused in its implication. The connotations of power are reminiscent of Pownall's Cohoes Falls or Jefferson's description of Harper's Ferry. But, like Bartram's hurricane at the head of the St. John's, the rapids at Louisville are a natural phenomenon caused by rocks that are characterized as being, not a part of, but opposed to "nature," identified by Imlay as by Bartram only with "genial" aspects of the landscape. Though temporarily interrupted by the turbulent falls, the river both above and below presents a scene of sensual and feminine tranquillity, lending its symbolic force to the loving mood of "Bellefont": "The winding river here presents itself in two directions, and on either hand the eye dwells with peculiar delight upon its fair bosom; and while the whispering breezes curl its limped waters, the azure veins seem to swell, as if they were enraptured with the soft dalliance of their fragrant sweets" (311).

"Bellefont" takes its name from a literal fountain, which is a sacramental sign of pastoral purity: "It gushes from a rock; and when its different pliant rills have joined at its base, they form an oval bason, about three hundred feet in diameter, which float over a bed of crystals, that eradiates its surface, and gives to it a polish more transparent than a mirror of glass" (312). Like the Ohio a symbol of fluvial union, the fountain (like the falls at Louisville) puts forth connotations of fruitfulness, becoming thereby a watery equivalent to the marriage of Caroline and Captain Arl——ton:

> *The water steals off in several directions, and in their meandering course moistens the flowery banks, which, as if to return the load, spread their blooming sweets on every side, and the soft gales gather their odours as they pass; and while they perfume the ambient air, the wanton hours dancing to the gentle harmony of sweet sounds, which the feathered songsters warble in modulate strains, love seems to have gained absolute and unbounded empire, and here in the couch of elegance and desire, to dally in all the charms of its various joys.* (312–13)

This pastoral intensity emphasizes Imlay's idealistic division between the corruptions of the old world east of the mountains and the "innocence" of the western regions—as in his topography, asserting the contiguity of the separatist, the secessionist, and the utopian impulses. "As the government of this district is not organized," writes Arl——ton to Ilr—ay, "it is my intention to form in epitome the model of a society which I conceive ought to form part of the polity of every civilized commonwealth" (295). Imlay's ideal state is ruled by a "house of representatives . . . who are to assemble every Sunday in the year, to take into consideration the measures necessary to promote the encouragements of agriculture and all useful arts, as well as to discuss upon the science of government and jurisprudence" (295). The quality of Imlay's Jacobinism is revealed by his version of a "Sunday meeting," an anti-ecclesiastical gesture that also betrays the subversive nature of his utopian society; for whatever else his community may be, it is obviously ruled by a government independent of the United States of America. The pan-Federalism of his topography has been set aside for an insular commonwealth, planted by the Falls of the Ohio, a notorious block to commerce in both directions, and thereby an impediment to the realization of national union. Thus the implicit separatism of the earlier book is made explicit in Imlay's novel.

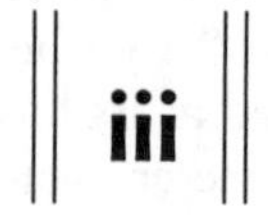

Even in 1793, Imlay's was a minority view in the United States, identified with the wilder fringes of Enlightenment idealism, and his affair with Mary Wollstonecraft, set against the backdrop of the French Revolution, could have done nothing to help promote his radically republican vision. Though many American intellectuals were sympathetic to the revolution at the start, enthusiasm waned as the guillotine, instead of the liberty pole, became its chief symbol. By 1798, paranoid Federalists like Jedidiah Morse were rushing to pulpits and into print with dire warnings against the danger to republican institutions of subversives who were immigrating from France as refugees but were in truth members of the Bavarian Illuminati seeeking to foment discontent. In that same year, Federalist legislators pushed through the notorious Alien and Sedition Acts in reaction to the perceived threat of subversion from abroad, an extreme reflection of a mood hardly friendly to Imlay's Jacobin schemes, especially when set forth in a novel concerning idealistic immigrants' removing to America in order to set up a republican utopia that observed the Sabbath in unusual ways.

More in tune with the reactionary mood as the Enlightenment darkened into the early stages of Romanticism were the works of Charles Brockden Brown. Brown began his literary career in sympathy with Jacobin ideals, was an admiring reader of William Godwin's *Political Justice,* and framed his earliest fiction in imitation of *Caleb Williams.* But by 1798 he had begun to illustrate Morse's paranoia in a series of determinedly anti-Jacobin novels, beginning with *Wieland* (1798), fiction that is centered by the activities of Godwinian hero-villains—who are often foreign-born or influenced by notions associated with illuminatism, and whose schemes, even when well-intentioned, wreak havoc on American soil. What Brown wrote in 1806 of the French Revolution, that "the revolutionary patriots whose intentions are admitted to be pure" were nonetheless guilty "of great precipitation, presumption, and imprudence," may be said of all his Godwin-inspired protagonists, whose activities parallel events in France under Jacobin rule, which he defined as "the miscarriage of a scheme of frantic innovation" (Clark: 235).

Charles Brockden Brown is seldom viewed as a geopolitician, but in 1803 he wrote two anonymous pamphlets that may be set against Gilbert Imlay's topography for their contrasting view of western regions. Though Brown regarded the Mississippi Valley as a region of great value, he pursued a rhetorical strategy that provides a key to his paranoiac fictional world. Arguing for the acquisition of French Louisiana, Brown proposed that conquest would be cheaper than purchase, and he attempted to stir up francophobia by claiming to have in his possession a secret paper outlining a French plot for masterminding a revolt in the United States by slaves and Indians, using Louisiana as a base for intervention in succeeding events. This "plot" (it was Brown's invention, warranted by the Genêt Affair and the recent insurrection on Santo Domingo) would realize the Federalist nightmare of geopolitical dissolution, setting state against state, section against section, each pursuing its private interests at the cost of national unity. Brown thereby promoted an image of the Southwest as a nursery of anarchy, not a hothouse of republican virtue. There was, as we shall see, an enduring tradition of secession in the Mississippi Valley (of which Imlay was not innocent nor Brown apparently unaware), but it was distinctly a homegrown movement with little active French encouragement, despite the efforts of the conspirators to win Napoleon's support. And fears of slave insurrections would endure until the Civil War. But Brown's anonymous pamphlets are chiefly of value here as illuminating his essential political paranoia, especially as evidenced by his unwavering hostility towards French-inspired meddling in American affairs.

In Brown's novels we have a reactionary alternative to Imlay's radical utopian vision, promoting the gothic, rather than the sentimental romance as a primary vehicle for literary landscape in America. Both writers are indebted to William Godwin, but in antithetical ways, the Reign of Terror having aroused in Brown considerable hostility concerning the effects and even the origins of Jacobin schemes for social improvement. In most of Brown's novels the brilliant irradiations of the Enlightenment are illustrated not by sunny vistas but by deeply shadowed ravines. Instead of promoting situations in which the New World landscape acts as a benign influence, inspiring well-reasoned plans for political utopias, Brown created natural settings that enhance his emphasis on irrational, antisocial behavior, an environment in which chance, not reasoned action, provides the chief agent of causation. Rather than emphasize admirable examples of virtuous activity, Brown preferred to foreground characters whose destructive impulses, though perhaps founded on benevolent ideas, result in an atmosphere of pervading evil. His landscapes, as a result, are melodramatic backdrops to episodes of terror and violence, in all ways antithetical to the sunlit scenery of Imlay's physiocratic drama. Instead of sparkling fountains, Brown prefers sinister caves, subterranean passageways leading to some perilous encounter with destructive forces in human form.

A key to the difference in America between the sentimental and the gothic modes as they relate to the Enlightenment heritage may be found by comparing pastoral Belleview in William Hill Brown's *The Power of Sympathy* to a similar setting in Brockden Brown's *Wieland*. The summerhouse at Belleview (where the "good" characters read aloud from Barlow and Dwight) is called "the TEMPLE of *Apollo,*" and is "elegantly furnished," which appointments include "a library and musick" (I: 28). Set in a pastoral spot notable for the "beauty of the prospect," which is enhanced by "the grandeur of the river that rolls through the meadow," the summerhouse is reached by an "avenue" whose gate is decorated by an allegorical figure "of CONTENT, pointing with one hand to the Temple, and with the other to an INVITATION, executed in an antique style, that you would think it done either by the ancient inhabitants of the country, or by the hand of a Fairy" (27–29). In *Wieland* a similar rural retreat has quite a different setting, placed

> *on the top of a rock whose sides were steep, rugged, and encumbered with dwarf cedars and stony asperities. . . . The eastern verge of this precipice was sixty feet above the river which flowed at its foot. The view before it consisted of a transparent current, fluctuating and rippling in a rocky channel, and bounded by a rising scene of corn-fields and orchards. The edifice was slight and airy. It was no more than a circular area, twelve feet in diameter, whose flooring was the rock, cleared of moss and shrubs, and exactly levelled, edged by twelve Tuscan columns, and covered by an undulating dome. My father furnished the dimensions and outlines, but allowed the artist whom he employed to complete the structure on his own plan. It was without seat, table, or ornament of any kind.*
>
> *It was the temple of his Deity.* (11–12)

As the landscape suggests, this deity is no pagan Apollo, but a punishing, Judeo-Christian God, derived from the "doctrine of the . . . Camisards," an apocalyptic-minded Huguenot sect. The elder Wieland, convinced that he has been divinely commanded "to disseminate the truths of the gospel" among the American Indians, suffers from the belief that his lack of success in this regard was equivalent to disobeying God (10–11). His divinely appointed mission is the original Puritan errand, and like preachers of a Matherish persuasion, old Wieland is convinced he will suffer a terrible penalty for his failure, an obsession which in time is realized: one night he is destroyed by spontaneous combustion, a horrid death that takes place in his temple.

The father's mental instability is passed on to his son, who converts the sanctuary to unlike purposes but with like results. Where the senior Wieland is a child of the Puritans, the younger is clearly an Enlightenment product: a devotee of classic authors, particularly Cicero—that apogee of reasoned discourse and balance—Wieland converts his father's retreat to uses much like that of the summerhouse at Belleview. Setting up a bust of Cicero on a pedestal, he adds a harpsichord, transforming the "temple" into "a place of resort in the evenings of summer. Here we sung, and talked, and read, and occasionally banqueted. Every joyous and tender scene most dear to my memory, is connected with this edifice" (24). So recalls Wieland's sister, Clara, the novel's narrator, but this idyll is changed with the appearance of the mysterious Carwin: thenceforth the summerhouse begins to take on dark implications, the temple of reason becoming increasingly associated with irrational acts, and finally with stark, raving madness.

Carwin is one of Brown's ambivalent, Godwin-inspired villain-heroes, who, though born in America (western Pennsylvania), has spent some time in Europe, where he came under the influence of the Illuminati. Though he is motivated by good intentions, Carwin's use of ventriloquism triggers the germ of paternal insanity in young Wieland, so that, like his father, he becomes convinced he is an agent of God. Misconstruing Carwin's ventriloquized messages as divine commands, he kills his wife and children and threatens the life of his sister. Throughout the ensuing melodrama, the domed Jeffersonian temple on the barren rock serves as a focal point, a scenic center to which much of the action refers. Clearly, in Brown's view, if the Puritan errand is clouded with psychosis, its Enlightenment counterpart is equally suspect: reason, under the right conditions, can become the agent of insanity. Where Joel Barlow, under the influence of the Enlightenment, could convert the Puritan tradition of prophecy to a celebration of man's capacity for improving both himself and his environment, the anti-Jacobin Brown treats the visionary mode as a matter of either madness or mere charlatanism. Thus the temple converted to a pleasure dome is haunted by the figure of Carwin, whose (literally) equivocal gift transforms him into something floating up from a madman's dream: "The face was turned towards me. Every muscle was tense; the forehead and brows were drawn into vehement expression; the lips were stretched as in the act of shrieking, and the eyes emitted sparks, which, no doubt, if I had been unattended by a light, would have illuminated like the corruscations of a meteor" (147–48).

In Carwin, it might be said, the Enlightenment tradition of benign prophecy in America receives a powerful antithesis, a seer whose warnings precipitate a tragedy. Symbolically, Carwin is first inspired to master his ambivalent art when, as a rural youth tending cows, he wanders into a gloomy ravine, a "vocal glen" that produces marvelous echoes. This becomes the boy's favorite spot, to which he repairs to read Milton's *Comus* and master the art of "biloquism" (252). Though located in western Pennsylvania, the dark ravine, in a metaphorical sense, leads Carwin to where the Schuylkill passes through the Wieland estate, and to "a bushy hollow on the edge of the hill" where stands the temple (200). Like the "vocal glen," this natural recess becomes Carwin's retreat: "The odour of its leaves, the coolness of its shade, and the music of its water-fall . . . converted . . . my sadness . . . into peaceful melancholy" (203). It is here that he first projects his voice to the confusion of Wieland, shouting warnings to "Hold!" that have a fatal effect on events. Brown thereby associates romantic riverine scenery not only with melancholy but with madness, promoting both a Gothic mood and an anti-Jacobin burden.

The identification of insanity with watercourses takes a wilder turn in *Edgar Huntly* (1799), where the American landscape is even more thoroughly gothicized, not only in terms of scenery but in terms of psychological symbolism as well. The region in which much of the action takes place is defined by "the forks of the Delaware" (though *which* forks Brown never makes clear), a fanciful region containing a township called "Solesbury," a wasteland district called "Norwalk," and a valley called "Chatasco," all depicted as a mountainous borderland between the settled regions of Pennsylvania and the Indian country beyond. What was for Brackenridge a social and satiric divide, and for Imlay a barrier between American innocence and European depravity, becomes for Brown a boundary region close to what Crèvecoeur's Farmer found along the headwaters of the Susquehanna and Delaware rivers. In pursuit of an Irish emigrant named Clithero Edny, a sleep-walking madman suspected of murdering Edgar Huntly's friend Waldemere (to whose sister the novel is written by Huntly as a lengthy epistle), the hero several times penetrates the mazelike countryside of "Norwalk," a region of "caverns and torrents": a surrealistic landscape of the mind, Norwalk is the antithesis of Imlay's sensual wilderness of the body, and is characterized by scenery that is anything but erotic.

Central to the place is a cave to which Huntly tracks Edny, a dark tunnel that leads up to a strangely symmetrical summit, rather like a ruined tower surrounded by a moat: "If you imagine a cylindrical mass, with a cavity dug in the centre, whose edge conforms to the exterior edge; and if you place in this cavity another cylinder, higher than that which surrounds it, but so small as to leave between its sides and those of the cavity, an hollow space, you will gain as distinct an image of this hill as words can convey" (102). From the central eminence there rushes a cataract, which, "dashing along the rugged pavement below" makes "a perpetual mist," so that "there was a desolate and solitary grandeur in the scene, enhanced by the circumstances in which it was beheld, and by the perils through which I had recently passed, that had never before been witnessed by me. A sort of sanctity and awe environed it, owing to the consciousness of absolute and utter loneliness" (103). To place this spot next to the falls of the Ohio in Imlay's novel is to epitomize the difference between the sentimental and gothic modes of landscape depiction.

Moreover, this awesome place corresponds to the temple in *Wieland,* the artificial structure here transformed into a sublime manifestation of natural landscape. Where the temple is manmade, the hill has all the characteristics of an artificial origin, being a wilderness equivalent to a Gothic ruin. Carwin lurks in a cave beneath Wieland's temple, and it is on the summit that Huntly first spies the crazed Clithero (a name echoing "Cicero"), who is seen standing on the central eminence (or tower), separated from him by the chasm and its cataract. Where Cicero is a symbol of reason, Clithero, now a half-naked wildman with "shaggy and tangled locks, and an air of melancholy wildness," is clearly a mad projection of the gothic scene (104). Huntly has already learned that Clithero's madness—like Wieland's—is the result of misplaced philanthropy, and much of the action that follows turns on the theme (frequently voiced by the hero): "How imperfect are the grounds of all our decisions!" (92). The region of Norwalk, with its maze of torrents, ravines, and caverns, is a wasteland that epitomizes the snares and delusions of the world, and the hero, disturbed profoundly by the spectacle of the mad Clithero and haunted by dreams of his dead friend, becomes a sleepwalker in that region himself, awakening in the bottom of a cylindrical pit located in the cave that leads down from the symmetrical summit where Huntly first sighted the madman.

Again, these architectonic touches evoke the ruins of the gothic novel, but that this is an American scene is emphasized several times over by Brown: having escaped the pit, Huntly encounters a panther, which he kills with a tomahawk, and when he finds his way out of the far end of the cave, it is only to stumble upon a band of marauding Indians who are holding a white girl captive. Huntly, having been orphaned by an Indian attack, is no American Rousseau, yet his abhorrence of violence is such that he must reason his way to killing an Indian sentinel so he can rescue the girl. Imlay's Arl——ton likewise is reluctant to kill the Indians holding Caroline captive, but where pacifism suits the utopian motive of *The Emigrants,* Brown is much more interested in challenging the Enlightenment attitudes of his hero by dropping him into situations where they simply are not operative. Huntly refrains from killing the other Indians, but when, as a consequence, they follow him and the girl, he is forced to kill them anyway. Unlike Imlay's Ohio Valley, Brown's wilderness tract is no place for philanthropy, nor, as the story seems to testify, is much of the rest of the world, where the most benevolent impluses often inspire disastrous actions. The difference between wilderness and civilization is merely one of degree: one goes to sleep in one's own bed and wakes up in a cave.

Brown's quarrel with Enlightenment attitudes aside, what is most interesting about the landscape in *Edgar Huntly* is its symbolic function. Thus the cavern serves as a corridor from the wilder aspects of nature found on the border of civilization into the heart of savage darkness. In following the lightless maze of passageways, Huntly is led first to his vision of the madman-wildman Clithero, then to his encounter with the pit and the panther, and finally, into the midst of the native American phenomenon that, having served the pious propaganda of the Puritans, would provide aspiring writers with a situation analogous to the imperiled heroines of romantic fiction—capture by Indians. Huntly, escaping with the unnamed white girl, has further encounters with her Indian captors, kills them, but is so badly wounded he is left for dead by a band of settlers. His subsequent wanderings through Norwalk bring him to the Delaware River, on the east side of which is the road to civilization and safety: "The opposite bank was five hundred yards distant, and was equally towering and steep as that on which I stood. Appearances were adapted to persuade you that these rocks had formerly joined, but by some mighty effort of nature, had been severed, that the stream might find way through the chasm. The channel, however, was encumbered with asperities over which the river fretted and foamed with thundering impetuousity" (214). At the start of his adventure, Huntly confronted the mad Clithero across a waterfall; toward the end he stares longingly at the road to home across an impassable river. The symmetry suggests a design, but not the kind accommodated by Enlightenment optimism.

In effect, the Delaware presents a picture equivalent to Jefferson's Potomac, but with antithetical implications. "Norwalk," as Huntly describes it, is a vale between two mountain ridges that "gradually widened as it tended to the westward," and whose southern ridge is bordered by the river, but as the Delaware tends eastward, "the mountain and river receded from each other, and one of the cultivable districts lying between them was Solesbury, my natal township" (211). Here again we have the theme of a two-sided river, a dividing line between the wilderness and civilization, but the Delaware is also a counterpart to the cavern, being a conduit out of, not into, the wilderness zone. Moreover, like so many features of Brown's landscapes, it does not lend itself readily to man's dispensations, and though Huntly has earlier declared that "a mountain-cave and the rumbling of an unseen torrent are . . . dear to my youthful imagination," by this point in his narrative he has presumably exhausted his "devotion to the spirit that breathes its inspiration in the gloom of forests and on the verge of streams" (22, 94). If, in one dimension, the novel is an antidote to Enlightenment optimism, in the other it presents a view of nature that is hostile to the emerging Romantic love of wildness—in effect, a throwback to an earlier, neoclassical aversion to landscape in its harsher aspects.

Where the rivers in Imlay's *Emigrants* put forth an erotic symbolism, Brown's Delaware (like Crèvecoeur's) is a disobliging, savage stream, and, having followed its course along a precipitous path, Huntly is forced by the sudden appearance of a man he assumes to be another marauding Indian to leap "from this tremendous height into the river," a plunge that serves as yet another example of the circumstantial pattern that has determined the haphazard course of the hero's adventures: "Had the depth been less, [the water's] resistance would not perhaps have hindered me from being mortally injured against the rocky bottom. Had the depth been greater, time enough would not have been allowed me to regain the surface. . . . As it was, my fate was suspended on a thread . . . [and] I only emerged from the gulf to encounter new perils" (221–22). Shot at from above, Huntly is carried along by the river between cliffs too steep to climb, and he is saved from drowning only by the chance advent of a pine protruding from a crack near the water. Seizing hold of a low branch, he manages to pull himself out of the river, but his situation is not much improved, for he is still on the wrong side of the stream—chance now operating against him. Only after much further wandering does Huntly hit upon a road that takes him to a ford in the river, enabling him to cross over into the regions of comparative safety.

"Few, perhaps, among mankind," muses Huntly toward the end of his lengthy letter to his sister, "have undergone vicissitudes of peril and wonder equal to mine. The miracles of poetry, the transitions of enchantment, are beggarly and mean compared with those which I had experienced: Passage into new forms, overleaping the bars of time and space, reversal of the laws of inanimate and intelligent existence had been mine to perform and to witness" (239). Yet the lesson apparently derived from the experience is hardly encouraging to the heroic impulse: "Disastrous and humiliating is the state of man! By his own hands, is constructed the mass of misery and error in which his steps are forever involved. . . . How little cognizance have men over the actions and motives of each other! How total is our blindness with regard to our own performances" (278). As his friend Sarsefield, a maximizing, Franklinesque character, adjures him: "Be more circumspect and more obsequious for the future" (292).

As an expression of Godwinianism, *Edgar Huntly* is an idiosyncratic production. Much more clearly in the Jacobin vein is Imlay's *Emigrants,* with its utopian conclusion and its insistence that good men and women, acting wisely, can control and direct their destinies—given the proper (western) environment. And yet, as Imlay's vision will disappear as a motive for fiction (but live on in the utopian impulse in America), so Brown's gothic territory will be the preferred frontier landscape of much literature yet to come, his cavern and river pointing the way into (and out of) a region, not of lovely pastoral promise, but of harshness and danger. There runs through Brown's fiction a Calvinist emphasis on man's innate capacity for doing evil—even if inadvertently—and he shares with the Puritans a tendency to portray the wilderness in adversarial terms; not, as in the novels of Cooper, as a terrain of adventure—an heroic zone—but as an arena of misadventure, from which the hero eventually retreats. Anticipating the darker aspects of Romanticism as they were manifested in America, Brown's landscape was conceived in the shadows cast by the rising glory of the Enlightenment imagination, resulting in the nightmare regions of Norwalk and beyond. It is a horrific territory and a tragic setting, and like Crèvecoeur's sin-burdened Susquehanna and Bartram's alligator-invested fountains, it calls into question the orderly systems rivers were essential to in the enlightened projections of republican plans.

And yet, it needs to be said that Brown's ideological differences with Imlay do not extend to a literal geopolitical dimension: his forbidding landscape is not created to promote anti-expansion propaganda, as were the captivity narratives written by the Puritans, for the wasteland atmosphere is chiefly in the service of his psychological (gothic) drama. As we have seen, Brown in 1803 argued for imperial expansion beyond the Mississippi, and during this same period he was a consistent advocate of uniting the republic "into one compact mass of contiguous provinces traversed by roads, canals, and rivers, and blended into one system of convenient and unrestricted intercourse" (Clark: 281). In urging the construction of a road between Philadelphia and Pittsburgh, Brown in 1806 sounded a millenarian note in harmony with republican schemes for like improvements:

> *To enlarge on the high importance of cementing the union of our citizens on the western waters with those of the Atlantic states would be unnecessary. Politicians have generally agreed that rivers unite the interests and promote the friendship of those who inhabit their banks; while mountains, on the contrary, tend to the disunion and estrangement of those who are separated by them. In the present case, to make the crooked ways straight and the rough ways smooth will in effect remove the intervening mountains, and, by facilitating the intercourse of our western brethren with those on the Atlantic, essentially unite them* in interest, *which is the most effectual means of uniting the human race.* (Clark: 231)

As we shall see, this is a virtual epitome of Washington's parting advice to his countrymen concerning interior communications and is in complete harmony with Jefferson's plans at the start of his second term for an ambitious program of public works.

Still, the net effect of Brown's use of the American landscape in his fiction is a negative one, in contradistinction to those paradisiac vistas promoted by propagandists like Imlay. But Imlay was alone, among many such promoters and projectors, in being able to sustain his vision in fiction, and his novel was a minority strain, a hybrid apparently incapable of reproducing its kind. Like his *Topographical Description,* moreover, *Emigrants* contains a subversive, secessionist bias, the implications of which were inimical to the Jeffersonian (and Barlovian) schemes. Here too, Brown was prophetic, for his imaginary French plot hatched in Louisiana was truer to conditions in the western valley than was Imlay's pacific scheme. As we shall soon find, there is an outlaw and renegade spirit literally resident in Imlay's utopia, and subsequent travelers along the Ohio belied his rhapsodic account with reports of deep divisions equivalent to those listed by Brown's imaginary Frenchman. The Jacobin ideal, when actually transported into the great western valley, proved indeed to be a disruptive and secessionist element, just such a force for separation as Washington and Jefferson feared, personified as agents of disunion incarnate. Historical figures worthy of fiction who, like Brown's Carwin, spoke in two voices, these agents of misrule could convert riverside temples of Enlightenment reason into very real ruins of Gothic disarray, while defining their actions as having been inspired by the most admirable intentions.

VII

Damsel with a Dulcimer

i

Among the first of the European writers to be influenced by Bartram's *Travels* was the Vicomte François-Auguste-René de Chateaubriand, whose epochal voyage to America took place the year Bartram's book was finally published. Since the young French emigré passed twice through Pennsylvania on his American voyage, it is likely he visited Bartram's gardens, for he had an interest in botany, and he may even have acquired a copy of *Travels* from the author himself. Certainly, marvelous coincidences are the stuff Chateaubriand's life and writings are made of, and his American journey may be said to have continued the French connection commenced by Jacques Cartier and extended by Crèvecoeur into the eighteenth century. Like Cartier, Chateaubriand was born in Saint-Malo, Brittany, and the year of his voyage to North America (1791) was the three-hundredth anniversary of Cartier's birth.

Like Cartier, moreover, Chateaubriand came over with the expectation of discovering the Northwest Passage, a compounded folly that was the last sortie of French imperialism north of Mexico: where Cartier's voyages served chiefly to inspire those explorers who followed him over, Chateaubriand contributed mostly to the course, not of empire, but of Romantic literature. If the record of his experiences was influenced by Bartram's *Travels,* the result has a dark quality that is in keeping with Charles Brockden Brown's gothicism: there is a distinctly erotic coloration to Chateaubriand's American landscape, but it has a tragic, not a pastoral-comic, implication.

Even before departing for America, Chateaubriand had plans to gather material for a Rousseau-inspired Indian epic, which eventually became his *Natchez,* but needing a more "practical" errand in order to obtain his family's permission, he hit upon the impossible search for a passage to India. In so doing, he read the travels of Charlevoix and Du Pratz, the Gallic equivalents of Daniel Coxe, whose writings and maps verged on pure fiction as they trended west; and though Chateaubriand abandoned his geopolitical errand about where Champlain ended his explorations nearly 200 years earlier—in the vicinity of Lake Erie—he persisted in gathering materials for poetry, thereby ushering in a new era of French imperialism, in which America was claimed as a territory of the Romantic sensibility. During Chateaubriand's stay in America, the French Revolution was moving from its moderate to its most radical phase, and it may be said that the Frenchman's use of his American experience effected a transformation of essentially Enlightenment materials to darkly Romantic uses, a metamorphosis entirely in harmony with that found in Brown's fiction.

Les Natchez was not published until 1829, and Chateaubriand's earliest writings were expressions of his disenchantment with the French Revolution and a reassertion of Christian pieties, writings that made the publication in 1801 of *Atala*—a fragment of the later work—preeminently successful in those parts of America where Federalist sympathies were strongest. It was translated into English in America by Caleb Bingham, a Connecticut-born Congregationalist minister and teacher, whose Dartmouth education and Jeffersonian brand of republicanism made him a uniquely qualified conduit of conservative French ideology. For, when read closely, Chateaubriand's pious little Indian story has a dark dimension, suggesting that savages and the Saviour are not a fruitful but a fated combination.

Chateaubriand's own travels did not take him further south than the mouth of the Ohio River, and the lush, tropical overture to *Atala* is heavily indebted to Bartram, though the style is very much Chateaubriand's own:

> *From the mouth of the Meschaceba* [*Chateaubriand preferred the antique usage for "Mississippi"*] *to its junction with the Ohio, one continued picture covers its surface. On the western shore, savannas open to view as far as the sight extends. Their waving verdure, as the prospect stretches, seems to reach the azure vault of heaven, where it wholly disappears. In these boundless meadows are seen, straying, droves of three or four thousand wild buffaloes. Sometimes a bison, borne down with years, cuts through the waves, and lands upon some island of the Meschaceba, to sleep quietly among the high grass. By his forehead, ornamented with two crescents, and his grisly beard, you would take him for the bellowing river god; who casts a look over the waters, and seems satisfied with the wild productions which its shores so abundantly yield.* (8)

The transformation of the buffalo into a pagan deity is the typical and definitive touch, for no such figure is provided by Bartram, whose own description of the Mississippi is relatively spare. Chateaubriand does, however, seem dependent on Bartram's account of the St. John's, his bellowing buffalo a symbolic equivalent of Bartram's regnant alligator.

Chateaubriand's geographies are in all ways an exotic blend, and the movement of *Atala* takes the reader from the Mississippi to the Chattahoochee, thence to the headwaters of the Tennessee, and from there into the Allegheny Mountains, where, by means of a recollected voyage, the reader visits Niagara Falls, a "spectacle of nature's mighty work," which the narrator regards "with admiration bordering on terror." The reader, however, is apt to be distracted by the unlikely "caracajous suspend[ed] . . . by their long tails [from] the ends of bending branches, to seize in the abyss the carcases of animals, killed by passing over the falls" (167–68). This scenic grand tour ends in the shadow of a great natural bridge, against which the climactic scenes of the story are enacted, a geological freak transferred for the narrative occasion from Jefferson's neighborhood (and *Notes*). The effect depends upon an eclectic (and exotic) enjambment of the continent's growing catalogue of natural wonders.

By so doing, Chateaubriand demonstrated to a new generation of American writers (and painters) the manifold possibilities of their native scenery, much as, with his Rousseauvean characters and romantic plot, he outlined an essential fable in which the union of civilization and savagery, even when overseen by benevolent Christian intentions, will prove ultimately tragic. Chactaw, the noble Indian brave; Atala, the beautiful but doomed half-breed; and Father Aubry, the long-suffering Christian missionary; all will have their counterparts in later American literature, as will René, auditor in *Atala,* hero of his own eponymous story, who is the spirit of Romantic melancholy adrift with his authorial namesake René Chateaubriand on the rivers of the New World. Halfway between Imlay's Arl——ton and Brown's Carwin, René represents the Romantic golden mean, which will inspire and nourish the genteel tradition that placed the dying Indian in a sentimental frame.

After Chateaubriand's brief passage through it, the American landscape would never be the same again, yet the first results were quite different from anything imagined by Cooper at the near end of the spectrum and Longfellow at the far. Imlay's *Topographical Description* was one of Chateaubriand's chief literary sources—despite the disparity of the two men in piety and politics—which resulted in a complex and even alchemical transformation. The most immediate by-product was a book by Thomas Ashe, who visited America in 1805 with

Atala in his pocket and Bartram's *Travels* in his portmanteau. Ashe was an Anglo-Irish Army officer of dubious reputation, whose epistolary account of a journey he claimed to have made down the Ohio and Mississippi rivers has never been taken as a matter of facts.

Yet the picaresque spirit would not, in future years, be a stranger to the rivers of western America, and Ashe's *Travels,* though discredited as a factual account, recommends itself to us because of its large fictional element. Though somewhat anomalous, like Imlay's *Emigrants,* it provides a salutary antidote to the other man's unqualified and unremitting idealism, while yet deriving from Imlay a distinctly utopian impulse connecting him with the most extreme version of separatism the republic was to experience during its early years. If it took a Frenchman like Chateaubriand to provide a broad survey of the darkly romantic potential of the American scene, then it was the Irishman Ashe who rendered the sharpest satiric delineation available of the disjunctive dimension of life along the Ohio River as the eighteenth gave way to the nineteenth century.

ii

From its first publication in 1809, Ashe's *Travels* was looked at askance, in part because of the author's easily detected fictions, but also because of his hostile attitude toward aspects of the American character as it manifested itself in western regions. Ashe's account of life in the Ohio Valley comports well with Brackenridge's picaresque version, yet he must be accounted the first of those British travel writers who drew the wrath of Americans by according their experiment something less than unqualified praise. The passage of eighty years somewhat mitigated the quality of American indignation, and, by 1889, Henry Adams could pronounce Ashe's "invented" *Travels* as "quite untrustworthy" but "amusing," going on to quote it at length as an authority on the frontier scene, perhaps because the Irishman's view of the South as a region of barbarism supported Adams' own (*History*: I: 43).

Thomas Ashe has proved less useful (if still amusing) to later historians, who have preferred to maintain a distinction between factual commentary and clever invention, and though Adams could be beguiled by the "something of the artist" in Ashe, literary scholars have apparently not found enough of the creative element to treat his *Travels* as fiction. Some doubt is shed on Ashe's reliability from the very start: his first letter is dated "October, 1806," and the year remains the same throughout the succeeding twelve months. Given a few salient facts derived from his text, we may assume Ashe meant to divide his account between 1805 and 1806, but the typographical error is definitive. That Ashe made the trip he narrates has even been questioned, though he most certainly passed through Cincinnati, traveling under the name "Arville," for he there made the acquaintance of Dr. William Goforth, a pioneering paleontologist, who had excavated some mammoth bones from a salt lick. These Ashe took in charge and conveyed to London, where they were displayed at some profit presumably to himself though certainly not to Dr. Goforth; and if the misappropriation of the bones testifies to the man's essential dishonesty, it also suggests his route took him down the Mississippi to New Orleans. The publication of his *Travels* was preceded by Ashe's *Memoirs of Mammoth and Other Extraordinary Bones* (1806), published to promote his exhibition, and the obvious hunger in England for American marvels seems to have been the main inspiration for his second book. In sum, what a London hack did to Carver's factual observations, Ashe willingly did to his own.

He therefore belongs to a line of mendacious travelers that begins with Mandeville, who start out in recognizable terrain and then move on through realms of possibility to the territory of the probable with occasional forays into sheer fantasy. He is, that is to say, a fabulator, his *Travels* ultimately passing into an America already rendered mythic by Chateaubriand. Seizing on those aspects of American character and scenery that had already proved fascinating to European readers, he welded them into a reasonably coherent narrative. While obviously working within the tradition of propaganda for settlement as established by Manasseh Cutler and Imlay, Ashe has also a deep satiric streak, which often gives way to an even darker impulse, both springing from his bias against the New World and its inhabitants.

Ashe's technique of amplification and exaggeration can be clearly demonstrated by his account of the rapids at Louisville. Where for Imlay they provided a picturesque adjunct to the pastoral scene, Ashe converts them into something more than faintly reminiscent of Pownall's Cohoes Falls and Chateaubriand's Niagara, for the prospect tends toward the recently discovered region called the "sublime," which involves, in descriptions of falling water, a certain predictable pattern. First, "the roaring of the falls which reached me at the distance of fifteen miles." Next, "the danger it announces to the mind," because of "the velocity of the water" and its "tremendous uproar," peril given particularly when Ashe finds himself and his flatboat very nearly drawn into "sudden and violent perdition" (212).

But the formula also calls for a more pleasurable dimension, and if the rapids from an upstream point of view connote danger, from below—as seen from Louisville—the prospect is much closer to that conveyed by Imlay: "Here the magnificence of the scene, the grandeur of the falls, the unceasing brawl of the cataract, and the beauty of the surrounding prospect, all contribute to render the place truly delightful, and to impress every man of observation who beholds it, with ideas of its future importance . . ." (213). But it is precisely here that Ashe introduces his characteristically blighting touch: ". . . till he inquires more minutely, and discovers a character of unhealthiness in the place, which forbids the encouragement of any hope of its permanency of improvement." So much for Imlay's utopia, doomed by the selfsame falls that guaranteed its separatist sanctity.

Ashe's picture of the American scene is complex, and his purposes are often obscure, but one thing is quite clear—his dislike of everything associated with Kentucky. Louisville is simply no exception. If Ashe's *Travels* is mostly known, in the phrase of a prominent bibliographer, for its "snarling asperity," much of the venom is expended on the "left-hand" side of the Ohio River: "Many of the small inns on the Virginia and Kentucky shore were held in solitary situations by persons of infamous character, driven from the interior and the head waters, by the gradual encroachments made on them by morals, religion, and justice" (88). On the "right-hand" side, by contrast, as at Marietta, "if justice be impotent on the opposite Virginian shore, and morals and laws be trampled upon and despised, here they are strengthened by authority; and upheld, respected, and supported by all ranks," thanks to the enforcement of "New-England regulations of church and magistry" (110). As a result, Ashe swivels back and forth between admiration and disgust, the latter mostly inspired by the emigrant population, for the Virginian becomes a Kentuckian, who subsists on a diet of pork and whiskey and who hurls himself repeatedly into gouging matches with his neighbors. Filson's Daniel Boone clearly had already moved on. The river itself retains a certain innocence, beautiful much of the time, yet with a definable element of constant danger, particularly to the navigators of flatboats descending to the Mississippi.

Scorning the advice of a pilot hired to guide him through the rapids, Ashe refuses to wait out an approaching thunderstorm: "Whenever I have determined on acting, I have not easily been turned from my intention" (214). The result is a highly theatrical transition from the upper to the lower Ohio:

> *The pilot and I governed the helm, and my passengers sat on the roof of the boat. A profound silence reigned. A sentiment of awe and terror occupied every mind, and urged the necessity of a*

fixed and resolute duty. In a few minutes we worked across the eddy, and reached the current of the north fall, which hurried us on with an awful swiftness, and made impressions vain to describe. The water soon rushed with a more horrid fury, and seemed to threaten destruction even to the solid rock which opposed its passage in the centre of the river, and the terrific and incessant din with which this was accompanied, almost overcame and unnerved the heart. At the distance of half a mile a thick mist, like volumes of smoke, rose to the skies, and as we advanced we heard a more sullen noise, which soon after almost stunned our ears. Making as we proceeded the north side, we were struck with the most terrific event and awful scene. The expected thunder burst at once in heavy peals over our heads, and the gust with which it was accompanied raged up the river, and held our boat in agitated suspense on the verge of the precipitating flood. The lightning, too, glanced and flashed on the furious cataract, which rushed down with tremendous fury within sight of the eye. We doubled the most fatal rock, and though the storm increased to a dreadful degree, we held the boat in the channel, took the chute, *and following with skilful helm its narrow and winding bed, filled with rocks, and confined by a vortex which appears the residence of death, we floated in uninterrupted water of one calm confined sheet. The instant of taking the fall was certainly sublime and awful. The organs of perception were hurried along, and partook of the turbulence of the roaring water. The powers of recollection were even suspended by the sudden shock; and it was not till after a considerable time that I was enabled to look back and contemplate the sublime horrors of the scene from which I made so fortunate an escape.* (214-15)

This is a remarkable passage, not so much in the manner of its telling—though that is effective—as in its advance over other, earlier descriptions of rapids and falls, as by Pownall and Jefferson, for where they struck postures of aesthetic observation, Ashe literally participates in the sublime element itself. Brown has his Huntly leap into the turbulent Delaware, but this is not quite the same: Ashe and his pilot guide themselves through the terrifying chute as the elements crash about them, taking mastery over the sheer force that is the point of earlier descriptions.

Where for Bartram a hurricane marks the upper limits of his journey on the St. John's—signaling his ultimate penetration into wildness—the thunderstorm serves for Ashe as a vehicle of transition, emphasizing the fact that he is the kind of proto-Byronic traveler who actually *courts* danger. Bartram, the Enlightened man, is caught by the storm, but Ashe, the Romantic, sails knowingly into it. His passengers are "two ladies and gentlemen, who had courage to take the fall out of mere curiosity, notwithstanding the great peril with which the act was allied," and at the end of the ride, the ladies are "exhausted" and "unnerved" by the experience. Yet they quickly recover, and the party "enjoyed a pleasant walk back to the town, and passed the evening with that serene delight which is only known to those who have experienced an equally extraordinary and eventful day" (215), an extreme version of the participatory aestheticism that pervades Ashe's *Travels* throughout.

Ashe's account of the Ohio Valley is obviously much less idyllic than are the descriptions of Filson and Imlay. He refers explicitly to Imlay's "dreams" in a derogatory sense, and at another place he describes "one mighty scene of endless mountains covered with ponderous and gloomy wood" as having been "described by Imlay and others, as a lawn producing shrubs and flowers, and fit for the abode of gods instead of men" (75, 168). "And others" obviously includes Filson, alluded to elsewhere by Ashe as "an author residing in Philadelphia," who was hired by "the first explorer of Kentuckey" (Boone) to write "an animated and embellished description of that country. The narrative was in a florid, beautiful, and almost poetical style: in short, the work possessed every merit except truth" (29). Frequent mention is made throughout of "the authors who have given descriptions of Kentuckey" for the purpose of luring settlers there, but who, according to Ashe, "either never see that state, or only would see a small portion of highly beautiful land which it contains in its centre, sixty miles long by about thirty miles broad" (151).

Ashe, by constantly stressing the negative aspects of the Ohio Valley, seems to be setting himself up as a reliable reporter, who may be depended upon to render an honest account of what he actually saw: "Had such writers been aware that their romance might occasion miseries in real life, I am willing to think that they would have controuled the fancy which produced it, and have given the world plain and useful truths, which would have served the unfortunate emigrant as a faithful and honest guide, in the place of offering him flattering and fallacious images, the pursuit of which winds up his history of calamity, disappointment and destruction; and he discovers the nature of romance at the price of his happiness and fortune" (168). That Ashe was absolutely right—as witness the fate of the French at Marietta—does not relieve us of the responsibility of squinting hard at his own account.

For it also needs to be observed once again that much of the negative comment has to do with the Kentucky side of the river, while the Ohio side tends to take on a very Imlay-like coloration. Ashe is particularly lyrical about the region between the two Miamis: "I . . . must candidly confess, I never beheld a tract of land so favoured by nature, and so susceptible of improvement by art" (204). Some suggestion as to his reasons occurs at the end of his travels along the two rivers, when he stops "to take breakfast with the hospitable judge Symmes, the original proprietor, after the extinction of the Indian title, of the whole of the country lying between the two Miamis" (208). Symmes provides Ashe an ideal pattern of Enlightened Settler, placing his own home in a situation that "cannot be equaled for the variety and elegance of its prospects. Improved farms, villages, seats, and the remains of ancient and modern military works, decorate the banks of the finest place of water in the world, and present themselves to view from the principal apartments of the house, which is a noble stone mansion, erected at great expence, and on a plan which does infinite honour to the artist, and to the taste of the proprietor" (208).

All along the way, Ashe takes pains to point out how the aesthetic dimension has been neglected by settlers in laying out their buildings and towns, as at Charlestown ("on the Virginia side"), which was "originally laid out with the best row [of houses] facing the river, and the intermediate space answer[ing] the purpose of a street, explanade [*sic*]*, and water terrace, giving an air of health and cheerfulness gratifying to the inhabitants, and highly pleasing to those descending the stream. However, owing to the avarice of the proprietor of the terrace, and a disgraceful absence of judgment and taste, he has sold his title to the water side, and the purchasers are building on it; turning the back of their houses immediately close to the edge of the bank, and excluding all manner of view and communication from the [r]est of the town" (82). Riverside towns would increasingly turn their backs to the beauties of the Ohio, and Ashe's point is well taken, yet there is something overdone, perhaps, about his contrasting enthusiasm for the aesthetic sense of Judge Symmes, who "has been studious to give the river sides a pastoral effect, by preserving woods, planting orchards, and diversifying these with corn-fields, sloping pastures, and every other effect incidental both to an improved and rural life" (208).

Ashe's "Symmes," however, can be distinguished from the historical John Cleve Symmes usually associated with the Miami Purchase, for to his English correspondent, Ashe attributes the Judge's "expression of

*The first edition is plagued with typographical errors throughout.

elevated judgment" to the "fact" that "the proprietor formerly resided in England, and after in New York," where he met and married his wife, "a lady distinguished by elegance of mind, and a general and correct information" (208). The historical Symmes moved from military exploits as a New Jersey officer in the Revolution to the western regions without a recorded stay in England, but by 1805 he was indeed married to a lady from New York, his third wife, the former Susanna Livingston. According to Ashe, the Judge's household also contains "a Miss Livingston, on whom they fix their affections; and whom they treat with parental kindness and respectful urbanity, the one being due to her intrinsic merit, and the other to her family, which is eminent for birth, property, and talent, in the state of New York" (208).

We may smile at this cosmopolitan chauvinism, yet Ashe's ideal settler clearly brings to the wilderness an English notion of well-ordered landscape, which (as in Imlay's novel) is given a definitive touch by a feminine presence. Where "the judge passes his time in directing his various works, the ladies read, walk, and attend to various birds and animals, which they domesticate both for entertainment and use" (209). Miss Livingston, in particular, "is much of a botanist—a practical one. She collects seeds from such plants and flowers as are most conspicuous in the prairies, and cultivates them with care on the banks and in the vicinity of the house" (209). Moreover, in a parklike group of native shrubs, this paragon has erected "a small Indian temple, where [she] preserves seeds and plants, and classes specimens of wood, which contribute much to her knowledge and entertainment" (209).

Still, despite its apparent unlikeliness, the genteel impress of the Symmes ménage on the Ohio landscape should not be dismissed as irrelevant to historical reality. Though Ashe has his differences with Imlay, his notion of the best way to settle a new country is quite similar, for both men stress the importance of an aristocracy (made up of officer veterans of the Revolution), which provides a stabilizing and elevating presence. The invisible but assumed analogue is Washington's Mount Vernon as described in 1789 by Jedidiah Morse: "He lives as he ever has done, in the unvarying habits of regularity, temperence, and industry" (*Geography:* 131). Moreover, Ashe's landscape aesthetic, derived from English neoclassic models, will have a long reach in American art and literature. In some ways a realization of Crèvecoeur's Enlightenment ideal as manifested by John Bartram, it points forward to subsequent accounts of cosmopolitan gentlemen rearing utopias of reasoned order in a new country, and particularly to the community set up in the New York wilderness by Cooper's Judge Marmaduke Temple.

Ashe also shares with Imlay a predisposition to draw unfavorable distinctions between the regions east and west of the Alleghenies. The opening pages of his *Travels* dismiss summarily the settled portions of the continent: "Those [states] to the north-east are indebted to nature for but few gifts . . . and bigotry, pride, and a malignant hatred to the mother-country characterize the inhabitants," while in the South, "nature has done much, but man little. Society is here in a shameful degeneracy" (5–6). By contrast, his first view of the Ohio Valley is a sight that "could not fail of gratifying and enchanting me; giving serenity to the mind, and gratitude to the heart; and awakening in the soul its most amiable and distinguished affections" (18). Where he differs with Imlay, chiefly, has to do with the character of Kentuckians and, equally important, with the place of the Indian in the landscape. Imlay, in his *Topographical Description,* advocates a policy of forceful assimilation, and in his novel assumes a posture of benign indifference, but Ashe throughout evinces a classically Rousseauvean attitude. At the start of his Ohio travels he is accompanied by his white servant, Mindeth, and a Mandan "mestee" (half-breed) called "Cuff." As the narrative progresses, Mindeth takes second place to the Indian, who (aided by his marvelous gift for imitating the sounds of various birds and beasts of the American woodlands) increasingly serves as an intermediary between Ashe and the wilderness through which he passes. If in Judge Symmes we are not far from Marmaduke Temple, Cuff puts us on the road to Uncas and Chingachgook.

For Ashe obviously views the American Indian through the sympathetic eyes of Bartram and Chateaubriand: he regards the red man as the "real proprietor" of the western lands, and spends some time in the camps of the "Mongos" and "Shawanese," where he engages in "very interesting conversation" with "Onamo" and "Adario," whose Italianate names suggest tribal connections with Chateaubriand's Atala (243, 203, 245). At the far end of his journey he even makes a "visit to the remains of the nation of the Natchez Indians, once the most powerful and enlightened people of all the continent of America," but now reduced by "the ravage of war, the small-pox, and spirituous liquors" to a "slighted and despised" remnant who persist in their ancestral worship of the sun (290). From the start of his excursion, Ashe's presentation of the American aborigine is an extended *ubi sunt* lament, as when, in making his first crossing of the Alleghenies, he happens across a deserted Indian village, a site that dramatizes the tragic facts of the conflict between white men and red, being "a last refuge in the hour of melancholy and despair . . . hid in the depth of the valley, amidst the profoundest gloom of the woods" (16).

Throughout Ashe's journey, evidence of Indian "proprietorship" of the Ohio Valley is mostly a matter of the earthen and other monuments they have left behind them. Bartram was similarly fascinated by

Florida's ancient remains, but for Ashe the theme becomes obsessive. Writers before him (on whom he obviously depends) described the mysterious mounds of the Ohio Valley, but in Ashe's account the prevalence and extent of "monuments of Indian antiquity" is disproportionately large (142). Mounds, barrows, tombs, forts, even natural catacombs filled with mummies abound, virtually littering the landscape with an Egyptian-like evidence of an earlier and relatively advanced culture. Periodicals of the day contain similar accounts, including reports of mummified remains found in the caverns of Kentucky, but Ashe's handling of this "factual" material is typically exaggerated to the point of pure fiction.

Ashe's method in these and other matters is to start with a bona fide fact, and then proceed into the realm of the imagination. One such "discovery" occurs at Cave-in-Rock, the cavern on the lower Ohio that served as the haven for numerous land pirates and river outlaws. By 1806, this place was already notorious, and Ashe rehearses (and expands) its grim history, though less as a hideout for bloodthirsty bandits than as "an object of terror and astonishment from having been the retreat of the remains of an Indian nation exasperated against the Americans, and resolved to put as many of them as possible to death" (225). Jonathan Carver reported the existence of a cave near the Falls of St. Anthony that was decorated with mysterious hieroglyphs, which Ashe likewise finds, along with "names of persons and dates, and other remarks, etched by former [civilized] inhabitants, and nearly by every visitor" (227). The modern graffiti in effect disqualified the cave as an exotic zone, but Ashe purports to have also discovered "an orifice in the roof of the cave," which, when he worked his way up through it, led into "an apartment of greater magnitude than that from which I had immediately ascended, and of infinitely more splendour, magnificence, and variety. . . . The roof, which was arched, the sides and natural pillars that supported it, seemed at first sight to be cut out and wrought into innumerable figures and ornaments, not unlike those of a gothic cathedral. These were formed by a thousand perpetual distillations of the coldest and most petrifying quality imaginable, and which besides exhibited an infinite number of objects that bore some imperfect resemblance to many different kinds of animals" (228). Beyond lay "another vault of very great depth," a "frightful chasm" into which he dared not descend, and he found a heap of human bones in another part of the cave—not ancient but modern remains, whose broken skulls and bones attested that "murder had been committed, and that the dreadful reports respecting the cave were neither fabulous nor exaggerated" (229). It is perhaps needless to add that these upper chambers of Ashe's gothic cavern were entirely fictitious, wonders that may have been inspired by reports of the recently discovered Mammoth Cave.

Captain Farrago in *Modern Chivalry* has an experience similar to Ashe's when he is taken to a cavern in the Allegheny mountains, whence issues a fountain with a marvelous "petrifying quality," so that the cave is not only full of stalactites but contains an inner "apartment" containing "a vast bed of human skeletons petrified, but distinguishable by their forms. . . . The dimensions of some of these skeletons bespoke them giants; that of one measured eight feet, wanting an inch" (276). At this point, Brackenridge's account approximates Ashe's marvels, but he goes on to reveal increasingly exaggerated "discoveries" within the cavern, including a "petrified grove," containing "trees in their natural position, with wasps nests on them, all petrified; and buffaloes standing under, in their proper form, but as hard as adamant. A bleak wind, with a petrifying dew, had arrested them in life, and fixed them to the spot; while the mountain in a series of ages had grown over them" (277). Included in this frozen tableau is "an Indian man reduced to stone, with a bundle of peltry on his back," while nearby "the skin of a wildcat . . . hung upon a stone peg in the side of the grotto," the last of which Captain Farrago takes along as a souvenir. The target of Brackenridge's ridicule is Ashe's sources, the accounts of Western wonders that earned the credulity of learned (Eastern) associations like the American Philosophical Society, who regularly published without question as "scientific" whatever vagabond information wandered over the mountains. The same sources inspired similarly exaggerated accounts by such as Ashe, resulting in a prophetic anticipation of the tall tale that would flourish along the rivers he descended, no more so than in the Irishman's account of the alligators he encountered on the lower Mississippi.

At the start, Ashe is clearly in Bartram's debt, as when his boat is attacked by "a huge alligator, at least twenty feet long, of proportionate circumference, and with a head containing one fourth of the length of the body!" (275). The exclamation is well justified: one doubts from the proportions that Ashe himself ever saw an alligator save for Bartram's drawings of one. Ashe likewise seems indebted to Bartram in claiming that alligators "roar as loud as thunder," but the further information that they kept him awake at night by "sobs, sighs, and tears, and moans of inexpressible anguish and length" has a less certain source (282). This seems to be extending the Power of Sympathy into an unlikely range, as when, having killed a giant alligator, Ashe finds it is the mother of a host of infants, who run "over and around it in great agitation, and whining and moaning, because they discovered it without animation, and destitute of all symptoms of life" (283).

Ashe may have been inspired here by the legendary propensity of crocodiles to weep, and he uses this myth in his customary backwards fashion: examining the eyes of the young alligators he has orphaned, he finds them dry "of any moisture whatever . . . though the plaints are piteous to the most distressing degree" (284). They rather resemble kittens by Ashe's account ("They have beautiful blue eyes, with an expression extremely soft and sensible" [283]), and he is so taken by them that he adopts three: "I should tell you that my own little ones thrive well, and take on all the airs of a pet. They take their food out of my hand, and by their voice express much satisfaction whenever they are bathed" (285). At junctures like this the reader definitely has a sensation of having a leg pulled, and looking down, finds himself confronting a sharp-toothed grin. By this point in the narrative, the guidebook character of *Travels* has completely given way to the increasingly unlikely marvels that Ashe claims to have encountered.

And yet one of the most marvelous episodes in his book is verifiable—his visit to "Bacchus Island" near the town of Bellepré on the Ohio, an encounter with all the earmarks of fiction, which records an undeniable fact:

> *The island hove in sight to great advantage from the middle of the river, from which point of view little more appeared than the simple decorations of nature; trees, shrubs, and flowers of every perfume and kind. The next point of view on running with the current on the right hand side, varied to a scene of enchantment; a lawn, in the form of a fan inverted, presented itself; the nut forming the centre and summit of the island, and the broad segment the borders of the water. The lawn contained one hundred acres of the best pasture interspersed with flowering shrubs and clumps of trees, in a manner that conveyed a strong conviction of the taste and judgment of the proprietor. The house came into view at the instant I was signifying a wish that such a lawn had a mansion. It stands on the immediate summit of the island, whose ascent is very gradual; is snow white, three stories high, and furnished with wings which interlock the adjoining trees, confine the prospect, and intercept the sight of barns, stables, and out-offices, which are so often suffered to destroy the effect of the noblest views in England.* (135)

The view is like that of Mt. Vernon from the Potomac, and this Ohio River mansion likewise is a veritable avatar of civilized order and southern hospitality. Having carried down with him from Marietta the lady and gentleman who reside in this marvelous place, Ashe is invited by them to spend the night, a "friendly importunity" that "flowed from hearts desirous not to be refused" (136). They arrive at teatime, which ritual is "conducted with a propriety and elegance which I never witnessed out of Britain," Ashe's usual standard for assessing the quality of American taste, and "the conversation" likewise "was chaste and general, and the manners of the lady and gentleman were refined without being frigid; distinguished without being ostentatious, and familiar without being vulgar, importunate, or absurd" (136). Ashe bestows similar encomiums on the gardens about the house, "which were elegantly laid out" in English style, and describes in like terms the events of the evening, including a fishing party and a picnic supper, "after which we returned to the house, where over a bottle of wine one hour longer we conversed on the pleasures of our rural sports, and retired to rest with that heartfelt ease and serenity which follows an innocent and well-spent day" (137). It was therefore "with difficulty" that Ashe tore himself away from "this interesting family," "emigrants of the first distinction from Ireland" (137).

Ashe begs to be excused from naming "the amiable couple," but by 1809 he did not have to bother. For "Bacchus" could be easily read by most Americans as a more elegant version of "Backus," the island in the Ohio on which an emigré Irishman named Blennerhassett had built an elegant retreat, associated less, however, with evidences of taste in the lives of the rich and famous than with a notorious scheme to realize the secessionist impulse feared by Washington and Jefferson and latent in Imlay's accounts of the western region. The presence of Blennerhassett and his lovely wife in Ashe's *Travels* is problematic—he and his house may have been intended as a complement to the description of Symmes' establishment—yet the very anomalousness of the Irishman's island retreat as described by Ashe asserts a disjunctive difference. And when regarded in its historical context, it conflates the neat division between life as it is lived on the northern and southern banks of the river in Ashe's dichotomized account.

iii

Harman Blennerhassett gained what measure of immortality he enjoys by association, for he was an important member of the Burr Conspiracy of 1806, the actual aims of which will never be known, but which was identified at the time with the setting up of an independent empire in Louisiana and the invasion of Mexico. The first item was disclaimed ardently by Burr as treasonous while he claimed the last was a patriotic act. Historians still differ about Burr's veracity, since he was at the time on trial for his life, and it does seem that in the minds of the conspirators, including Blennerhassett, the invasion of Mexico was but part of a much larger geopolitical scheme. It was in any event a grandiose design of tragic, not epic, dimensions, bound up with personal ambition and weakness and thereby doomed to fail. At one point in danger of being hanged as traitors, Burr and Blennerhassett lost their fortunes and their reputations, and the two ended their relationship permanently estranged.

And yet at the start, the plan seemed sure of success to all parties concerned, particularly to General James Wilkinson, who first proposed the scheme to Burr and at the last betrayed it to Jefferson. Wilkinson had long been associated with secessionist sentiments in the western regions, and was likewise known for his subversive dealings with Spanish authorities in New Orleans before the Louisiana Purchase secured that important port for the United States; if the plan was truly a conspiracy, then its darker dimensions took its shadows from the general's hinder parts. But there can be no doubt that secessionists could still be found in the Ohio Valley in 1806, nor was Wilkinson alone in his lust for the conquest of Spanish territory, as the subsequent career of Andrew Jackson attests. Burr was probably less interested in conquest than in recouping his financial and political fortunes, and as always, Burr's part was bereft of ideology while heavy with personal interest. The resident ideologue was the master of Backus Island, who loaned the conspiracy a Jacobin dimension, albeit with an Imlay-like profile, which soon took a Charles-Brockden-Brown twist.

A man of considerable means, Blennerhassett was nonetheless a radical by temperament and had spent some time in revolutionary France before deciding to emigrate to North America. He may well have had his Jacobin idealism pointed in that direction by Gilbert Imlay's writings about the Ohio Valley: the kind of life he designed for himself and his young wife on Backus Island was very similar to that enjoyed by the hero and heroine of *The Emigrants*. Whatever the specifics of inspiration, the Blennerhassett admired by Ashe was Imlay's ideal settler in the flesh, a paradigm of enlightened physiocracy establishing an orderly oasis of culture in the heart of the western wilderness. That Blennerhassett seems to have been drawn to Burr's scheme in part out of the boredom that such a life predictably produced distinguishes him from that other exemplary settler described by Ashe, Judge Symmes, who was motivated not by utopian idealism but by a desire to get rich. Whatever his impulse, Blennerhassett dramatized the tendency of the separatist ideal to turn secessionist when it is converted to political ends.

Moreover, Symmes and Blennerhassett had in common an aristocratic ethos. It is this that recommended them to Ashe, whose interest in Indians was likewise rooted in a desire to establish (even manufacture) the prior existence of a wealthy and advanced culture in the Ohio Valley. It is against such standards that the backwoods culture of Kentucky is implicitly measured, and it is precisely such standards that lend Ashe's book a dark, even tragic, dimension. Though lured to America by Jacobin ideals, Blennerhassett elected to buy an island on the Virginia side of the river so that he would be able to keep slaves and thereby maintain his aristocratic way of life. Typically, under the charming influence of Burr he quickly shed his radicalism when tempted by a vision of a brave new world that would be organized along feudal lines, in effect projecting and intensifying plantation life as a basis for New World empire.

Ruined by his association with Burr, Blennerhassett ended in abject cringing and obsequious poverty, and spent the rest of his life in search of that kind of employment by which impoverished aristocrats hope to disguise their essentially menial positions. He was unsuccessful even in this degrading quest and, having spent a short period on the Isle of Jersey in 1827, he made his home on Guernsey and, suffering a series of paralytic strokes, died and was buried there in 1831. These were islands far different from the one he had settled on in 1796, yet they brought him once again into the neighborhood of Gilbert Imlay. One wonders if these two exiles ever met—Imlay the propagandist for a pastoral utopia in the Ohio Valley, Blennerhassett the man who realized the dream but then moved on under the much more dangerous influence of Aaron Burr. If in Imlay the libertinism concealed in the concept of personal liberty is revealed, then in Blennerhassett we find the reactionary impulse hidden at the heart of utopian pastoralism. His last surviving son, Joseph Lewis, enlisted and died in the service of the Confederacy.

Speaking for the prosecution at the trial of Burr and Blennerhassett, William Wirt put the dapper, elegant New Yorker at the heart of the conspiracy that has always borne his name, figuring the Blennerhassetts as a modern Adam and Eve, lured from their Eden by a Satan in human form: "In the midst of all this peace, this innocent simplicity and tranquillity, this feast of the mind, this pure banquet of the heart, the destroyer comes," who is able to "find his way into their hearts by the dignity and elegance of this demeanor, the light and beauty of his conversation, and the seductive and fascinating power of his address" ("Speech": V: 403). But if Burr can be seen as a Miltonic serpent, "winding himself into the open and unpractised heart of Blennerhassett," it is because he evinced just those tasteful qualities Ashe found so praiseworthy in the homes of Judge Symmes and the Blennerhassetts, qualities associated with Old World manners and values, not with American innocence. By Wirt's account, Burr can also be read as an equivalent to those illuminati-like heroes who appear in the fiction of Charles Brockden Brown: Europeans like the cynical, opportunistic and amoral Ormond in the novel of that name, a Jacobin-inspired rakehell who commits murder, robbery, seduction and rape while planning to establish an enlightened utopia in some far-western zone.

But Burr perhaps more closely resembles Carwin, his arguments for western empire serving to trigger, not cause, Blennerhassett's irrational ambition for more extensive domain upon which he could impose his orderly plans. Motivated at least in part by the indebtedness he had incurred by building his island home, which was later ransacked by debtors seeking compensation, then ravaged by mobs of casual vandals, Blennerhassett in all ways symbolizes the dark side of Jacobin schemes, ruined by the base human impulses they never seem to accommodate in utopian plans, including their own. By such transformations are the neoclassical architectonics of the early republic given the ragged, irregular shapes of the gothic scene.

It was the peculiar fortune of Thomas Ashe to pass down the Ohio River just ahead of Aaron Burr, and to visit (if indeed he did) the Blennerhassetts as they seemed to be enjoying unqualified felicity. Nothing of the grim sequel is hinted, yet Ashe's constant emphasis on the dark side of life in the Ohio Valley, his dramatic division between the orderly settlement on the north side and the disorder on the south side of the great river, his reiterated descriptions of Indian ruins, all serve to give a fatal and potentially tragic cast to the pattern of settlement there. Though written to discountenance Imlay and Filson, Ashe's arbitrariness makes of Blennerhassett's island a paradise equivalent to the Kentucky garden evoked by the other writers, yet he surrounds it with signs of disorder and disarray that are equivalents to the somber (if covert) symbolism found in Bartram and Chateaubriand. At a late point in his travels, Ashe tells us, he delivered to "Col. Bruin . . . at Payeu [*sic*] Pierre . . . a letter from his friend Burr," the only mention made in his book of the chief figure in the conspiracy even then unfolding (286). But such a letter would have been carried by Ashe the entire route of his downriver journey, along which Burr himself would soon move, and it provides thereby a key to the cipher, by means of which we can decode the whole.

VIII

Measureless to Man

i

There is no more sizeable evidence verifying the Great Man theory of history than the circumstances of the Louisiana Purchase, made possible by the whim of a self-anointed emperor and the pen of an American president who had never writ so large. Seeking merely to negotiate for navigation rights on the Mississippi, Jefferson's embassy to Napoleon was handed half a continent, and Jefferson, in accepting it, strategically revised the outlines of the American empire, enlarged the powers of the President, and altered the balance of geopolitical power in the United States forever. Before 1803, the Mississippi was chiefly viewed as an avenue of commercial exports pointed southward, permitting the flow (when allowed by Spain) of commodities out of western regions to New Orleans markets. Afterward it served, at least symbolically, as a frontier dividing East and West, the crossing of which had powerful metaphorical and geopolitical associations. Where before it was seen as the southern part of the Ohio River, it was now regarded as a trunk whose chief branch was the Missouri River, permitting national expansion toward the West.

Jefferson, who had written so well on behalf of American independence, had been less successful in convincing Spain that rivers like the ocean "are free to all men," extending parietal rights to nations as well as to individuals (*Report*: I: 253). But after 1803, virtually all American rivers were free, at least of geopolitical obstructions to navigation, releasing a westward flow of imperial energies, of which Aaron Burr was but the most radical symbol. Yet what was set in motion in part by Napoleon's fears concerning the activities of a handful of revolutionaries in Haiti aroused similar fears in New England; for what Napoleon had regarded as a potential burden, New Englanders saw as a counterweight that would soon shift the balance of geopolitical power to the South and West. In the eyes of New England, which (in 1803) were identified with the eyes of the Adamses, the purchase of Louisiana was a bargain equivalent to buying a pig in a poke, from which came growling sounds.

Writing in 1803 to his fellow citizens of Massachusetts, John Quincy Adams acknowledged that the inevitable shift of "the centre of power" south and west was "founded in nature" and could not be resisted, but as much as it was not anticipated by the Founding Fathers, it set in motion "an immense force" whose advantages were entirely conjectural, none of which would benefit the eastern states: it was not, he concluded, a time for men of New England to celebrate but rather to fear (*Writings*: III: 59). Loyal to his section, Adams as a member of the Senate voted against the purchase of Louisiana, but then, acknowledging the inevitable, he voted for its incorporation as a part of the United States. In doing so, he also acknowledged the convincing arguments of Senator Breckinridge of Kentucky, the articulate spokesman of western interests, who was able to couple them to eastern concerns.

Is the Union, Breckinridge had asked, "more in danger from Louisiana, when colonized by American people under American jurisdiction, than when populated by Americans under the control of some foreign, powerful, and rival nation?" a question that aroused the old New England bugaboo, Separatism, and reawakened from its abbreviated sleep the ghost of Western Secession. "Is the goddess of liberty," asked Breckinridge, "restrained by watercourses? Is she governed by geographical limits? Is her dominion on this continent confined to the east side of the Mississippi?" (Ogg: 563–64). What could the Adams say to this who, ten years earlier, had celebrated the Fourth of July by observing that the Revolution had sown "the seeds of liberty" afar, that "however barren the soil of the regions in which they have been received, such is the native exuberance of the plant, that it must eventually flourish with luxuriant profusion" ("Oration": 108). The regions Adams had specifically in mind were in Europe, not west of the Mississippi, but consistency demanded that he anticipate with as much gladness "the fair fabric of universal liberty ris[ing]" in an opposite direction from that which he had anticipated (109).

John Quincy Adams' grandson saw the matter with a serer eye. Writing at the end of a century of expanding domain and centralized power, Henry Adams set down the hard lines of impossible alternatives: "Either Louisiana [having been purchased] must be admitted as a State, or it must be held as territory. In the first case the old Union was at an end; in the second case the national government was an empire, with 'inherent sovereignty' derived from the war and treaty-making powers" (*History*: II: 113). Henry Adams saw this undeniable "dilemma" as the jaws of a logical vise that would squash "Virginian," i.e., Republican "theories," but today it is easier to see the dilemma as the horns of a monolithic, even apocalyptic, beast, in which Union and Empire were one and the same. And if the imperial urge has less to do with the lust to conquer, as one modern sage has observed, than with the love of order, then the beast had its coefficient in the man regarded by Henry Adams as the statist equivalent of the Antichrist, who had engineered the purchase and who saw the territory so acquired as a vast opportunity for the exercise of Enlightenment reason in all its manifestations.

Louisiana in a sense was for Jefferson what Florida had been for the Bartrams, a region crying out for the dispensations of orderliness; and having acquired his immense prize, he ensured possession by mapping it. In 1792, Jefferson had regarded the region beyond the Mississippi as an extension of Virginia, and as he set about in 1803 to lend order to Louisiana, he saw to it that Virginians would have that responsibility. In a letter to General George Rogers Clark, written late in 1783, a document that serves as a strategic addendum to his *Notes on Virginia,* Jefferson begins by thanking Clark for some shells and seeds the General has sent from the West, and then goes on to restate "the hope of getting for me as many of the different species of bones, teeth & tusks of the Mammoth as can now be found" (*LL&CE*: 654). But in virtually the same breath, Jefferson, the Enlightenment man of purely scientific curiosity, went on to reveal some imperialistic plans: "I find they have subscribed a very large sum of money in England for exploring the country from the Missisipi to California. They pretend it is only to promote knolege [*sic*]. I am afraid they have thoughts of colonising into that quarter. Some of us have been talking here in a feeble way of making the attempt to search that country. But I doubt whether we have enough of that kind of spirit to raise the money. How would you like to lead such a party?" (654–55). This is the first tender sprout of the seed planted thirty years earlier by James Maury, and it would take another twenty years to mature as the Lewis and Clark expedition, in which the party of the second part would not be George Rogers but his younger brother, William. Like the English "subscribers" before him, Jefferson maintained at first that the Expedition was "only to promote knolege," when much more was at stake, yet we cannot deny the Enlightenment motive, which from shells and seeds to mammoth bones was for Jefferson a version of intellectual imperialism.

Clark declined the opportunity to undertake yet another non-lucrative errand for the sake of the "Publick Interest," but he did caution that any group assembled for the purpose should not be large: "They will allarm the Indian Nations they pass through" (656). Following Clark's logic to its extreme, Jefferson went on to engage the services of the Yankee adventurer John Ledyard in a solitary voyage of exploration. Ledyard, in his characteristically eccentric way, proposed to walk from St. Petersburg eastward to Pittsburg, but Catherine the Great plucked him up in mid-Siberia and set him back on the Polish border facing west. The French naturalist André Michaux was Jefferson's next choice, through the powers vested in him as president of the American Philosophical Society, but Michaux got no further than Kentucky before it was discovered he was an agent of the French Republic, sent over to raise an expeditionary force quite different from that proposed by Jefferson, for the purpose of mounting an invasion of Spanish territory beyond the Mississippi in which George Rogers Clark was not entirely innocent of involvement.

Once again, the Enlightenment espousal of science was a convenient match with imperialism, and the fine print of Jefferson's instructions to Michaux suggests that the United States as well as France had something more than purely scientific knowledge to gain from the expedition: "If you reach the Pacific ocean & return, the [American Philosophical] Society assign to you all the benefits of the subscription before-mentioned. If you reach the waters only which run into that ocean, the society reserve to themselves the apportionment of the reward according to the conditions expressed in the subscription. If you do not reach even those waters, they refuse all reward, and reclaim the money you may have received" (*LL&CE*: 671). News of mammoth bones and tidings of whether the Peruvian llama "is found in those parts of the continent" were all well and good, but the route to the Pacific Ocean was the main item in the contract.

The presidency of the United States gave Jefferson much more power to promote inquiry than did presiding over the American Philosophical Society, and the Lewis and Clark Expedition was conceived and launched even before Louisiana was dropped entire into the national lap. Here again, Jefferson demonstrated his talent for the oblique, and in attempting to wring funds from a penurious Congress he pointed out the commercial advantages of opening a route to the Pacific while at the same time explaining to French and Spanish embassy officials that (as reported by Carlos Martinez de Yrugo) "in reality" the expedition "would have no other view than the advancement of the geography. He said they would give it the denomination of mercantile, inasmuch as only in this way would the Congress have the power of voting the necessary funds" (4).

As with his ornate brief for opening the navigation of the Mississippi, Jefferson's scientific disinterestedness did not impress the Spanish, whose ambassador ended his report with a very acute précis of the Virginian's character: "The President has been all his life a man of letters, very speculative and a lover of glory, and it would be possible he might attempt to perpetuate the fame of his administration not only by the measures of frugality and economy which characterize him, but also by discovering or attempting at least to discover the way by which the Americans may some day extend their population and their influence up to the coasts of the South Sea" (5). The French were more taken by Jefferson's stated motive, and the naturalist Bernard Lacépède sent him a flattering note in which he observed that one seldom sees "the chief magistrate of a great nation unite the enlightenment and solicitude which the well-being of his fellow-citizens demands with the boundless knowledge and the labors of a celebrated philosopher," sentiments that suggest he was no less shrewd than de Yrugo in comprehending the character of the man he so fulsomely addressed (46).

Yet the French, as the evidence of Chateaubriand's intentions attests, sincerely shared Jefferson's belief that "an easy communication by river, canal and *short* portages," in Lacépède's words, might be established "between New Yorck [*sic*] . . . and the town which would be built at the mouth of the Columbia," and the French naturalist saw the opening of such an extended route as a "great means to civilization. . . . Hitherto, the movement of enlightenment has been from east to west. The inhabitants of the United States, if they do not reject their destiny, will one day halt and reverse this movement" (47). Such were also the sentiments of republican bards and Manasseh Cutler, and we may enlist Jefferson's Lewis and Clark expedition as an outreach of the Enlightenment intended to heighten America's rising glory, a nimbus specifically identified with regions in the West.

ii

As an extended arm of the Enlightenment (and Jefferson's Virginia) the Lewis and Clark expedition had a symbolic leader in the person of Meriwether Lewis, not so much as a lesser or younger version of his patron, but as a creation and agent of the older man. As Jefferson was the student of James Maury, so Lewis was his prize pupil, even though in assigning his private secretary the responsibility of heading the mission, as Jefferson admitted to Robert Patterson, professor of mathematics at the University of Pennsylvania, he was aware that Lewis fell somewhat short of the ideal mark: he would have preferred "a person perfectly skilled in botany, natural history, mineralogy, astronomy, with at the same time the necessary firmness of body & mind, habits of living in the woods & familiarity with the Indian character" (*LL&CE*: 21).

In sum, what he needed was an Enlightenment polymath with the physique and talents of a frontiersman, a Jefferson crossed with a Daniel Boone, not necessarily an impossible combination (John Bartram in his prime came close to the mark); but since no one with all these qualities was close at hand, Jefferson decided that Lewis would do. Certainly his secretary knew his way through the woods—and so that he could distinguish the trees (as it were) he was put through what amounted to a crash course in natural history in Philadelphia, a choice of location both convenient and symbolic as the training ground for the man who was to be the avatar of American civilization, much as the same place had provided Lewis Evans the first meridian on his American map.

By 1803, Jefferson had returned to his earlier idea of mounting a relatively large-scale operation, in terms of both men and equipment. Like his binary rationale for sending out the exploring party, the details of planning and logistics look forward to more than a century of government-backed scientific expeditions, the licensing of inquiry by means of commercial considerations matched by a mixture of theoretical anticipations and practical needs based on prior experience. The Enlightenment rage for utilitarian order dominates the documents recording the preparations, which contain lists of equipment gauged to meet any possible contingency—military, commercial, scientific, or diplomatic. Such lists seem to leave no margin for error, and are truly impressive when compared to the suggestion made, in 1789, by Henry Knox, then Secretary of War, that a party about to engage in a similar operation would find pocket compasses "necessary to their success, and Pencils and papers to assist their remarks" (662).

Knox also suggested that canoes would be sufficient for such an operation, while Lewis and Clark were to be supplied with a keel boat. Yet bearing in mind that such a ponderous craft could not be lugged through the Rocky Mountains to the navigable waters of the "Oregan" river, Captain Lewis put his Enlightenment mind to work devising a solution, the details of which delayed his departure from Harper's Ferry, where the government arsenal was located that was to supply him with arms and ammunition for the expedition. Lewis' invention was a portable cast-iron canoe frame, which was to be covered (once they arrived on the "Oregan") with the "bark and wood" he anticipated would be available "in all seasons of the year, and in every quarter of the century, which is tolerably furnished with forest trees" (39–40).

Lewis devoted considerable time while at Harper's Ferry to the "experiments necessary to determine it's dementions," a misspelling not without coincidental meaning. What seemed a commonsensical, even ingenious, invention at the Philadelphia end of things turned into something else beyond the Great Divide. Not all the lists devised by men nor instruments fashioned in Philadelphia could anticipate conditions beyond the anticipation and, occasionally, the control of men: that is, once outside the perimeters of prior experience, encounter took the form of parameters in which the element of the unexpected became a constant.

As an epic action (and it surely qualifies as that) the Lewis and Clark expedition was to a large degree quixotic, yet another in a long series of misconceived errands in the New World undertaken in the long shadow cast by the Spanish conquistadors and French cavaliers. After Captain John Smith, the English in America did not waste much time looking for the northwest passage, and it was only in the wake of the French and Indian War that Anglo-Americans, in taking possession of the French territory, awoke to La Salle's old hope of finding a water route to India. Even then, it was a minor, even eccentric, adjunct to Anglo-American empire, until the pupil of James Maury took charge. "The object of your mission is single," Jefferson wrote Lewis after the leader of the expedition had arrived in the Ohio Valley; "the direct water communication from sea to sea formed by the bed of the Missouri & perhaps the Oregon" (137). At last all the glorious obfuscation of Enlightenment inquiry blew away, revealing the hard line of an imperial diagram, yet at the far end of the journey undertaken by Lewis and Clark, Jefferson's neoclassical vector would be swallowed up by a labyrinthine maze.

For Jefferson's instructions were based on the received and mistaken idea that the continent of North America put forth a symmetrical and convenient shape, that the Rocky (or "Shining") Mountains matched the Alleghenies in height and extent, that the Missouri was a far-western equivalent to the Ohio, its sources in the mountains interlocking with the headwaters of a western equivalent to the Potomac. On Jefferson's imaginary map, the twin, pyramidal ranges of the Alleghenies and the Rockies looked not unlike a great suspension bridge, for from the mountain heights hung a reticulated web of rivers that held together the otherwise disparate parts of the continent, anchoring them to the Atlantic and Pacific coasts. These montane streams were coefficients and tributaries to the sublime system that fed and sustained the Mississippi, both as river and as a waterway for commerce and national expansion. As a collective idea, this map showed a harmonious, unified, and above all else beautifully proportioned plan, lovely to an eye trained to admire the neoclassical ideal.

Like many of Jefferson's other schemes, however, the expedition was forced to revise itself as it went, and took its final and definitive shape from circumstances beyond his expectation and control—in this case the lay of the land. Lewis and Clark were successful in that they brought back a wealth of information, concerning flora and fauna (including Indians) as well as geographical data, but their expedition was a failure in direct and ironic proportion to the accuracy of the map that resulted, for that document put to rest forever the hope of finding a water-borne passage to the Pacific. It also destroyed the symmetrical diagram that sustained it, for if America, as Dr. Thomas Cooper had observed, was "otherwise," it got more so as the continent stretched west, and, as westerners would in later years be fond of pointing out, it got bigger also, bringing about just the kind of geopolitical imbalance that John Quincy Adams feared (Imlay, *TD*: 185). Yet Adams, once again, was the kind of reasonable man that the New England brand of Enlightened thinking produced, quick to accept change and an adept at converting it to the status quo. When his turn as Secretary of State came around, Adams' was the hand that drew the northern boundary of the Louisiana Territory, and if it lay somewhat to the south of the meandering path of the Lewis and Clark expedition, it was a line that coincidentally became the Oregon Trail.

Heroic errands with a defining quixotic dimension often verge on the picaresque, and there is, particularly in the original journals on which the official account of the expedition was based, a decidedly (albeit unintended) comic vein. As literature, the journals kept by Lewis and Clark belong to that class of American adventures in which a certain tension is provided by an uneasy partnership, a marriage of cosmopolitan and vernacular values that expresses the disjunction inherent in a republican society. In a metaphorical sense (permitted by chronology) the expedition of Lewis and Clark is a logical extension of the literary journey of Captain John Farrago and Teague O'Regan, which ends in the western regions of Pennsylvania, the place where the expedition began.

Captain Meriwether Lewis provides the cosmopolitan, aristocratic, genteel element—albeit with a somewhat rustic profile. A hunter by inclination and a military man by training, Lewis may be placed in terms of character as well as birthplace in the neighborhood of Washington and Jefferson. By contrast, his companion officer, William Clark, was both a simpler and a more pragmatic person, in outlines more closely resembling Daniel Boone. As a boy he had migrated with his family from Virginia to Kentucky, and if Lewis belonged to that class of Virginians one generation removed from the Tidewater gentry—then Clark was representative of those who poured through the Cumberland Gap to settle Virginia's western lands.

As a writer, Meriwether Lewis was capable of graceful prose periods, and he brought to the wild West not only his quickly acquired knowledge of natural history but an aesthetic sensibility. By circumstance, Captain Lewis was superior to his partner, Captain Clark, in rank, and though the other was equal in authority (a situation that at times produced friction), Clark played symbolic squire to Lewis' questing knight, a commonsensical Sancho Panza to the other's Quixote. The quality of difference is clearly defined by comparing the journals kept by each man, for Clark's (at times audible) vernacular observations provide a balance to the often self-conscious compositions of Lewis. Of the two, Lewis was clearly more typical of Enlightenment Man, while Clark was the kind of Natural Man whom the men of the Enlightenment liked to celebrate. It was a marriage very much of convenience, and like many such had its tensions, for in Clark's eyes Enlightenment inquiry tended to hamper the onward march of the expedition. He had no great admiration for Lewis' cast-iron canoe, which he wryly christened "The Experiment," for the boat was never able to fulfill its intended function, due to the absence of pine pitch on the westward slopes of the mountains.

In appraising the expedition as literature, we can assign it a dual hero, that Lewis-and-Clark compound with which its record is always associated. At the start, the line of inquiry is associated with Lewis, the Enlightenment Man traveling out of Philadelphia to the arsenal at Harper's Ferry, where the expedition took on expansionist stamp, because of its critical location at the juncture of the Potomac and Susquehanna Rivers. But it was at Louisville on the Ohio that Clark added the final, definitive dimension: joining the expedition at that symbolic point, he brought with him a number of other Kentuckians, who carried with them the frontier spirit. All in all, it would be difficult to conceive of a purely poetic (i.e., fictitious) arrangement of people, places, and incidental appurtenances that would match those of the Lewis and Clark expedition for symbolic potential.

The ratification of the Louisiana Purchase gave a new emphasis and momentum to the expedition, transforming Lewis' errand from a tacit to an explicit national mission. It guaranteed the expectations of experienced public men like Albert Gallatin that American settlement would follow the course of the Missouri no matter to whom the river belonged. Lewis and Clark thereby became both surveyors *and* ambassadors, projecting George Washington's errand in 1753 to a further range. Though their chief assignment remained the search for a Northwest Passage to the Pacific, they were also to secure alliances with the Indian tribes along the route, to maintain friendly relationships with the British fur-traders who were highly active along (and protective of) the upper reaches of western rivers, and to determine, if only by hearsay, the headwaters of streams like the Platte and the Colorado, which defined the very vague borders of the Louisiana Territory.

As harbingers of empire, the two captains were bringers of order, in terms both of the land they passed through and the men who already inhabited it. To secure the goodwill of Indian tribes along the Missouri, on which the fate of future commerce and settlement along the great river and its tributaries depended, they were instructed to give out presents and presidential medals, making peace not only between red men and white but between red men as well, for they were supposed to settle local grievances and rivalries in order to smooth out the imperial path. Captain John Smith had attended to similar business while exploring the rivers of the Chesapeake Bay for a Northwest Passage two centuries earlier, and a romantic adjunct to his peacemaking efforts was provided by Pocahontas, the Indian princess who loaned a friendly, even loving, presence to the virgin land. For their part, Lewis and Clark had Sacajawea, wife of Toussaint Charbonneau, the French-Canadian guide who provided the expedition its most picaresque accoutrement.

Sacajawea was the useful half of the family, and though neither princess nor maiden, she was instrumental at a critical point in the journey, when the expedition reached the mountain stronghold of her people, the horse-trading Shoshoni. Despite their differences, therefore, Pocahontas and Sacajawea are joint symbols of mediation, testifying to the ubiquitous yet ambivalent presence of the Indian in the epic of New World empire, at once a necessary adjunct and a potential adversary to the heroic action. Like the rivers they lived along, which bore Indian names, the native people were both essential to the westward progress of civilization and resistant to it. Lewis' many observations about the shifting, treacherous waters of the Missouri are coupled with his remarks concerning the equally unstable and undependable nature of the Indian. It is significant in this regard that Sacajawea (like Pocahontas) is not accorded much space in the published history of the expedition: symbols of mediation tend to emerge after the martial aspects of the imperial process have subsided.

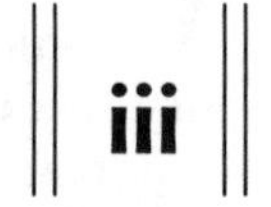

Although the Missouri provided the main axis of westward movement for the expedition, and held out the chief hope for a passage to the Pacific, the facts of actual encounter with the river, including collapsing cutbanks, oxbows, sawyers, and sandbars, gave dramatic evidence that the Missouri would be a violent and moody ally to the westward advance of personified Enlightenment. This instability was epitomized by the melting away in one night of the sandbar the explorers had pitched their camp on. The remains of Indian towns and fortifications were likewise a constant reminder of the "unsteady movements and tottering fortunes" of a nomadic people, who lived in a constant state of warfare, which had caused the disappearance of entire tribes in the recent past. Thus the zone defined by the winding, turbulent course of the Missouri was one of constant flux, upon which the expedition as an arm of the Enlightenment attempted to impose an element of order, mapping the river and its environs and through treaties hoping to stabilize Indian relationships—a familiar pattern of imperial design with predictable results.

In order to better impress the Indians with United States might, Lewis and Clark in their negotiations along the Missouri took a firm line and gave it a suitable backdrop of ceremony and martial strength. Shortly after arriving at the village of the Mandans, the Americans met with the chiefs in council, and "that the impression might be more forcible, the men were all paraded, and the council was opened by a discharge from the swivel [gun] of the boat. We then delivered a speech which, like those we had already made, intermingled advice with the assurances of friendship and trade" (*History:* I: 182). One old chief became impatient and interrupted the ceremony, but he "was instantly rebuked with great dignity by one of [his fellow] chiefs for this violation of decorum at such a moment," and the proceedings continued with the ritual handing out of presents, the amount and quality of which were adjusted to the rank of each chief.

This carefully staged pageant was a kind of imperial theatrics, lending the business of power play a certain aesthetic grace, a combination of forms understood and appreciated by the Indians, yet one that was intended by the Americans as an extended instrument of republican law and order. As symbolized by the clasped hands that appeared under the profile of Jefferson on the medals given to each head chief, the presidential arm had a very long reach, an Enlightenment extension whose intention was likewise displayed in the iconography on medals given to lesser chiefs, "representing some domestic animals and a loom for weaving" and the "impressions of a farmer sowing grain" (182). Agrarian and pacific symbols, these icons posited the conversion of Indians from a nomadic hunting culture to a more advanced and stable way of life; in effect projecting a physiocratic ideal onto the western region and its inhabitants. The reality of the encounter between civilized and savage peoples is better illustrated by another, perhaps unintended icon: early in the voyage, the Americans planted the national ensign on the great funeral mound of the powerful Indian chief Blackbird, who, according to Meriwether Lewis, "seems to have been a personage of great consideration" (72).

The projection westward of imperial aesthetics took a number of forms during the upriver voyage, an orderly dispensation that informs Lewis' description of the cliffs along the Missouri, which looked to him like "elegant ranges of freestone buildings, with columns variously sculptured, and supporting long and elegant galleries, while the parapets are adorned with statuary. On a nearer approach they represent every form of elegant ruins—columns, some with pedestals and capitals entire, other mutilated and prostrate, and some rising pyramidically over each other till they terminate in a sharp point" (339). Lewis fancied these natural approximations to "productions of art" as having a "romantic" (because ruinous) appearance, but they notably resemble not Gothic but classical examples of architecture; in effect, a westward transference of the same Palladian ideal that was giving shape to the new national capital. "Clark," notes Dr. Coues, "is Doric, Lewis is Corinthian" in style, giving by their journal entries a neoclassical impress to the land (260n).

The expedition spent the winter encamped near the Mandan village, and the next stage of the upriver voyage brought Lewis and Clark to the great falls of the Missouri, a movement into hitherto unexplored territory that culminated in an episode of sheer power and beauty central to the aesthetic dimension of Lewis' narrative. This part of the book is given added tension by the arrival of the explorers at a twin fork in the Missouri, and the necessity therefore of determining which was the main channel to the river, a decision on which rested the success of the expedition, posited as it was on the proximity of the headwaters of the Missouri to the sources of the Columbia ("Oregon") River. Lewis and Clark chose the right river for their assignment but the wrong direction for their purposes, for it was the other (Maria's) river that led to the shortest way through the mountains. Yet, as Elliot Coues observed, the "worst possible" direction from which to reach the Columbia led the two explorers through unmatchable scenery. In terms of utility, therefore, their decision was wrong, but in terms of beauty—and literature—it was right.

Led to the Great Falls of the Missouri by a Mosaic cloud of "spray, which . . . arose above the plain like a column of smoke," Meriwether Lewis "hurried with impatience" to the spot, where he "seated himself on some rocks near the center of the falls" to "enjoy the sublime spectacle of this stupendous object, which since the creation had been lavishing its magnificence upon the desert, unknown to civilization" (II: 365). Edmund Burke had instructed Lewis' generation on what to do when confronted by a natural phenomenon of this magnitude, and by seating himself near its center, Lewis became not only an audience of one but the necessary perspective figure in the picture he renders for us. Without a spectator, as Lewis observes, even the grandest manifestations of landscape are wasted. In John Bartram's words, an unexplored paradise remains a desert.

Having enjoyed this sublime manifestation of wilderness power, Lewis continued to explore beyond it, and next came upon its Burkean counterpart, "one of the most beautiful objects in nature," another and contrasting waterfall that "with an edge as straight and regular as if formed by art, stretches itself from one side of the river to the other for at least a quarter of a mile. . . . The scene which it presented was indeed singularly beautiful, since, without any of the wild, irregular sublimity of the lower falls, it combined all the regular elegances which the fancy of a painter would select to form a beautiful waterfall" (368–69). Here again the binary theme of the riverine landscape, and, as before, the aesthetic is neoclassical, determined by a linear, hence regular, ideal, yet another instance of natural Palladianism. To this aesthetic diagram Lewis added the requisite expansionist view, for, ascending a nearby hill, he obtained an approximate Pisgah prospect, "a delightful plain, extending from the river to the base of the Snow [Rocky] Mountains," and through which "the Missouri stretches to the south in one unruffled stream of water, as if unconscious of the roughness it must soon encounter, and bearing on its bosom vast flocks of geese; while numerous herds of buffalo are feeding on the plains which surround it" (370).

Descending to the plain, Lewis shot a buffalo for his supper, and having neglected to reload his rifle, was surprised by a large bear, which he escaped by running to the river and jumping in, thereby giving his scenic excursion a nearly tragic but finally a comic conclusion. He was next menaced by a "brownish-yellow animal," probably a wolverine, and then, "as if the beasts of the forests had conspired against him" he was charged by three buffalo bulls (372). Lewis was successful in facing them down, and continued his return to camp as night fell, "reflecting on the strange adventures and sights of the day; which crowded on his mind so rapidly that he would have been inclined to believe it all enchantment, if the thorns of the prickly-pear, piercing his feet, had not dispelled at every moment the illusion" (372).

Of the several agents acting to dispel aesthetic enchantment, it is the bear that adds most meaning to the epic framework of the expedition.

Among the American fauna first recorded by Lewis and Clark was the grizzly, which by its various colorations at first confused them, appearing in brown, white, and truly grizzled coats, but which was finally classified by its concealed testicles and by its seldom concealed hostility to their presence, as if reluctant "to yield their dominion over the neighborhood" (378). Grizzlies became more numerous as the expedition approached the Great Falls, and Lewis' nearly fatal encounter testifies to their acute sense of territorial imperatives, providing a Rocky Mountain equivalent to Bartram's thundering alligator—here putting to flight Jefferson's agent of empire, who as Enlightenment Man was anxious to classify the grizzly bear for science, but who had no desire to share the embrace of the wilderness spirit incarnate.

The episode of the sublime and the beautiful falls and the terrible but comical bear occurs at a critical point in the narrative. Thenceforth the expedition would be following unnavigable waters, and as the distance lengthened and the terrain increased in roughness, the inherent fallacy of the pyramidal theory of North American geography became apparent. Passage to the Pacific is possible through the Rocky Mountains, but it would be made by other means than water carriage. Though the falls provided Lewis his most intensely aesthetic moment of the voyage, the accommodation of the wilderness to civilized notions of order is clearly illusory, as all things seemed to conspire against the advance of the expedition into the mountains. Yet even as Lewis and Clark continued to pursue their increasingly dubious quest, the spirit of the party became if anything even more expressive of both a nationalistic and an Enlightenment élan.

Following the Missouri to its ultimate source, the explorers gave the various headwaters contributing to the main river names that reflected their errands, a series of triads that lend the narrative a consciously symbolic dimension. The first three forks of the Missouri were named "Jefferson," "Madison," and "Gallatin"; while the tributaries to the greatest of these streams were called "Philosophy," "Philanthropy," and "Wisdom"; qualities associated not only with Jefferson and the "virtue-bent" Enlightenment Man, but with the ideals of Freemasonry. Yet all of this highly civilized (even politicized) name-giving is overshadowed by the terrain through which the explorers had just passed—the formidable "Gates of the Rockies," "a most sublime and extraordinary spectacle. . . . Nothing can be imagined more tremendous than the frowning darkness of these rocks, which project over the river and menace us with destruction" (425). Like Pownall's Hudson and Jefferson's Potomac, the Missouri "seems to have forced its channel down this solid mass," the rock having "given way . . . reluctantly."

The Jefferson River led Lewis and Clark to what they presumed was the ultimate source of the Missouri, a "chaste and icy fountain" from which they drank a sober but triumphant libation, feeling themselves "rewarded for all their labors and their difficulties" (484). Then, crossing over the ridge of the Great Divide, they drank "for the first time the waters of the Columbia." Yet the subsequent exploration of the Great River of the West only further verified the truth already discovered, that the heads of navigation of the Missouri and the Columbia were not near but very far apart; and though Clark persisted in mapping out the most direct overland route between the two rivers, the great hope that had inspired the expedition proved chimerical. The narrative continues to maintain its epic coherence, but the original purpose of the expedition having no longer any validity, the account loses its heroic dimension. Having reached the navigable part of the Columbia, Lewis delayed the expedition as he struggled to float his cast-iron canoe, the failure of his "experiment" foreshadowing much that was to come, including the hardships of the long, dismal winter in Fort Clastrop, at the mouth of the Columbia River. Lewis and Clark continued to explore and map the territory on the far side of the mountains, gathering the requisite data concerning the wildlife, the minerals, and the people they found there. Yet this last part of the narrative loses the exuberance of the earlier sections, and, taking its tone from the misery endured by the expedition while on the Pacific coast, it is notably lacking in Enlightenment optimism.

Failing to encounter the trading vessels from which they expected to obtain the beads that passed among the Indians for currency, the Americans (like the English at Jamestown two centuries earlier, at the very start of the process that the Lewis and Clark Expedition ended) found themselves increasingly at a disadvantage when bargaining for food and the souvenirs that were for them of scientific interest. Though continuing to act as peacemakers, the ragged group of half-starved soldiers lost much of their impressiveness as agents of empire, and the Flathead Indians with whom they dealt correspondingly lacked the good looks and dignified bearing of the Plains tribes. They did, however, turn out to be very sharp businessmen. Being experienced at trade, they drove hard bargains, and did not hesitate to make the best of their advantage. Where the Great Falls of the Missouri provided a sublime spectacle, untenanted by humans, the falls of the Columbia, where the Indians periodically met for purposes of conducting trade, was virtually a marketplace.

Among Jefferson's Enlightenment theories set in rhyme by Joel Barlow was the idea that commerce could affect peace and harmony between nations, but this utopian notion was posited on an equal balance of supplies and needs that was lacking in the dealings between the Lewis and Clark expedition and the Flathead Indians. As the winter wore on, the supplies were mostly in the hands of the Indians, the Americans had all the needs; and as the balance of trade shifted, tensions began to increase, exacerbated by thefts and sharp dealing by the Flatheads. There were occasional outbreaks of violence, but an uneasy truce was generally maintained, and throughout their stay the soldiers frequently consorted with Indian women—though seldom for free. The Americans even developed a liking for that staple of Indian diet, dog meat, which they had at first found repugnant to their civilized palates, and despite Lewis' industry in gathering and classifying scientific data, one gets a very definite sense that the Enlightenment nimbus surrounding his enterprise was gradually dimming. At times a very thin line distinguished white men from red, civilized from savage beings, much as the Indians themselves demonstrated an almost seamless progression from the intelligent and mannerly Shoshoni—the horsemen of the mountains—to the Yahoo-like Chinooks who lived nearest the ocean.

In terms of Enlightenment imagery, the Oregon of Lewis and Clark may be said to lie on the far side of the Shining Mountains, in the dark shadow of primitivism whose reality always gives the lie to Rousseauvean notions of nature's noblemen. It is an equivalent experience to that had by Edgar Huntly's penetrator of similar terrain, an experience that belied a number of Enlightenment assumptions. Wilderness life, as suggested by John Colter's decision to leave the expedition for the life of a mountain man (recalling Edny's transformation to a wildman) is more infectious than law and order, which depends for its authority on force beyond persuasion—life in the wilds draws upon deep, dark springs in the soul. Whether as the bear, that manlike personification of the forest, or the man of woods and waters himself, the Indian, what lies beyond the Great Divide in the American psyche mocks the orderly dispensations of Enlightenment Man.

Still, the arguments of rifles and swivel guns are strong ones, and the bear and Indian eventually yielded dominion to the Americans who soon followed the expedition's lead: the first parties of fur traders were met by Lewis and Clark as they returned down the Missouri in the spring of 1806, an experience equivalent to Washington's when he returned from the Ohio Valley a half-century earlier. If Coulter remained in the wilderness instead of returning to civilization, it was for the purpose of exploiting the fur-bearing wealth it contained.

iv

Meriwether Lewis did not live to witness the commercial (and literary) consequences of his labor. Like earlier explorers associated with empire in America, he subsequently slipped from grace. Columbus died impoverished in prison; John Smith ended a tatterdemalion old adventurer still futilely peddling his talents in London; LaSalle was murdered by his men as he attempted to find the mouth of the Mississippi; and Lewis died by his own hand while traveling eastward from the Louisiana Territory to Washington, D.C., in 1809. The man who had endured so much hardship in the wilderness and who had persisted in giving the terrific abundance of the western regions a semblance of order—at least during his passage through—was finally unhinged, not by the chaos of wilderness, but by the messy affairs of men.

Lewis was rewarded by Jefferson for his exploits by being named General James Wilkinson's successor as the governor of Louisiana, a dubious reward at best, for the political and bureaucratic tangles that were inherited with the great wealth of Louisiana—the kind of labyrinth a Wilkinson could comfortably inhabit with Minotaurean grace—proved to be Meriwether Lewis' undoing. Although he performed a herculean labor of housecleaning, bringing (literal) law and order to the workings of the new territory, Lewis was called to account for some unauthorized expenditures by Congress, a small but humiliating matter that triggered deep depression. He was on his way east to explain matters, when, in an isolated ordinary on the Natchez Trace, he put a sudden termination to a brilliant and still promising career by blowing out his enlightened brains.

The memorial raised to Meriwether Lewis on his lonely grave was not erected until 1848, a moment well beyond his anticipated life span but a date sacred to the eventual meaning of the Oregon Trail. Placed alongside a wilderness trace that had lost its chief function with the coming of the steamboat, erected even as the United States was rushing to fill in its continental borders, the monument provides a suitable counterpart to the reverse side of the Great Seal of the United States, that glorious expression of Manifest Destiny. An unfinished column on a truncated pyramid of seven steps (one for each lustrum in Lewis' life), it symbolizes a tragically interrupted career, not an empire yet to be completed. As we contemplate the divinely sanctioned pyramid on the

Great Seal, a towering witness of technological superiority and imperial unity through architectonic order, we might think of that unfinished pyramid with its broken neoclassical column, an expression of the human cost that empire demands, set out in forms recalling Palladian plans.

Interlude

Milk of Paradise

i

The "Washington of the West" he was called, but General James Wilkinson deserves the sobriquet only if we give particular stress to the regional modifier, for "the West" during his period of ascendency in that trans-Mississippi region was a fractious geopolitical amalgam of rival ambitions, given unity by expansionist energies and by a rampant mood of opportunism, of which the general was a prime example. Where the "Washington of the East" stood for personal duty and national unity, Wilkinson was a symbol of self-aggrandizement and secession. Washington sought to strengthen trans-Allegheny ties between East and West; Wilkinson connived to establish connections between western territories and the Spanish authorities in Louisiana.

Washington was not without self-interest, regionally defined, for the union he sought to establish was a principle aligned with the Potomac River, but Wilkinson may best be associated with the Mississippi, that turbid flood that draws tribute from the entire western region, only to destroy the lands it passes through on its way to New Orleans. He is a classic example of the overreacher, undoing and undone by greed, and though Colonel Aaron Burr and his aborted scheme for empire may have set in temporary motion the danger feared by Washington and Jefferson, from the first, the dashing Easterner was the cat's-paw of Wilkinson, who hatched the scheme and who, when he realized its folly, aborted the seed of embryonic empire by betraying Burr and his pitifully small band of adventurers.

During the trial of Burr and Blennerhassett, Wilkinson was the chief (if not very useful) witness for the prosecution, and was regarded by many at the time as a living caricature of vaingloriousness. But from here we can see the extent to which Wilkinson was the chief figure in the action, being both the tempter and the betrayer of Burr and hence the true author of events. Burr emerged from his aborted adventure the type and symbol of prideful ambition, styled by William Wirt a Satan tempting the Blennerhassetts from their temporal Eden; but in truth he was merely a Faustus, led on by the regnant Mephistopheles, General Wilkinson, who ranked the little colonel in so many, mostly unpleasant, ways. Sheltered by Jefferson (for reasons only the Virginian knew) from the punishment he richly deserved, Wilkinson seems at times a shadowy projection of Jefferson's idealism, a dark counterpart to Meriwether Lewis, and part and parcel of those aspects of his administration that Jefferson's detractors emphasize and his champions must always try to explain away. But Wilkinson is perhaps best defined by the title he aspired to, the Washington of the West, as an avatar of all that the Washington of the East abhorred.

Though like Washington the son of a planter, Wilkinson was born on the opposite side of the Potomac, in Maryland, and he introduced into the western regions what can only be called the sot-weed factor, in terms not of tobacco but besottedness, the effect of rum and addiction to political power in large doses. As George Washington stood for all that was admirable east of the Alleghenies, and loaned by his presence a measure of national stability equivalent to the Constitution, so Wilkinson stood for all that was contemptible west of the mountains, fostering instability to the point of chaos. Though his plans never came to anything save personal dishonor, he managed to leave his fingerprints all over the emerging map of the Mississippi Valley, as testified by the place names that still endure. Washington as a monumental man was carved from marble, but Wilkinson was shaped from soapstone; there is a latent Caesar in the first General, but the second was every inch (save his profile) a Cassius.

If the first half-century of the republic had a resident Vice and an emblem of misrule, then it was General Wilkinson, virtually a resurrection of the Thomas Morton who plagued the Puritans' sensibilities with his phallic maypole and threatened their lives with his easy way with Indians, including trading rifles for furs. Rampant sexuality was never Wilkinson's game, however; only lust for wealth and power, but he was a very Machiavel for intrigue, compulsive double-dealing covered with a great deal of bluster and lies. He was blessed—perhaps by the devil—with a remarkably long life, a symbolic span that coincides with the rise of American empire, having been born in 1757, at the start of the Seven Years' War, and dying in 1825, a year that preceded the fiftieth anniversary of the republic and the deaths of Jefferson and John Adams. Wilkinson first rose to public view during the Revolution, having survived the siege of Boston, then joined Benedict Arnold in the siege of Quebec in time for the American retreat, an association (and action) that symbolized his subsequent military career. Serving thenceforth under Washington, at the battles of Trenton and Princeton, Wilkinson now first emerged as the great general's shadow, his anti-type. He joined in the Conway Cabal, and, the intrigue discovered (thanks to Wilkinson's bungling), was forced to resign his commission; but instead of creeping away into the darkness of anonymity, Wilkinson persisted in seeking both reputation and wealth by whatever means, usually oblique and secretive and more often than not dishonest. Following the end of the war he got a leg up in society by marrying the sister of a Biddle, and stood for (and won) election to the Pennsylvania state assembly.

But Wilkinson's fate and future lay in the West, as his subsequent activities show. He soon removed to Kentucky, where from real-estate speculation he enlarged his sphere of self-aggrandizement to double-dealing on an international scale, attempting to profit from the prolonged dispute between Spain and the United States over mutual boundaries and navigation rights to the Mississippi. While pretending to act in the interests of his country, he did what he could to prolong the dispute to which his prosperity was so closely allied. Indeed, the warrant for his title of "Washington of the West" was based on Spanish promises should Wilkinson succeed in leading a massive defection of the western regions from the United States, thereby personifying a massive contradiction in terms. He was rather an un-Washington, the perfect antitype of the man with whom he sought comparison.

Falstaff, perhaps, is the most convenient analogue, for the general was a prime specimen of *Miles Gloriosus,* whose activities lend themselves to mock-heroic rather than epic forms, being a parody of the virtuous, self-sacrificing, and courageous actions we associate with true heroism. Notably, Wilkinson was burlesqued as General Jacobus von Poffenberg in Washington Irving's satiric account of the political wrangles of the early republic, disguised as a comic history of New York. And yet, like Falstaff, Wilkinson was not without companions and admirers, for like all great confidence men—and he was undeniably great in that regard—the general had considerable powers of charm and an ability to convince, resulting in an astounding record of survival. No one can rise as high as Wilkinson did (and stay there) without having, at least superficially, qualities that earned his contemporaries' respect.

Nor did he lack consensus. Gilbert Imlay was one of his admirers, and among the expansionist documents assembled in the 1793 (American) edition of his *Topographical Description* there is a copy of the petition addressed in 1788 from "the people of Kentucky in convention" to the United States. The apparent work of General Wilkinson, who was a very busy delegate to that meeting, espousing aloud patriotic opinions while working as always in quiet for himself, the document is resonant with region feeling. Imlay was of the opinion that the petition contained language and sentiments "so pure and so manly" that it could be taken as a perfect expression of "the disposition and manners" of the Kentucky settlers. Though Imlay was undoubtedly exaggerating for his own purposes, the statement was issued by all present at the convention and it certainly is a small masterpiece of expansionist rhetoric. We may accept it, therefore, as an epitome of what the Mississippi River meant to General James Wilkinson and his fraternal band in 1788. Moreover, whatever may have been Wilkinson's subversive ambitions concerning the future of the territory watered by the Mississippi, his published declaration was hardly at odds with the hopes and fears of George Washington himself concerning the western regions.

The petition begins with a long account of the settlers' ordeals, particularly their bloody struggles with the Indians, the victorious conclusion of which attested that "'no human cause could control that Providence which had destined this western country to be the seat of a civilized and happy people'" (xviii). Such sentiments were completely in accord with Manasseh Cutler's pacifist imperialism, but where Cutler envisioned the ongoing and orderly expansion westward of a divinely favored civilization, Wilkinson put his emphasis on a major impediment to further progress and prosperity in the West, "a narrow policy [that] shut up our navigation [on the Mississippi] and discourag[d] our industry":

> *"In this situation, we call for your attention—we beg you to trace the Mississippi from the ocean—survey the numerable rivers which water your western territory, and pay their tribute to its greatness—examine the luxuriant soil which those rivers traverse. Then we ask, can the* GOD OF WISDOM AND NATURE *have created that vast Country in vain? Was it for nothing that he blessed it with a fertility so astonishing? Did he not provide those great streams which enter into the Mississippi, and by it communicate with the Atlantic, that other nations might enjoy with us the blessings of our prolific soil? View the country, and you will answer for yourselves. But can the presumptuous madness of men imagine a policy inconsistent with the immense design of the* DEITY*? Americans cannot."* (xix–xx)

The logic of this paragraph is inexorable: to deny Americans access to the entire length of the Mississippi is tantamount to blasphemy, being a denial of sacred writ, not in the form of biblical scripture but in the handwriting of God upon the land, whose fertility and rivers are a divine kabala prophesying empire. Manifest Destiny as a journalist's term had not yet been invented, but the idea it expressed was current wisdom in Kentucky in 1788.

If Wilkinson resembles, in his addiction to misrule and rebellion, Thomas Morton of Merrymount, so his language in the petition also evokes the sensual diction of the author of *New English Canaan,* suggesting that Imlay himself may have had a hand in its composition, erotic hyperbole being the stock in trade of the author of *The Emigrants,* as well. Deriving (or imposing) sexuality from (or on) the landscape is often an outlaw and alien theme in early American literature, and in the 1790s Imlay was its chief practitioner: it is a minority expression that springs up only occasionally in a landscape otherwise characterized by the kind of adversity that produces a hard-faced virtue, as in Crèvecoeur's antithetical view. Prominent among the settlers in Imlay's libertarian utopia is "General W——," a good natured, avuncular figure clearly patterned after Imlay's Kentucky patron, whose role in the novel, however, hardly approximates his real activities. Yet his presence is symbolic, and the fact that Wilkinson named his subsequent home in St. Louis "Belle-Fontaine" cements the connection between the incipient secessionist gestalt in Imlay's book and the activities of the corpulent general.

Much as Ashe's *Travels* puts forth a tantalizing shadow of the Burr conspiracy, which would attempt to realize the latent outline of secession in Imlay's *Topographical Description* and novel, so we catch there also a brief glimpse of the meddlesome general, referred to briefly at the end of Ashe's narrative. The peripatetic Irishman noted as he passed down the Mississippi on his way to New Orleans that Wilkinson was busy at Fort Adams, below Natchez, "collecting troops to drive the Spaniards beyond the Louisiana line, from the limits of which the domains of the United States would extend to the Florida gulph and the isthmus of Dawen [Darien]" (293). This was not quite Wilkinson's accomplishment during the ensuing military inaction, nor was it his official assignment (to say nothing of his real intention), but it was an essential part of the scheme launched by Wilkinson and Burr on western waters, even as the general's maneuvers reported by Ashe were essential to Burr's betrayal.

ii

Wilkinson's operation against the Spanish forces in Mexico (now Texas) was but one of several spurious gestures he was making in a desperate last-minute attempt to play all ends against the middle as the

Burr conspiracy rushed toward its disastrous conclusion. The frantic efforts to save his unsavory hide were in a sense successful, for if Wilkinson failed in his hopes to advance his personal fortune, he did manage to scramble clear of Burr's sinking scheme. Moreover, during the same period that he was engineering his hydra-headed plan of conquest, Wilkinson was sponsoring exploratory parties; though purportedly mounted in the spirit of the Lewis and Clark expedition, they were enlisted more in the service of Wilkinson's lust for personal empire than in the hope of extending an orderly republic westward.

Moreover, as Captain Meriwether Lewis was the creation and extension of Thomas Jefferson's brighter side, so Wilkinson had his agent also, in Captain Zebulon Pike; for if Lewis was an immaculate avatar of the Enlightenment spirit in western regions, then Pike was to a very large degree Wilkinson's creature, a personification of opportunism and personal self-aggrandisement, which was one of the several forms assumed by Liberty when (in the words of Senator Breckinridge of Kentucky) she leapt the Mississippi and sped on toward the Sabine River. Where Meriwether Lewis borrows a measure of fame from his association with Jefferson, Pike gained only obloquy from Wilkinson's patronage, much as his expedition into the Southwest remains as cloud-hidden as the peak that marked its northwestern terminus.

Pike was Lewis' antithesis in a number of ways, not the least of which was his ancestral origins in New England. He lacked the Virginian's intellectual brilliance and aristocratic background, and instead of electing to serve in the Army because such service was in accord with the aristocratic élan of Virginia gentry, he did so because it was both a family tradition and a way an ambitious man without inherited property could get ahead in the world. He was in pursuit, therefore, of what Lewis already possessed, a difference that helps to distinguish between these two undeniably brave instruments of empire. Having spent twenty years awaiting his destiny on a number of frontier posts, Pike was certainly qualified to carry out the tasks assigned him. But he was therefore hot iron for Wilkinson's hammer and anvil, and such was his gratitude (or ingenuousness) that he never abandoned his stated faith in his benefactor's sterling character.

The makeup of Pike's two expeditions likewise presents a number of defining differences when compared to the Lewis and Clark party, for Pike set out on both adventures lacking the equipment, planning, and prestigious backing enjoyed by the other men. His expeditions were distinctly of the cut-rate kind, and, as a result, the last of the two very nearly had a tragic termination. Moreover, where the Lewis and Clark errand was a collective enterprise, a depersonalized national mission given even more objectivity by a dual leadership, Pike's were very much extensions of his own personality. Neither Lewis nor Clark was averse to becoming famous, but Pike seems to have been assiduous in his pursuit of glory, and where Lewis' journals languished in the care of others for years, Pike pushed his own as quickly as possible into print. This gave him the only margin he enjoyed over the competing expedition, but it also resulted in hasty and careless composition, which eventually put Pike's narrative at a disadvantage when compared with Nicholas Biddle's polished text. Still, Pike's account makes better reading than some critics have allowed, and even the obscurity of references may be seen as integral to the opacity of his second and most important errand. We can never be sure whether Pike is vague by accident or intention.

Most important, we cannot deny the man his hard-sought recognition, even if we are unwilling to call it fame. If his missions were unsuccessful, they were also, given his lack of resources and information, virtually impossible, and may be compared with those several Arctic and African explorations that symbolized imperial enterprise somewhat later in the century, in that failure becomes a very relative term in the face of the great, even magnificent, effort merely to survive. As Pike and his party made their way north along the upper reaches of the Mississippi in the winter of 1805–1806, they verified Lewis' wisdom in waiting for springtime before voyaging up the Missouri, but they also demonstrated qualities of endurance and ingenuity, fashioning toboggans when all other means of transportation failed them. Caught likewise a year later by the approach of winter on the headwaters of the Arkansas, Pike and his men suffered greatly from the cold, yet they persisted in carrying out their assignment, the men presumably because of their loyalty to Pike, Pike because of something more than merely a love of fame. He seems, even in his letters, to be a man for whom glory is not a goal but an inner necessity, and if he refused to admit that Wilkinson was anything less than a stainless blade of heroic temper, it was perhaps because to do so would have dimmed his own luster.

In a certain sense, Pike's futile because unsuccessful heroics take the meaning from his failure. Where Lewis and Clark passed through the mountains and came out on the Pacific side, Pike in both his northern and western advances came up against a terminus, best symbolized by the peak that he discovered, a pyramid shape standing for limits, not growth. But Pike's importance returns ever again to his mountain, as the mountain remained so long in view during his travels, and until the publication of the account of the Stephen Long expedition some ten years later, Pike's journals served to introduce to the American imagination the idea that the far western prairies were a wasteland, a zone best left as a permanent geopolitical barrier. Antedating the narrative of the Lewis and Clark expedition by some four years, Pike's book undoubtedly provided a bleak backdrop to that lengthy account of western hardship as well.

The mountain as terminus is the more meaningful because Pike's most ambitious undertaking (in both personal and geopolitical terms) was not defined, like Lewis's, by the westward-trending courses of rivers, but by the invisible barrier line determined by their sources. Though his travels on the Mississippi were, during the early phase, by water, that errand was not one of crossing over some great divide, but of establishing the location of a headwater that would define the northernmost limit of United States domain. And most of his second journey was overland, bisecting as he went the headwaters (he assumed) of the Platte, the Kansas, the Arkansas, and the Red River of the South, which last turned out, to his genuine surprise or mock mortification (and it is on this pivot that the ambiguity turns), to be the Rio Grande del Norte, leading directly south into the heart of Spanish territory. There was certainly a tacit element of military reconnaissance involved in Lewis' mission, but the element was both larger and more tacit in Pike's case, a subversive dimension that points beyond the newly established bounds of United States dominion. It is this dimension that reveals the ubiquitous hand of General Wilkinson, the plump equivalent of Jefferson's long, lean fingers, both with an imperialistic reach, but the one signifying appetite, and the other, intellect.

As Wilkinson's agent, Pike was a suitable exemplar of qualities best described as physical. Though the appendixes to his journals indicate that he was second to none in his avidity for collecting data, Pike on the march seems to have been happiest when killing something. Lewis himself was a skillful hunter, but Pike displays an almost savage joy when pitting his skill against "big" game, as when, early on his voyage up the Mississippi, he "left the camp with the fullest determination to kill an elk, if it were possible, before my return," for the simple reason that he had never before "killed one of those animals" (38). The first day's hunt resulted in two elk wounded (both escaped), three deer, ditto, and the conclusion that Pike's balls (musket) were too small. His second sortie, the next day, resulted in several more wounded deer before one elk finally fell to Pike's marksmanship: "This was the cause of much exultation, because it fulfilled my determination, and, as we had been two days and nights without victuals, it was very acceptable" (40). Clearly, to Pike's thinking, glory was more important than a full gut. Still, to be fair to the man, he was satisfied with having bagged his one elk, and a week later, he "passed several deer and one elk, which I might probably have killed; but not knowing whether I should be able to secure the meat, if I killed them, and bearing in mind that they were created for the use, and not the sport of man, I did not fire at them" (42).

A Pathfinder, especially during the Mississippi voyage, Pike served also as Deerslayer, being the chief provider while on the overland march. One of those men for whom leadership means total control, he seems by his own account to have been ubiquitous if not invaluable to the success of the expedition: "Never did I undergo more fatigue, in performing the duties of hunter, spy, guide, commanding officer, &c. Sometimes in front; sometimes in the rear; frequently in advance of my party 10 or 15 miles; that at night I was scarcely able to make my notes intelligible" (55). But of his hardihood there can be no doubt, and if Captain Lewis may be regarded as a personification of the inquiring spirit of the Enlightenment, Clark the exemplification of the natural man as hunter-guide, then Lieutenant Pike embodies both qualities: "the searching spirit" as he called it, not for the mere "trade" to which he applied the phrase, but, as with the glory to be gained, purely for the love of the quest. Against this role, his function as peacemaker—"Peace has reigned through my mediation"—seems largely incidental and, as it proved, ineffective (79).

Like Lewis, Pike had a literary flair, to which he added what seems to be a self-conscious effort to convey the impression of being well read, dropping occasional references to the books he carried along with him, neoclassic authors like Shenstone and Pope. He also had something of an erotic streak—if with a domestic bias—rejecting the offer of a nice warm squaw at one far point on the Mississippi voyage for the sake of the "civilized fair" who awaited him at home, a note missing from the published version of Lewis' journals. As with Gilbert Imlay, this eroticism can invest the landscape with sensual analogies: watching with avidity for the first signs of spring along the upper Mississippi, Pike compared the anxiety of his men about the breakup of the ice—which would signal their return—to that of "a lover [waiting for] the arrival of the priest who was to unite him to his beloved" (88).

Divorced from the lengthy, data-filled appendix, Pike's Mississippi journal conveys the impression of almost boyish exuberance, a result not only of the excitement natural to a junior officer suddenly given his first important assignment, but of the relative simplicity of his task. This is not to gainsay the difficulties he encountered, and here, too, his errand may be compared with Washington's embassy to the French in the Ohio Valley, both being singleminded endeavors, dutifully executed. But Pike's second journal (like Washington's also) borrows complexity from the ambiguous assignment: Pike's ebullience seldom breaks through, and the narrative increasingly conveys a note of bewilderment and even confusion, and thereby resembles Washington's account of his Fort Necessity debacle. Like Washington's second foray across the mountains, Pike's southwestern expedition ended with a hastily thrown up fortification, and with capitulation (perhaps anticipated) to the Spanish authorities. And, like Fort Necessity, Pike's little blockhouse on a tributary of the Rio Grande was an outpost of the empire that would follow. On this second tour of exploration, also, the hazards were far greater; for, unlike the upriver voyage on and along the Mississippi, there were no physical guides, and it was a venture into a virtually unoccupied and unknown territory, without the reassuring (if ambivalent) presence of fur traders and Indians who were friendly or even obsequious to Americans passing through.

Despite these dangers, Pike's second journal displays, if anything, a stronger aesthetic dimension, inspired by the remarkable territory through which the party traveled: as Pike moved into the foothills of the Rockies he was able to obtain vistas worthy of being called "sublime," an experience denied him while traveling up the Mississippi. Along with his constant admiration for purely animal beauty, whether of the wild horses he encountered or an occasional Indian, we can place an aesthetic capacity comparable to Meriwether Lewis', which reached an apogee of excitement at the end of his southwestern adventure. After climbing a hill near his little fort on the west branch of the Rio Grande, Pike "had a view of all the prairie and rivers to the north of us," a scene that, like Lewis' description of the headwaters of the Missouri, was

> *one of the most sublime and beautiful inland prospects ever presented to the eyes of man. . . . The main river bursting out of the western mountain, and meeting from the north-east, a large branch, which divides the chain of mountains, proceeds down the prairie, making many large and beautiful islands, one of which I judge contains 100,000 acres of land, all meadow ground, covered with innumerable herds of deer; about six miles from the mountains which cross the prairie, at the south end, a branch of 12 steps wide, pays its tribute to the main stream from the west course. Due W. 12°. N. 75°. W. 6°. Four miles below is a stream of the same size, which enters on the east; its general course is N. 65°. E. up which was a large road; from the entrance of this down, was about three miles, to the junction of the west fork, which waters the foot on the north, whilst the main river wound along in meanders on the east. In short, this view combined the sublime and beautiful; the great and lofty mountains covered with eternal snows, seemed to surround the luxuriant vale, crowned with perennial flowers, like a terrestrial paradise, shut out from the view of man.* (193–94)

Like Lewis' vision of the two contrasting falls, Pike's is an imperialistic epiphany, a fit of rapture after having pierced to the center of the unknown. And yet, as indicated by the presence of the road, which gives a civilized dimension to the composition, this territory was not unknown to the Spanish, and the particulars of Pike's astronomical coordinates add a military note to his aestheticism that is missing from Lewis' description. For Lewis, moreover, the falls of the Missouri provided a true (albeit metaphorical) center, a midpoint in his western traverse, a focus of his emotions at a moment when he still thought himself near the opposing headwaters of the Columbia River. For Pike, it was a turning point, or pivot, from which he headed, not further west but south; not into a wilder range but into the already civilized parts of Spanish America. Yet the coincidental manifestation of an artistic sensibility in both explorers at a critical juncture in their narratives testifies to the quality of the expansionist mentality in the opening years of the nineteenth century, certifying the ubiquity of an imperialistic aesthetic that lent order other than geographical and geopolitical to the land.

Both episodes, moreover, had a concluding descent: Lewis drops from the sublime heights of contemplation into the Missouri River while being pursued by a bear, while Pike drops into the hands of the Spanish authorities. Thenceforth, Lewis would find few occasions for aesthetic rapture, and Pike would become chiefly a tourist, for as a prisoner he was hospitably treated by gallant Army officers and eagerly sought by literate, amiable priests hoping to lure young Protestants into their fold. The world entered by Pike was not much different from the French Canada into which Puritan prisoners were carried a century earlier, but where the most famous of those redeemed captives, the Reverend John Williams, thought of himself as in the heart of New Babylon, Pike tended to regard Mexico as a potential New Canaan. Even as Catholic priests sought his conversion, Pike portrayed a territory ripe for takeover, being in the care of noble dons resentful that France should "usurp the government of Spain," and who were therefore ready for "a revolution or separation" from European control (229). The Mexican newspapers Pike read were filled with "rumors of colonel Burr's conspiracies, the movements of our troops, &c. &c.," and at least one discontented Mexican officer, held back forty years in rank, was convinced that "we would invade [his] country the ensuing spring" and was therefore delighted to receive from Pike a certificate "addressed to the citizens of the United States, stating his friendly disposition and his being a man of influence" (235; 218–19). Though Pike claimed he was not an *agent provocateur* in the service of Wilkinson, he certainly portrays Mexico and the Mexicans as ready and willing for annexation by the United States.

The point to make, however, lies beyond mere conjecture, and turns on the distinctions to be made when confronted by the process of exploration in the America of Zebulon Montgomery Pike and Meriwether Lewis. Lewis is the type and symbol of the process at its most transcendent, idealistic stage, irradiated with the bold, inquiring spirit of Enlightenment enterprise in America during the first expansive stage of United States history; Pike, because of his entanglement in Wilkinson's Laocoön group, has his lower parts enfolded in serpentine coils. Yet it is impossible to separate the two men from the expansionist spirit they both, in different ways, expressed, much as General Washington and General Wilkinson are but two sides of the same imperial coin.

Such ironies are compounded by tracing the destinies of Meriwether Lewis and Zebulon Pike to the end. The noble and stainless protégé of Jefferson died by his own hand, his career abruptly ending in disgrace. Pike, having vigorously (and successfully) defended himself against all charges of complicity in the Burr conspiracy, died a bona fide hero during the War of 1812 (in contrast to General Wilkinson, who did not). If there is something in Zebulon Montgomery Pike that looks forward to John Charles Frémont, there is that also which anticipates George Armstrong Custer, the military man who is the type and symbol of vainglory exercised in the name of American empire, yet who, in the end, preserves a measure of true heroism, giving a peculiar grace to an otherwise questionable career. It is, finally, in these anticipatory qualities that Pike's true meaning lies.

General Wilkinson, the great hydra whose coils beslimed so many men and reputations, ended his days in that object of his heart's desire, Mexico City, in yet another futile pursuit of favor and wealth. Wilkinson's bones lie unidentified in a common tomb, his only monument a handful of place names left on the map, traces of his once ubiquitous presence. Captain Lewis has his broken plinth, ordering the wilderness round in Tennessee, but Captain Pike has his mountain, that enduring symbol of the American West, less terminus than pivot to empire. He also has his county in Missouri, which in time became associated with a people transformed by the western experience into a perpetually restless breed. For them, no mountain, however large, was a barrier to their further passage, however brief; and we should think of the Pikes (as they were called) as the explorer's children, and think of them also in terms of the rivers Pike explored, stretching out westward from the Mississippi like the fingers on an extended hand, the skeletal remains of General Wilkinson's imperial reach.

IX

Midway on the Waves

i

The contingency of Barlow's *Columbiad,* Fulton's *Clermont,* and the return of the Lewis and Clark expedition make 1807 a marvelous year in the chronology of the young republic, a high point, surely, in the emergence of the Enlightenment in America, matched only by the Constitutional Convention of twenty years earlier. All three events, moreover, can be seen as resulting from lengthy gestation, for much as Barlow's poem was an expansion on his earlier *Vision* and the expedition had long been a favorite project of Jefferson's, so Fulton's steamboat was not only the last of several prototypes he had developed, but was indebted (in concept if not details) to the earlier models produced by Rumsey and John Fitch. Like the Constitution itself, therefore, all three events more properly belong to the decade of the 1780s, rather than to the year in which they occurred. Of the three, finally, only the steamboat was an unqualified success, transcending its late eighteenth-century origins and moving out of the recent past into the future. The main accomplishment of the expedition was to lay to rest forever the old dream of a river-borne passage to India, and Barlow's *Columbiad* was largely ignored by posterity, being old-fashioned in genre and poetics virtually before it was published.

The Columbiad is dated also in a way that ill accords with its prophetic character, given Barlow's great faith in scientific and technological progress, for his account is limited (as in his earlier *Vision*) to the accomplishments of savants like Rittenhouse and Franklin, though the first had been dead for more than ten and the other for nearly twenty years. Most peculiar is the lack of any mention of the steamboat in Barlow's poem: as the representative of the Ohio Company in France, Barlow was certainly familiar with Cutler's prophecy, and, much more to the point, he had been Fulton's close friend and associate while both were living abroad, and had even contributed a boiler of his own design to an early prototype of the *Clermont* that was launched on (and sank into) the Seine.

The secrecy Fulton was working under after he had returned to America might have justified a tactful silence concerning the steamboat, but the *The Columbiad* also contains no mention of the young inventor's work with the submarine and torpedo, inventions that he had long been promoting and that were conceived as instruments of the universal peace Barlow's poem ends by celebrating. Fulton was helpful in seeing *The Columbiad* through the press, even executing a portrait of the poet for the frontispiece, but Barlow's gratitude was chiefly (if generously) expressed by the dedication page, though there is an offhand allusion in the book to the "Panorama" Fulton had displayed with great profit in Paris, a combination of the inventor's painterly and technical skills.

A more important (though less specific) acknowledgement by Barlow of Fulton's emergence, after Franklin's death, as the regnant American Prometheus may be found in the expanded section concerning canals in *The Columbiad*. For Barlow's friend and protégé had in 1796 published in England an ambitious and influential plan for extending a system of small canals throughout the island, using inclined planes in the place of locks. Fulton's treatise was addressed to the British Board of Agriculture, but the young American took care to include a letter to the governor of Pennsylvania, suggesting the usefulness of his plan to commercial exchange in the United States: "Canals will pass through every vale, meander round each hill, and bind the whole country in the bonds of social intercourse; hence population will be increased, each acre of land will become valuable, industry will be stimulated, and the nation, gaining strength, will rise to unparalleled importance, by virtue of so powerful an ally as canals" (142). "Canals careering climb your sunbright hills," wrote Barlow in *The Columbiad*, "Vein the green slopes and strow their nurturing rills," a satisfactory (if compressed) translation of Fulton's sentiments (*Works*: II: 700).

Though associated thenceforth chiefly with his steamboat, Fulton would even after 1807 remain an enthusiast for canals, yet he was hardly even in 1796 a solitary prophet in this regard, and Barlow in his *Vision* of 1783 had cited canals as instruments of progress. The expanded reference in 1807 might have been made had Fulton not written his treatise at all, for both men were indebted for their enthusiastic view of artificial waterways to the canal "mania" (as it was called) then sweeping Great Britain, a result of the successful and highly profitable building of a canal by the Duke of Bridgewater between his coalmines at Worsely and the mills of Manchester. Both Jefferson and Washington were early advocates of canals as necessary to the Potomac–Monongahela system (Rumsey, we should recall, when not tinkering with his boats, supervised the building of canals bypassing the rapids on the Potomac), and an Irish engineer named Christopher Colles had in 1785 put into print the Virginians' fears concerning the ease of running a canal between the Hudson and the Great Lakes. So symbolic had artificial waterways become as instruments of technological progress, that Barlow and Fulton had planned, while still living in Paris, to collaborate on a didactic poem called *The Canal,* which in four books would interlock in couplets recent advances of technology and human progress, using the canal as a central symbol of both. The existence of this project may be the best explanation of Barlow's omission of recent inventions in his epic celebration of progress, holding back the submarine, torpedo, and steamboat for the work then in progress.

"Mount on the boat," invited Barlow in the opening stanza to *The Canal,* "and as it glides along, / We'll cheer the long Canal with useful song" (232). Work on the project, for which Fulton was supposed to supply the "ideas," Barlow the "poetry," did not, however, glide past the first book. Like many real canal projects of the day, the proposed union of "physical science" and "political economy" did not get very far before it was abandoned. By 1802 it had been set aside forever, Fulton having submersed himself in his "plunging boat" and other scientific projects in the service of personal not political economy. Barlow did not entirely give up his hopes for the poem until Fulton's marriage into the wealthy and powerful Livingston family of New York and his successful launching of the *Clermont* carried the inventor out of the Barlows' circle into other spheres than poetic. The part that was completed celebrates not only artificial waterways, but the benefits of damming and diking, climate control, and (by controlling rainfall) alleviating the ravages of yellow fever in Philadelphia, demonstrating in a scant 290 lines not only the benefits of canals but the virtual impossibility of bending poetry to such uses, suggesting that "useful song" was a contradiction in terms.

Until Walt Whitman in 1871 ushered the Muse of Poetry past the "drain-pipes, gasometers, artificial fertilizers" displayed at the New York Exposition and "install'd" her "amid the kitchen ware!" no major poetic champion of technology emerged in America after Joel Barlow. The visionary genre remained, in terms of scientific advance, relatively fixed, along with its Jacobin ideology, in the ebullient but highly generalized and abstract mood of the late 1780s. As it was the role of poetic visionaries during the early years of the republic to foresee great advances in communication, so it was the responsibility of their engineering successors to develop and build them, a division of labor that revealed not so much a dichotomy as a complexly integrated notion of what constituted "art" at the turn of the centuries. No better exemplification of that notion can be found than the man whose name appears on the dedication page of *The Columbiad*. If the Lewis and Clark expedition drew the baseline for an expanded American empire, then Fulton's pioneering work with canals and steamboats would provide the means by which imperial expansion would be accelerated, in terms that put forth the union of commerce and culture as an aesthetic as well as a technical idea.

Commencing his public career as a painter, Fulton eventually turned to invention (and promoting his inventions) as a more promising direction. His treatise on canals contains a telling if somewhat pretentious paragraph, in which he suggests that engineering is much like poetry (in the eighteenth-century view of poetics), in that both involve "improvements" rather than original "inventions": "Therefore the mechanic should sit down among levers, screws, wedges, wheels, &c., like a poet among the letters of the alphabet, considering them as the exhibition of his thoughts; in which a new arrangement transmits a new idea to the world" (x). Such analogies were made possible by the neoclassical notion of art, which, with its emphasis on "improvement," made possible a very generous definition of the word "artist." Moreover, public works like viaducts and aqueducts were often given forms inspired by classical models (though good republicans like Fulton and Thomas Paine preferred to work with cast-iron rather than stone), paradoxically dignifying the latest creations of engineers with the appearance of antiquity.

The Canal by Barlow and Fulton remained unfinished, but within twenty years it was realized—not as a poem but as an artifact, for the Erie Canal was a feat of engineering celebrated at its completion as a magnificent work of art challenging the achievements of Rome. The rhetoric it inspired, however, was but a somewhat more euphoric version of the prose generated by promoters and politicians of the early republic. They shared with Barlow and Fulton the conviction that canals were a means of perfecting the God-given design of the American map, an extended version of landscaping that would create a more perfect union and guarantee the universal spread of material prosperity and cultural improvement. Poets like Barlow may have envisioned a better, even millennialist future for the United States, but it would be the "art" of the inventor and engineer that would guarantee the certainty of the poetic vision.

In his *Treatise,* Fulton spelled out the inexorable logic of much rhetoric that would follow:

> [1] *Agriculture and commerce will improve, and happiness spread, in proportion as the facility of conveyance increases.*

> [2] *In proportion as the difficulty of communication is removed, the spirit of enterprize increases, and neighboring associations begin to mingle, their habits and customs assimilate, each transmits its improvements to the other, and each feels the beneficial effects resulting from the union.*
>
> [3] *Thus an easy communication brings remote parts into nearer alliance, combines the exertions of men, distributes their labours through a variety of channels, and spreads with greater regularity the blessings of life.* (11–15)

Fulton was talking about canals, but his syllogism could be applied to the steamboat also, for in his view the canal was an "improved machine" like the "looms of the drapier or hosier," which reduced "the labour" while multiplying "the produce," and thereby rendered "the necessaries, and conveniences of life more abundant," a cyclical yet progressive movement that seems, once set in motion, to know no bounds. "The easy means of procuring the accommodations of life increases the population of a country, and population, creating a greater demand, proceeds to further improvement" (19). Such a view was, if anything, more suitable to a New than an Old World, a New World in which space seemed an infinite extension westward of available land, needing only a system of articulated waterways to give it both population and coherence.

Nothing happened in the thirty years following the publication of Fulton's *Treatise* that would qualify the simple truth of his grand syllogism—years that saw not only the invention of the paddle-wheel steamer and its introduction to the Ohio and Mississippi, but the completion of the Erie Canal, which connected Fulton's Hudson to the rivers on which his steamboat already moved. If Washington may be seen, from the Apollonian height his posterity assigned him, as pointing to the incomplete parts of the great design imposed on the North American continent by Providence, then Fulton may be seen as the artisan who provided the means by which that design would be completed. From a painter of miniatures he matured into a designer of panoramas and moved on to ever grander schemes; for if epic is associated with empire, then it must be said that the man who was Barlow's friend was the author of the imperial design that projected the vision of *The Columbiad* upon the land. Whether as steamboat or canal, Fulton's marvelous mechanisms were the means by which the United States would become a beautiful machine.

Fulton's hegemony in this regard has not gone unchallenged. Certainly where canals were concerned he played a relatively modest role: prior to 1807 as an ambitious young man eager to build his reputation as a man of projects on the successes of others, afterwards as a rich and famous inventor willing to lend the authority of his name to worthy schemes. Because of the precedents set by Rumsey and Fitch, the last of whom ran a working (if cranky) steamboat on the Delaware River during the summer of 1787, Fulton has been denied by historians all but the popular fame that still associates him with his best-known invention. The steps (and men) that pointed the way for the Erie Canal will be discussed in subsequent chapters, but Fulton and his famous steamboat need to be here detached from that counterpart mechanism, not only because of chronological necessity—ten more years would elapse before construction began on the Erie Canal—but because the *Clermont* (and Fulton) are so contiguous to both *The Columbiad* and the Lewis and Clark expedition—and so discrete. Once again, both poem and exploratory mission were in some sense failures, but the steamboat built by Fulton was not just a success, it was a triumph from the start, and was quickly developed by the inventor and his successors into a machine that dominated the inland waters of North America as a symbol of both culture and commerce, a vehicle instrumental to material progress that was also a thing of beauty.

It is in this regard that Fulton, the artist turned inventor, is so important. Though his originality as an inventor may be disputed, as a promoter of innovation he was unmatched in his day. Washington dominates the pantheon of American heroes who emerged during the latter half of the eighteenth century, obliging decadal decorum by dying in 1799, and Fulton emerged early in the next century as a successor archetype. Virtually without rival, he reigns over the rapid advances in technology associated with the first twenty years of the nineteenth century, and, even after his early death, in 1816, his ghost, like Washington's, was often evoked in connection with internal improvements and invention. Like Washington, also, Fulton had his antitype, his Wilkinson, for in John Fitch, the man who twenty years earlier anticipated Fulton's success with the *Clermont,* we are given a dark alternative.

Where Wilkinson is a rascally burlesque of Washington's Virtuous Man, Fitch is a tragic figure: a man of flawed genius, lacking the polish of a Rumsey or Fulton and possessed by self-destructive compulsions, Fitch played the crackbrained inventor to Fulton's suave promoter. Most important, perhaps, he follows Imlay and Wilkinson, those other shadowy, troublesome spectres, in seeking his fortune beyond the Alleghenies. Though born on the Connecticut River, he was buried on the banks of the Ohio, symbol of the peripatetic Yankee celebrated in Irving's humorous *History of New York* (1809), though there was nothing funny—in the sense of hilarity—about John Fitch: quite the reverse. His is one of the earliest lives to call into question the emerging faith in progress that will promote the American myth of the self-made man. Grim alternative to the bright figure of Robert Fulton, John Fitch is a lean and hungry shadow cast by that corpulent corporation of many talented men in one body, Ben Franklin. He recommends himself to us as an incarnation of western failure, a Yankee equivalent to Daniel Boone; and like the westering long-hunter, the emigrating peddler-inventor is inextricably linked to the waterways of the great western valley.

ii

It is a nice coincidence that one of the earliest accounts of regular steamboat travel on the Hudson River is an entry in a journal written by DeWitt Clinton in 1810, kept to record the details of a trip he took from New York to Albany for the purpose of surveying, as a member of a commission, the route of what would become the Erie Canal. "The weather was warm," wrote Clinton, "and the boat crowded" (*Life*: 29). Such a blasé entry suggests how quickly New Yorkers had adapted themselves to an invention only three years old: clearly, steamboat travel was rapidly becoming part of everyday life in the Northeast, something one simply did, without further comment, save on those occasions where rhetoric of a public nature was demanded.

Three years earlier, on the first day of the first trip by steamboat to Albany, the mood was hardly so sanguine, though Fulton maintained his usual rigidly calm demeanor. The use of steam for navigation was still regarded by many in 1807 as chimerical, and previous experiments by Fulton had not always gone well. There was also considerable hope for profits invested in the voyage, most notably (and tangibly) by Chancellor Robert Livingston, the man who engineered the Louisiana Purchase, who was the chief financial backer of the scheme. Moreover, there were those like Manasseh Cutler who felt that steam navigation was no longer theoretical, let alone chimerical, and that it was only a matter of time before someone hit upon the right combination of boiler, piston, and wheel; and another inventor across the Hudson in New Jersey was already in the field. The monopoly held by Fulton and Livingston for exclusive navigational rights on that river was contingent upon the successful development of a steamboat, and it was good only for a specified period of years.

Whatever fears and apprehensions Fulton may have concealed beneath his calm manner, he hid them well, and wore for the occasion his customary formal attire, patterned after the dress of an English nobleman. His intended passengers were also elegantly decked-out for the event, which in terms of costume resembled a garden party more than a serious experiment on which private fortune and public well-being rested. Yet Fulton's formal wear and that of his passengers convey the aura surrounding the inventor that made his eventual success possible: he needed the New York State gentry who accompanied him on that day, for social sanction as much as for the necessary capital, and the successful end of the trial was celebrated symbolically by the announcement of Fulton's engagement to the daughter of one of the Livingstons—no longer a presumptuous gesture, thanks to the success of the Albany run.

Oral tradition down through the years has associated the maiden voyage of Fulton's boat with terrified countryfolk fleeing the approaching craft. Matters of record are more demure, for once the boat was well under way, the passengers quickly relaxed, and celebrated the great moment by singing the sentimental strains of a Scotch ballad, "Ye Banks and Braes o' Bonny Doon." The mood was one of polite jubilation, not universal consternation, and recognizing the importance of having his invention accepted by citizens of the politer classes, Fulton spent considerable money fitting up his steamboats with elegant appointments and posted regulations dictating comportment, in all ways fashioning them into socially acceptable vehicles, a tradition that would last as long as paddle wheels turned on American waters.

Although the palatial steamboats associated with flush times on the Mississippi River were many years in coming, from the *Clermont* to the *Robert E. Lee* the ugly boiler and furnace were concealed from sight. Steamboat design was derived from aesthetic principles similar to those found in Benjamin A. Latrobe's pumphouse in Philadelphia, which adapted the classical dome to a use not foreseen by Palladio: a tasteful and proper combination of neoclassical proportions, combining the square, rectangle, cylinder, and dome, the pumphouse adapted Jefferson's formula for structures of state to utilitarian ends, in effect installing a steam engine in a Roman temple. But where in the original a resident god would have been displayed to advantage, in Latrobe's pumphouse, classical architecture is used to conceal the building's true function. It would be many years before the beauty of the steam engine was appreciated as a marvel of clean lines expressing pure function—notably by Horatio Greenough, a sculptor trained to appreciate neoclassical linearity—and steamboat design never yielded to an aesthetic of mechanical utility, but increasingly expressed a version of Victorian prudery, putting forth a marvel of fretwork and gingerbread that belied the mechanism within, even to covering the paddle wheels with decorated skirts.

Like the invention of the steamboat, this aesthetic of concealment may be credited to Robert Fulton, at least in symbolic terms, and equally symbolic is the fact that Fitch, not he, invented the steamboat. Yet as appearance was so essential to the promotion of steamboat travel in its infancy, so Fitch and his prototype invention had an essential ugliness that provides stark contrast to Fulton and his beautiful machine. Luck likewise played a role in granting Fulton success where Fitch had failed: "Quicksilver Bob" was Fulton's nickname as a boy, a tribute to his mercurial personality, and Fulton's rise was in all ways evocative of the god sacred to science, commerce, travel—and thieves. Fitch was hardly so blessed, and his life seems to have been haunted by Poe's Imp of the Perverse: "Poor John Fitch" he habitually called himself, echoing Edgar's "Poor Tom," and evoking an image of near madness and starvation in rags.

For Fitch had the kind of genius so closely allied to madness that he was shunned by just those people essential to the promotion of any invention, both in the eighteenth century and now; unlike William Rumsey, his exact contemporary, or the much younger Fulton, Fitch lacked the polish and charm that recommended the first man to George Washington and the other to Chancellor Livingston. Yet, like the other two, he was driven by a powerful force of conviction, and shared with them a desire to be rich and famous. Unlike Rumsey, he developed a workable steamboat; unlike Fulton, he failed in his efforts to promote it. The success of Fulton in this regard and Fitch's failure have often been used to illustrate what it takes to be a success (and, conversely, a failure) in America, a truth not unlike that which resides in the contrast between the exterior appearance of the steamboat and the inner reality.

As early as 1850, Louisa Tuthill, a writer for children, put her index finger on the problem: "Genius John Fitch possessed, every one must acknowledge, but genius uncontrolled by common sense, and united with an irritable temperament, which he took no pains to control; yet he was a man of integrity and strict veracity, and generous, even to a fault" (49). Fulton, by contrast, according to Mrs. Tuthill, "displayed *industry, perseverance, indefatigability, presence of mind, and command of temper*. Who would not be sure of success with such available means?" (94). In fairness to Fitch, who possessed more than sufficient industry, *Perseverance* was the name of the boat that finally resulted. It was "presence of mind and command of temper" that he lacked, those Franklinesque virtues Fulton evinced, which, along with his elegant manners and clothes (not stressed by Mrs. Tuthill), allowed him to command the friendship and considerable wealth of Chancellor Livingston. And yet, as even Mrs. Tuthill tacitly admits, it is Fitch, with his "integrity, veracity, and generosity," who remains the most attractive of the two men. Fulton's rise from relative poverty to wealth and power seems a reprise of Franklin's similar progress, demonstrating the opportunities open to people of talent in America, and Mrs. Tuthill portrayed him as not only a builder of boats but a fashioner of self. Fitch, by contrast, whose autobiography was a perfect antithesis to Franklin's, being a record of an endless series of insults, bad luck, and reversals, is a member of the dark pantheon in America, men of genius and sterling worth who ended their lives in (and, in Fitch's case, from) undeserved poverty and neglect.

From our present perspective, moreover, we can see Fitch as being essential somehow to the truth of the matter. Like Meriwether Lewis, Fitch ended his miseries by taking his own life, and both men thereby challenge the easy assumptions of the Enlightenment while anticipating the darker aspects of literary Romanticism. Moreover, "little Johnny Fitch," as he called himself when in a defiant mood, was cut from the American grain, and the details of his life reveal a pattern in all ways contradistinctive to Fulton's cosmopolitan progress. From his impoverished, horribly abused childhood in Windsor on the Connecticut River, to his captivity among Indians in the unexplored wilderness of the (then) far West, to his residence in the Philadelphia of the *Philosophes* and, briefly, in the Paris of the Terror, to his besotted death in Bardstown, Kentucky, Fitch's miserable pilgrimage qualifies him as a genuine American anti-hero, a mechanic version of Johnny Appleseed and a Yankee version of Daniel Boone, providing a dark and tragic thread in the fabric of homespun heroics.

John Fitch, however, inspired no legends, was the occasion of few contemporary anecdotes, and is chiefly known through matters of undisputed record, including his own account of his wretched life, leaving little room for creative improvisation. There is in his story no touch of saving grace, nothing transcendent or hopeful, and he has served chiefly as either a bad example of genius blighted by antisocial behavior or as a sad example of genius unrecognized or neglected by contemporaries. Yet the thread of his story has a fascinating continuity, for like many another Yankee, Fitch was a jack of many trades and a master of most he attempted: from button-molding, to silversmithing, to clock- and map-making, all of the crafts he put his hand to, at least by his own account, made him considerable money. But his painful childhood had given a definite twist to his personality, a deep sense of inferiority that paradoxically would not allow him to rest content with modest attainments, but impelled him to realize some great goal, at first vaguely conceived. This is the eccentric bias that drove him on until a repeated series of reversals left him with no answer to his inner compulsion save self-destruction.

Part Franklin, part Boone, John Fitch combines the two great archetypal strands that characterize the emerging American character, incorporating in his eccentric pilgrimage some of the most important aspects of contemporary life, albeit in a self-contradictory manner. Born in 1743, Fitch had a symbolic birth at the epicenter of the Great Awakening then inspiring the natives of the Connecticut Valley, but his own enthusiasm was of a far different sort, aligned—as always, with an eccentric bias—with the Enlightenment forces then gathering strength in America. Fitch's subsequent attempt to found his own religion upon atheistic principles—his optimistically named "Universal Society"—reveals both his modernity and his perversity, for though partaking of the Enlightenment spirit of free inquiry, it was essentially an antisocial gesture, bound to fail.

Fitch's early life was spent acting out a sequence of roles that put him in the mainstream of American history, yet each time a trick of the current shot him off into some backwater or bay. His promising early career practicing Paul Revere's trade was interrupted by the Revolution, and after a very brief and abortive try at soldiering, Fitch set up business for a time at Valley Forge supplying Washington's men. With his profits, he determined next to survey and invest in western lands, but having crossed the Alleghenies, he decided to increase his capital by floating a load of flour down the Ohio on a flatboat, a venture that soon went aground. Losing his cargo and his money, Fitch and his companions were taken captive by a party of Delaware Indians, and though the obvious parallel is with Daniel Boone, Fitch's account of his experience retains the strong lines of the traditional Puritan captivity narrative.

The forceful removal of civilized men and women into the heart of darkness has been seen as both traumatic and transformational, a ritual of sorts (as indeed the Indians often conceived of it) by which an experience of savage life has a lasting impact on the captive. Some whites were brought over to adopt Indian ways, but many were not, and those New Englanders who wrote accounts of the experiences stress the means they used to maintain and fortify their commitment to Christ and western civilization. The southern tradition is somewhat different, for from Captain John Smith to Daniel Boone, captives tended to use the experience for personal advantage, learning something more about the terrain and the aborigines who occupied it—the better to dispossess them. Like Boone, they became somewhat Indianized in the process, a halfway covenant symbolized by the buckskin costume and moccasins they adopted.

John Fitch, as always, remained his own man: instead of emulating Boone's cunning by cooperating with his captors, he stubbornly refused to do the Indians' bidding, risking death or torture. Somehow he emerged whole, even somewhat admired by the Delawares by the time the party reached the end of the long trek to the British fort at Detroit. While being held prisoner there, Fitch, with typical Yankee industriousness and persistence, fashioned some crude tools and set up as a button maker to the British soldiers, and he even, so he claimed, supplied the fort with wooden clocks. Where his Puritan predecessors in captivity found solace in the scriptures, Fitch, the atheist son of the Enlightenment, found it in fashioning clockworks. Throughout his autobiography, Fitch often pauses to look back on his experience among the Indians as an episode of idealized simplicity, and in the midst of his struggles to perfect and promote his steamboat, he regarded with nostalgia the relatively happy and peaceful days in the midst of a simple people.

Fitch's enforced pilgrimage across the Ohio Valley is a quintessential frontier experience, a later equivalent to Washington's embassy to the French and Boone's own captivity experience, yet it was an interlude that provides a paradoxical contrast, not a vital continuity, with his subsequent career as a visionary inventor. It is his stay among the British that maintains continuity, the restless, ingenious Yankee relieving himself of boredom (and the British of cash) by setting up his primitive manufactory. Purveyors of the dominant American myth have insisted upon the transformational power of wilderness ordeals, but Fitch proves otherwise, demonstrating that the Yankee type, by 1780, had developed boilerplate immunity to the lure of wilderness life. Still, given his eventual return to the Ohio Valley, his ordeal there had symbolic meaning, and it would be a while before he gave up hope of realizing profit from speculation in western lands. One of the first fruits of his enforced sojourn was a map of the region he had traversed, which he drew, engraved, and then printed by means of a cider press, providing a handy, pocket-sized diagram designed for intended settlers.

Although the sudden Rousseauvean flash of inspiration that he associated with his desire to build a workable steamboat occurred after his return to the East, it was from the beginning associated with the need for improved transportation on the western rivers, on which his own flatboat voyage had come to grief. Much of what happened subsequently transpired in Philadelphia, and Fitch's story becomes thereafter essentially an urban tale, cast in the shadow of Ben Franklin himself. But that earlier circuit through the wilderness, and Fitch's eventual act of self-immolation and suicide on the banks of the Ohio in 1798, certifies the essentially western dimension of his two-stranded life.

Unfortunately, the decade in which Fitch tried to promote his invention was one in which any improvement in navigation on western waters would benefit Spain more than the United States; and when, in desperation, Fitch tried to interest the Spanish in his steamboat, he was nearly sucked into the still-invisible maelstrom being stirred up by General Wilkinson and other whirling dervishes of international intrigue. Even his successful demonstration of the *Perseverance* was staged in the wrong place, for the Delaware valley had sufficient alternative means of overland transportation, while the Hudson, where Fulton a full seventeen years later made his first successful run, was flanked by mountains. By 1807, moreover, the Louisiana Purchase had opened the Mississippi to American navigation, toward which Fulton's invention sped like the muse of technology itself. Still, it was Fitch's precedent that encouraged Fulton and others to follow him, developing the device that, as both Filson and Cutler prophesied, meant the acceleration of civilization's advance on western waters, beating back and ultimately displacing the Indian way of life. Daniel Boone in his buckskins may be celebrated as a mythic symbol of wilderness accommodation for the sake of imperial advance, while John Fitch, adamant in ragged broadcloth, is the symbol of unyielding civilization. For Fitch was very much a man of the Enlightenment, being a bringer of order, whether as craftsman, surveyor, mapmaker, or as the inventor of the machine that would realize Barlow's vision of American rivers as a vast system promoting the advance of civilization through the western garden. If Fulton was the mechanic Messiah, Fitch was his Evangelist, yet another St. John of the American woods, fated to announce but not to bring forth a millennial machine.

iii

From the start, Fulton, like Fitch, understood the importance of his invention to commerce on the western waters, and wrote to Joel Barlow, in 1807, that the steamboat "will give a cheap and quick conveyance to merchants on the Mississippi, the Missouri, and other great rivers which are now laying open their treasures to the enterprise of our country, and although the prospect of personal emolument has been some inducement to me, yet I feel infinitely more pleasure in reflecting on the numerous advantages that my country will derive from the invention" (Todd: 234). Fulton at this point seems to be writing for future publication, and the plain truth of the matter is that he spent much of the time left to him doing what he could to hold onto the monopoly that promised to make him a fortune. Thus the first steamboat built in Pittsburgh, only four years after the success of the North River prototype, was constructed from plans designed by Fulton, thereby extending by proxy his influence (and monopoly) over the Allegheny mountains.

And yet the maiden voyage of the boat that resulted from Fulton's plans, the *New Orleans,* was in character as well as duration a much different journey from the one taken in 1807 up the Hudson to Albany. Because of several catastrophic coincidences and the general nature of navigation on the Ohio and Mississippi, the initial trip of the *New Orleans* to its eponymous city is a paradigm of that wilder image associated with the western parts of North America. First of all, in 1811 the Ohio Valley was visited by a catalogue of ominous portents, including floods, prodigious flocks of passenger pigeons, and a spectacular comet. The only passengers (besides the crew) aboard the *New Orleans* were the builder, Nicholas Roosevelt, and his pregnant wife, Lydia Latrobe, daughter of the architect-engineer whose fortunes had increasingly become involved with Fulton's; and the only detailed account of the "days of horror" as she called them, was a letter written by Mrs. Roosevelt that was later amplified by her recollections (in 1871). Yet the facts alone were literally overwhelming, and though the exact design of the *New Orleans* is not known, it may be assumed she had a deep-draft hull, increasing the danger presented by the destruction the steamboat encountered along the route downriver. Setting out in the wake of the great comet, the boat was met en route by the even greater earthquake that devastated settlements and altered the landscape all along the Mississippi. The steamboat terrified Indians and not a few white settlers who thought the comet had plunged into the water and was headed downstream.

Though crowded with incident and near disasters, the voyage of the *New Orleans* from Pittsburg to the Gulf of Mexico was viewed as a blessed event, a technological equivalent to the child born to the Roosevelts at Louisville, and the marriage of the captain of the boat to Mrs. Roosevelt's maid in New Orleans put a final seal of promise on the trip. The motif of cosmic convulsion remained associated with western navigation, however, commemorated by the names of boats launched on the Ohio shortly afterwards: *Aetna, Vesuvius,* and *Comet.* When Lieutenant Stephen H. Long launched his exploration of the Rocky Mountains in 1819, the last of the epochal surveys of the Louisiana Territory sent out during the early years of the republic, he headed up the Missouri in a paddlewheeler with a dragon's head that spouted steam and smoke, with the intention of impressing the tribes along that river much the way the Chickasaws had been terrified by their first sight of the "fire dragon," as they called the *New Orleans*. Though this phenomenal vessel apparently had little effect on the Indians, it mightily impressed such eastern champions of technological improvements as Hezekiah Niles, editor of the *Niles' Register*. Aside from its dragon's head, the *Western Engineer* did signify yet another advance in technology, for it was the first sternwheeler ever built, giving it advantages in the shallow western rivers. Even the indifference of the Indians to the *Western Engineer*'s appearance is testimony to the speed with which the steamboat evolved and multiplied on western waters.

As with the easy acceptance in the east of Fulton's invention, the paddle-wheel steamer soon became a familiar sight in the West, inevitably coupled with the rapid spread of civilization. By 1824 the sight of a steamboat plowing western waters had become sufficiently accommodated to the scene to be a symbol for the Commerce it served. In the halls of Congress, Henry Clay of Kentucky, in arguing for a protective tariff, could describe the benefits of the "manufacturing arts" by means of what was to many of his auditors a familiar image:

> *The difference between a nation with, and without the arts, may be conceived, by the difference between a keel-boat and a steamboat, combatting the rapid torrent of the Mississippi. How slow does the former ascend, hugging the sinuosities of the shore, pushed on by her hardy and exposed crew, now throwing themselves in vigorous concert on their oars, and then seizing the pendent boughs of over-hanging trees: she seems hardly to move; and her scanty cargo is scarcely worth the transportation! With what ease is she not passed by the steamboat, laden with the riches of all quarters of the world, with a crowd of gay, cheerful, and protected passengers, now dashing into the midst of the current, or gliding through the eddies near the shore! Nature herself seems to survey, with astonishment, the passing wonder, and, in silent submission, reluctantly to own the magnificent triumphs, in her own vast dominion, of Fulton's immortal genius!* (*Papers*: III: 711)

And yet, Fulton's name at the time his steamboat was first introduced on western waters did not inspire universal praise and gratitude. In 1814, a champion of western progress grumbled about "the overwhelming patent of Fulton and Livingston," a monopolistic barrier to free enterprise on western waters equivalent to Spanish control over the

Mississippi River prior to the Louisiana Purchase, but within ten years the restraint had been removed, thanks to Fulton's early death in 1816 from overwork and to Daniel Webster's forensic powers (*Cramer*: 253). Thenceforth Robert Fulton could become a secular saint to emerging Whigs, whether of the East, like Webster, or of the West, like Clay.

From the start, as we have seen, writers familiar with western rivers anticipated the need for steam-powered navigation, even before the earliest models had been perfected. It was Fulton's invention that literally sped the course of progress along the Ohio and Mississippi, setting loose the spirit of Mercury with his wings and rod. No more fervent champion of progress could be found in the western regions in 1814 than the man who grumbled about Fulton's monopoly, Zadoc Cramer, a New Jersey–born Quaker who ran a publishing firm and bookstore in Pittsburgh until his premature death by tuberculosis, also in 1814, the penultimate year of his annual publication, *The Navigator,* by which his name is chiefly known. A cross between Morse's *American Geography* and Ames' almanac, Cramer's little book began to appear at the turn of the century and was definitely the century's child. A cumulative anthology of standard authorities on western topography and a constantly revised guide to changing navigational hazards along western rivers, the annual volume was also fleshed out with anecdotes and miscellaneous information designed to divert the emigrant during long stretches of inactivity while drifting downstream toward New Orleans. Though originally intended as an aid to the amateur navigators of flatboats and scows, Cramer's book made an easy transition to the age of steam-powered travel, in whose spirit it was compiled throughout.

Dependent upon earlier authorities like Thomas Hutchins, Cramer, like Jedidiah Morse, added new material on western regions when it became available. In 1808, Cramer provided one of the first published accounts of the Lewis and Clark expedition, and likewise early supplied his readers with a digest of Lieutenant Pike's travels to the source of the Mississippi. By 1814, the appendixes of borrowed information made up a good third of the entire edition, and by that year also his book was filled with notations concerning improvements in navigation, both under way and projected, along western rivers. For, like the town of Nashville as Cramer described it, he and his *Navigator* were animated by "the spirit of beginning," a faith in progress that characterized the regions bordering western waterways as the visions of republican poets began to take on specific shape.

From shipyards to shot-towers, from glass factories to floating gristmills, the burgeoning industry of the West is depicted by Cramer as the nucleus of a self-contained civilization, a citadel of republicanism "where each man is a prince in his own kingdom, and may without molestation, enjoy the frugal fare of his humble cot; where the clashing and terrifick sounds of war are not heard; where tyrants that desolate the earth dwell not; where man, simple man, is left to the guidance of his own will, subject only to laws of his own making, fraught with mildness, operating just on all, and by all protected and willingly obeyed" (28). Cramer's western man is Crèvecoeur's Farmer transplanted beyond the Alleghenies, recalling the Frenchman's euphoric description of the Ohio Valley in Manasseh Cutler's pamphlet, and evoking Cutler's sermon depicting the possibilities of life in a newer world, as the inhabitant of a peaceable kingdom: "Far removed from the clang of war, or the turmoils of the wicked and ambitious, with a most healthful and invigorating climate, what can prevent the Tennesseans from becoming a noble, a generous, a stout and hardy race of men, willing to enjoy the liberty they happily possess, and ready at a moment's warning to defend it with life, whenever invaded either by a foreign or domestic enemy" (279–80). Here, too, are the citizen-soldiers of New England, transplanted from the banks of the Concord to the Tennessee River—the very same Tennesseans who, with sundry Kentuckians, Creoles, Negro slaves, and some pirates, would successfully defend New Orleans against the British in 1815.

By Cramer's account, the Ohio River is a vast mechanism of commerce, its "very appearance" inviting "trade and enterprise," for by means of tributaries the river reaches out into the remotest settlements (33). It is also a mechanism of plenty, for, properly understood (with the help of *The Navigator*), the river takes care of its own, providing both avenues of trade and sustenance along the way: "Turkies, pheasant and partridges, are numerous on its banks; these, with the opportunity of sometimes shooting bears and deer swimming across the river, afford much pleasure to the navigator, and form sumptuous meals to the boat's crew" (28). The waters of the Ohio not only sustain commerce, they have remarkable philoprogenitive powers, instanced by a Mr. Charles Wells, Sr., "resident on the Ohio river, fifty miles below Wheeling," a gentleman "sixty-eight years of age, and truly a hale, healthy looking man, who has had by two wives . . . *twenty-two children,* sixteen of whom are living" (227–28). Mr. Wells confided to Cramer "that last year [1811] within a circuit of ten miles around him, ten women had born to them twenty children, each having had twins," although the credit for these additional feats of progenitive power apparently belonged to others.

> *The banks of the Ohio seem peculiarly grateful to the propagation of the human species, and perhaps, stronger evidences could not be produced than the anecdotes just related. Indeed, an observation to this effect can scarcely be missed by any person descending that river, and calling frequently at the cabins on its banks. Children are the first object that strikes the stranger's eye on mounting the bank, and their healthful, playful noise, the last thing that cheers his ears after leaving the, not unfrequently ragged looking premises. . . . Such would make Buffon stare, when he ungenerously asserts, as well as several other classical writers of Europe, that "animal life degenerates in America."* (228)

Yet when we notice that all of these instances of a full score and more infants fathered by men of four-score or more years were made possible by two (successive) wives, something of the price to be paid for peopling the Ohio Valley is suggested. Cramer's own premise seems a bit ragged, for if "animal life" does not degenerate in America, it seems to assert its vitality at the cost of the mothers.

Similar lights and shadows characterize Cramer's account of the Ohio Valley, unintentional counterpart to Ashe's pointed contrasts: at Pittsburgh, where "the prospect from the top of the Coal hill is extremely beautiful and romantic," the town is enveloped "in thick clouds of smoke" from foundries and fireplaces, a contrast between the beautiful and something other than the sublime (50, 49). Ever the meliorist, Cramer notes that "all this might be prevented by some additional expense on the construction of the chimnies," for there is little along the Ohio by his account that cannot be improved (69). From internal communications, to the landscape itself, the hand of man is productively transformational: "The country watered by Charties creek, a considerable stream which enters the Ohio a few miles below Pittsburgh, is extremely beautiful; much of the soil is admirably adapted to cultivation, and from the picturesque variety of the surface, it is susceptible of all those improvements which men of fortune and taste might be disposèd to bestow." (71). Here again, the Horatian formula is the operating principle: beauty alone does not suffice on western waters.

By means of his book, Cramer himself makes a few strategic improvements along the Ohio River; while he acknowledges unfavorable aspects of life on its banks—we hear about swindling boat-builders, counterfeiters, collapsed land schemes, deserted towns, disastrous floods and earthquakes—there is a stunning silence concerning the contrast associated by Thomas Ashe and others with the northern and southern sides of the Ohio. The good Quaker Cramer decries the continued use of whipping post and pillory on the Virginia side, but he makes no mention of an equally archaic and painful institution visible there. Having described the thriving industries of Frankfort, Kentucky, like the "bagging manufactory, in which about 25 hands, blackmen and boys, are busily engaged," Cramer goes on to extoll "the improvements of interior America, whose inhabitants begin to feel and act like the citizens of an independent nation, possessing an extent of country capable of producing, from the luxury of its soil, and variety of climate, *every thing* which ought to make a people happy, and independent of all the venomous combinations of maddened Europe" (264–65). All this is thanks to the labors of "blackmen and boys," otherwise not described.

Whatever its causes and consequences, Zadoc Cramer is a champion of commerce as a vehicle and beneficiary of enlightened progress, which, like the rivers that sustain it, "finds its own level, makes its own roads, and wants nothing but time to mature systems and open communications, which at present seem filled with insurmountable difficulties" (278–79). In 1814, chief among the agents of opening communications was the steamboat, that marvelously innovative machine with the capacity to greatly facilitate commercial exchange on western rivers, keeping up with the rapidity with which natural resources "throw themselves into our view" (279). By 1814, steam navigation had already begun on the lower Mississippi, and Cramer in his deathbed edition of *The Navigator* foresaw the day when "the citizens of Nashville will see a steam boat winding her course up the Cumberland, in all the majesty and nobleness of her internal and secreted power, without the assistance of poles, oars, or sails" (275–76). As early as 1811, Cramer had recognized the importance of Fulton's invention to commerce on the western waters, and wrote a long and revealing passage of mostly retrospective prophecy:

> *It will be a novel sight, and as pleasing as novel to see a huge boat working her way up the windings of the Ohio, without the appearance of sail, oar, pole, or any manual labour about her—moving within the secrets of her own wonderful mechanism, and propelled by power undiscoverable!—This plan if it succeeds, must open to view flattering prospects to an immense country, an interior of not less than two thousand miles of as fine a soil and climate, as the world can produce, and to a people worthy of all the advantages that nature and art can give them, a people the*

> *more meritorious, because they know how to sustain peace and live independent, among the crushing of empires, the falling of kings, the slaughter and bloodshed of millions, and the tumult, corruption, and tyranny of all the world beside. The immensity of country we have yet to settle, the vast riches of the bowels of the earth, the unexampled advantages of our water courses, which wind without interruption for thousands of miles, the numerous sources of trade and wealth opening to the enterprising and industrious citizens, are reflections that must rouse the most dull and stupid. Indeed the very appearance of the placid and unbroken surface of the Ohio invite to trade and enterprise, and from the canoe, which the adventurer manages with a single pole or paddle, he advances to a small square ark boat, which he loads at the headwaters with various wares . . . and starts his bark for the river traffic, stopping at every town and village to accommodate the inhabitants with the best of his cargo. This voyage performed, which generally occupies three months, and the ark sold for half its first cost, the trader returns doubly invigorated, and enabled to enlarge his vessel and cargo, he sets out again; this is repeated, until perhaps getting tired of this mode of merchandising, he sets himself down in some town or village as a wholesale merchant, druggist or apothecary, practicing physician or lawyer, or something else, that renders himself respectable in the eyes of his neighbors, where he lives amidst wealth and comforts the remainder of his days—nor is it by any known that his fortune was founded in the paddling of a canoe, or trafficking in apples, cider-royal and peach brandy, whiskey, &c. &c. &c. From the canoe, we now see ships of two or three hundred tons burden, masted and rigged, descending the same Ohio, laden with the products of the country, bound to New Orleans, thence to any part of the world. Thus, the rise and progress of the trade and the trader on the western waters; and thus, the flattering prospects of its future greatness through the channels of the Ohio and Mississippi rivers.* (32–33)

This is a passage with considerable complexity, which reveals much more than seems intended. Cramer begins by evoking the promised event of the steamboat, but adds the qualifying phrase, "if it succeeds," and then goes on to devote most of his description, not to the glorious future, but to the primitive (if industrious) past, a description of man- and wind-powered navigation of the Ohio, precisely the kind of vessels for whose pilots Cramer's *Navigator* was designed. Commencing with a canoe, Cramer (and his ideal navigator) moves up to a flatboat, and then, at the end of the passage, to "ships of two or three hundred tons burden."

Most curious perhaps is Cramer's repeated emphasis on the hidden springs of commercial success, whether "the secrets of [the steamboat's] wonderful mechanism," its "power undiscoverable," or his observation on respectability, picturing the prominent merchant, doctor, lawyer (or whatever) living "amidst wealth and comforts the remainder of his days" without anyone suspecting his fortune came from "trafficking" in apples or brandy dealt out from a canoe or flatboat. Whatever Cramer's intention, the passage puts a curious cast on the character of "the rise and progress of the trade and the trader on the western waters." Certain it is that gentlemen encountered whilst traveling along the Mississippi and Ohio are not always what they seem, much as steamboats pushing forward the boundaries of commerce often carry cargo invisible to the unheeding eye.

X

Stately Decree

i

Though George Washington as a national icon was a virtual personification of duty, the several stages of his public career may be seen as moving from a relatively active manifestation of dutifulness to something much more passive. At the start, with his embassy into the Ohio Valley and then his first trial of arms, Washington fits the standard notion of heroics, but his subsequent military career was increasingly characterized by that patient, even monumental, Fabian persistence that Leutze ennobled in his epical depiction of the Delaware crossing. As President, Washington labored to present both his country and himself in terms of strength, not so much exercised as held in reserve, duty being at that critical moment in the republic's history identified with the need to assert even while implementing national unity. If duty for Washington in 1753 was determined by a heroic errand, by the time he stepped down from the presidency in 1796 he associated it with passive restraint, and the rhetorical strategy of his farewell address is thoroughly consolidational, in effect a hardline Federalist sermon warning against foreign entanglements.

As a young man, Washington expressed the westering urge that periodically characterized American enterprise, but in old age he championed stability, not as stasis, but as the result of mutual efforts aimed at further perfecting the Union outlined by the Constitution: "Your union," he informed his countrymen as he returned to private life, "ought to be considered as a main prop of your liberty," but it is also that single principle in which "every part of our country . . . feels an immediate and particular interest" (*Messages*: I: 216). It was important, therefore, to avoid "*geographical* discriminations—*Northern* and *Southern, Atlantic* and *Western*—whence designing men may endeavor to excite a belief that there is a real difference of local interests and views" (216).

Moreover, in order to further cement national union and eliminate "geographical discriminations," the nation must look to "the progressive improvement of interior communications by land and water," and as coastal navigation already bound the North and South in mutual commercial interest, so the East and West should be linked by roads, rivers, and canals: "The *West* derives from the *East* supplies requisite to its growth and comfort, and what is perhaps of still greater consequence, it must of necessity owe the *secure* enjoyment of indispensable *outlets* for its own productions to the weight, influence, and the future maritime strength of the Atlantic side of the Union, directed by an indissoluble community of interest as *one nation*" (216).

In his private correspondence, as we have seen, Washington associated the vital commercial link between East and West with the Potomac River, but this does not gainsay the honesty of his conviction that the solidarity of national union was indissolubly linked to the perfection of a system of "interior communications." Washington's generation had given the United States both independence and the constitutional system that provided the basis for strong, confederated union, but there was much work to be done before that union became a geopolitical reality. Because he died in the closing days of the last year of the eighteenth century, Washington was fixed forever in the American memory in a time when national union was a reality and a more perfect union a realizable hope. He stands on a threshold, one hand pointing back to the accomplishments of the Fathers, the other pointing ahead to the work yet to be carried out by the Sons. "'Twice the saviour of his country,'" Washington came to be called: "'After conducting her to liberty, he opened her the way to prosperity by new roads and canals'" (Watson: *History:* 6). As National Deity, Washington was part farmer-soldier, part planner-projector; his spirit could be evoked by federalists and republicans alike. Wherever one turns in the promotional literature of the early nineteenth century espousing improved waterways and canals, the name of that most Roman of the forefathers appears, immaculate as the shining marble with which his form soon became associated, hovering far above the party and regional divisions he so abhorred, as pure as principle and as impossible to emulate.

Adams, Jefferson, and Madison were not so fortunate. The perilous political balance of the 1790s collapsed into party and faction, both identified with sectional rivalries, and foreign entanglements soon followed. If 1807 can be seen as a golden threshold to the kind of future Barlow foresaw, the imperial adventure of Lewis and Clark having carried the spirit of the Enlightenment like an Olympic torch all the way to the Pacific Ocean, even as Robert Fulton gave the Enlightenment faith in technological progress manifest form, 1807 was also the year of

Jefferson's ill-fated embargo. And yet Jefferson's administration had its bright side: anticipating an eventual surplus in national revenues, Jefferson at the start of his second term in office instructed his Secretary of State, Albert Gallatin, to prepare a survey of the present state of and desired improvements to interior communications. The result of Gallatin's labors, his epochal report on roads and canals, did not appear until 1808, as the hope of surplus revenues began to fade; but it was, nonetheless, as Joel Barlow called it, a "luminous" document. A record of the terrific industriousness of individual states as they labored with scant resources to implement "the progressive improvement of interior communications" within their borders, it was also a stately decree reflecting the importance of a national system of roads, canals, and waterways to the health of the expanding republic.

Like *The Federalist,* Gallatin's report was at once magisterial and urgent in tone, for if the prospect of surplus revenues seemed to warrant the mounting of an ambitious scheme of internal improvements, Burr's abortive expedition had theatrically demonstrated the strategic importance of improving the connection between eastern and western sections of the country. Finally, increased British depredations of American coastal shipping dramatized the need for developing what would eventually be called the "inland waterway" of protected sea-lanes. A mixture of sectional efforts and national goals, Gallatin's book is made up chiefly of reports by regional bodies that had already made significant achievements in promoting internal improvements, and it was given a prologue by Gallatin and an epilogue by Robert Fulton, the both designed to provide an overview of the benefits to be derived by the nation as a whole from the national funding of present and future projects.

In his opening remarks, Gallatin observed that though the general necessity of roads and canals was so universally recognized as to need no further argument, "public improvements" in America were often thwarted by "mistaken local interests" (*Roads*: I: 724). And yet the chief problem facing a national system of internal improvements was not so much sectional bias as lack of local capital permitting the funding of such projects, especially when they offered no immediate prospect of profit. Another problem resulted from the size and unsettled character of the country: unlike Great Britain, where short canals (as Fulton had argued) could be very useful and profitable, America needed to think of canals chiefly as artificial links in a system made up largely of natural (and already available) waterways. Canals were not so important in themselves, but as the means of opening "a communication with a natural extensive navigation which will flow through [those] new channel[s]" (725). Though a number of local efforts had already been undertaken, Gallatin observed, many were so inconsequential as to be unprofitable and others had been either abandoned or left on paper, "because their ultimate productiveness depends on other improvements, too extensive or too distant to be embraced by the same individuals. The General Government can alone remove these obstacles" (725).

Having reached that pivotal conclusion, Gallatin's argument tends to pick up headway: not only does the "General Government" have "resources amply sufficient for the completion of every practicable improvement," not only can the General Government "complete on any given line all the improvements, however distant, which may be necessary to render the whole productive, and eminently beneficial," but the danger to the republic of "a vast extent of territory" can best be averted "by opening speedy and easy communications through all its parts." In sum, "good roads and canals" are in the best national interests, for they will "shorten distances, facilitate commercial and personal intercourse, and unite, by a still more intimate community of interests, the most remote quarters of the United States. No other single operation, within the power of Government, can more effectually tend to strengthen and perpetuate that Union which secures external independence, domestic peace, and internal liberty."

Views of a similarly generous and accommodating nature distinguish the many surveys and reports incorporated in Gallatin's great anthology, in which projects of a clearly local nature are associated with (even as they are literally connected to) national well-being. "It is a circumstance of no small consequence to the happiness of a society," wrote Joshua Gilpin, son of Thomas Gilpin, the Philadelphia merchant who was also "the grandfather of the Chesapeake and Delaware Canal," "for people of various and distant districts under one Government to be constantly and extensively mingled together for the purpose of traffic and interchange of their respective arts and productions, so as to polish the local habits or prejudices of different parts, and unite them in one general sentiment of respect and affection for each other, and for the Government under which they live" (754). From Louisville, on the other side of the Alleghenies, came an observation from Jared Brooks concerning the proposed canal around the Falls of the Ohio, that "this project . . . includes the perfection of the navigation and the supply of water for manufactures to an immense extent; and it is evident that this operation will advance the national interest in a rate of progression that must infinitely exceed the most sanguine calculation" (822).

From Philadelphia came a bundle of pamphlets pertaining to the Susquehanna and Schuylkill Canal, undertaken in the 1790s, the first link in a projected chain of artificial and natural waterways intended to connect that city with the "great waters to the westward of the mountains" (829). Among these was "an historical account of the rise, progress, and present state of the canal navigation in Pennsylvania," which pointed out "the future prospects of commerce in the United States by means of canals and rivers joining the tide waters of Delaware, Susquehanna, Potomac, Hudson river, &c., with the Ohio, Mississippi, the great Western lakes, and perhaps the South Sea itself" (830). The author of this visionary essay seems to have been Robert Morris, the celebrated financier of the Revolution, who, having subsequently invested too heavily in the western lands through which the Pennsylvania canal was to pass, suffered the same fate as the project itself: his money ran out and he "was arrested." The once-wealthy merchant and friend of George Washington was thrown into debtors' prison for more than three years, and then was left to die in poverty in 1806, which denied him the consolations of Gallatin's report. There was no more vivid reminder of the high cost of canals when funded at a local level than Morris, yet none of his contemporaries matched him in championing internal improvements for the sake of national well-being: "Every improvement, and every new communication with the Western territories, promoted by any of the United States, by which the trade of the lakes, the Ohio and Mississippi waters can be drawn to our seaports, is a benefit to the whole Union" (837).

Where Morris and his generation of entrepreneurs went wrong had more to do with technology than ideology. Though Morris devoted a footnote to the opinion of the famous British engineer James Brindley that the purpose for which rivers were created was "to feed navigable canals," most of the schemes promoted by Morris and others during the late eighteenth century involved the use of canals to bypass falls and rapids in otherwise navigable rivers. By 1808, the school of experience had amply illustrated Brindley's terse truth. In the words of Benjamin H. Latrobe, the brilliant architect and engineer who had recently been appointed surveyor of public buildings in the District of Columbia by Jefferson: "Unfortunately those of our canals which have been cut to pass the rapids and falls of our rivers, partake, in a great measure, of the inconvenience of the rivers themselves; some wanting water when the river is low, some incapable of being entered excepting at a particular height of the water in the river; some subject to constant accumulation of bars, and all of those with which I am acquainted much less useful than the money expended on them ought to have made them" (*Roads*: 912). Latrobe's work on the Chesapeake Canal as well as similar projects had made him so sensitive to the problems resulting from the dependency of canals on natural forces that in a postscript to his report he suggested that railroads might in the long run provide the best means "of accomplishing distant lines of communication" in North America (916–17). Latrobe was (in 1807) speaking of horse-drawn railroads, but his observation was prophetic, and would be realized by engineers who learned their profession from him.

More in keeping with contemporary thought was the contribution by Robert Fulton, who imbued the canal idea with a traditional glory as a universal bestower of unmitigated blessings. Canals, he argued, would be of immense value to the republic: first, by raising the value of public lands through which they passed; second, by "cementing the Union, and extending the principles of confederated republican Government," so recently endangered, he noted, by the "intrigues" mounted by Aaron Burr and his friends "to sever the Western from the Eastern States." With the building of a trans-Allegheny system of canals, therefore, "the United States shall be bound together by . . . cheap and easy access to market[s] in all directions, [and] by a sense of mutual interests arising from mutual intercourse and mingled commerce" (920). Fulton, the astute student of Adam Smith, was also a canny geopolitician, and in putting forward the Federalist position that a happy nation is a tightly knit nation, he also proposed that the strongest bond holding the states together was "the people's interest," which he defined as the liberty enjoyed by "every man to sell the produce of his labor at the best market, and purchase at the cheapest" (921). To this paraphrase of Smith, Fulton added a direct quotation from Hume: "The Government of a wise people [is] little more than a system of civil police; for the best interest of man is industry, and free exchange of the produce of his labor for the things which he may require" (921). Until his premature death a decade later, Fulton did not abandon his faith in the canal as an engineering equivalent to the Constitution as a benevolent machine promoting law and order by means of mutual intercourse and commercial exchange.

"I am happy to find," concluded Fulton, "that through the good management of a wise administration, a period has arrived when an overflowing treasury exhibits abundant resources, and points the mind to works of such immense importance" such as those anthologized in Gallatin's report, and Fulton hoped they soon would become "favorite objects with the whole American people" (921). Unfortunately for Fulton's hopes, "the good manager" in question had already mounted his embargo, and the wisdom of his administration was no longer much admired. Though Gallatin's report remains a signal and visionary document, it suffered a fate similar to that of the many pamphlets and schemes for internal improvements collected within. Even in its conception, moreover, it was a curious and paradoxical creation, reflecting the disjunctiveness inherent in the president who authorized it.

If decentralization was a Jeffersonian ideal, it was more often honored in the breach than the observance; moreover, if a national system of internal communications was urgently needed, the urgency was compounded by the recent Louisiana Purchase, a massive instance of Jeffersonian abrogation of Constitutional limitations on the powers of his high office. Thus the anger felt by New England Federalists toward Jefferson because of the Embargo Act was quickly transferred to Gallatin's carefully depoliticized report. Any such program of internal improvements was characterized (to borrow John Adams' phrase) as "Gallican"—i.e., associated with Jefferson's republican love of all things French—just the sort of foreign entanglement Washington had warned about, even as he called for just such a program of public works as Gallatin proposed. So it was that Washington's advice concerning the importance to the Union of improved roads and waterways produced a document that served chiefly at the time to promote factionalism and sectionalism.

In any case, the noble sentiments and worthy projects in the report were doubly theoretical, because as both Jefferson and his Secretary of the Treasury knew, the national funding of internal improvements was not provided for by the Constitution. If in 1808 Washington's hope for "progressive improvements" for the sake of a more perfect union was to be realized, it would have to be accomplished by the efforts of the separate states. Yet, as the cumulative evidence of Gallatin's report clearly demonstrated, such efforts had by and large failed. By 1809, it might be said, a large number of influential and informed people agreed with Robert Fulton that public improvements were a very good thing, but few agreed among themselves as to the way they should be funded.

ii

With characteristic foresight, Benjamin Franklin was the first to raise the issue of internal improvements during the Constitutional Convention of 1787. He moved that Article I, section 8, which gave to Congress the power to "establish Post Offices and pose Roads," should be expanded to include the "power to provide for cutting canals where deemed necessary" (Farrand: II: 615). James Madison agreed but suggested that Franklin's motion be enlarged "to grant charters of incorporation where the interest of the U.S. might require & the legislative provisions of individual States may be incompetent" (615).

Madison added that his chief aim was "to secure an easy communication between the States which the free intercourse now to be opened, seemed to call for," as "the political obstacles [having been] removed, a removal of the natural ones as far as possible ought to follow." Madison thereby coupled the beautiful mechanism of the Constitution, which provided for a more perfect union of states, with the creation of devices that would more perfectly articulate the connection between the individual parts of that union.

The particulars of Madison's suggestion were unfortunate, however, associated as they were, not only with notions of monopoly and privilege, but with the issue of states' rights versus federal authority. Nor was the simple beauty of Franklin's logic apparent to all the representatives. Thus, although Franklin's motion had been seconded and supported by James Wilson, his fellow Philadelphian, who was conscious of the importance of roads and canals to the welfare of the one state whose territory was bisected by the Allegheny mountains, Roger Sherman—who perhaps saw little benefit accruing to Connecticut, which already had numerous good roads and bridges and anticipated no immediate need for extensive canals—objected to the motion, pointing out that while the expense of cutting canals "will fall on the United States . . . the benefit [would only] accrue to the places where the canals may be cut" (615). Wilson countered by proposing that the canals financed by the government might pay for themselves by becoming a source of revenue, and it was this suggestion, made to calm sectional jealousies, that inspired Madison's motion granting Congress the power to set up corporations, thereby creating agencies independent of both individual states and the national government.

But Madison in effect raised a new problem. Rufus King anticipated that any such provision would hinder the adoption of the Constitution, for in Philadelphia and New York it would be considered tantamount to "the establishment of a [national] Bank, which has been a subject of contention in those cities. In other places it will be referred to [compared to] mercantile monopolies" (616). Wilson dismissed King's objections as unfounded, and stressed "the importance of facilitating by canals the communication with the Western Settlements," and on George Mason's suggestion ("he was afraid of monopolies of every sort"), "the motion [was] modified as to admit a distinct question specifying and limited to the case of canals," and as such was defeated, the vote being split fairly cleanly along sectional lines: Pennsylvania, Virginia, and Georgia voted for the motion, while the delegates from New England voted as a solid bloc against it.

The delegates from New York did not vote on the motion, all three having earlier left for home. The antifederalists, Yates and Lansing, departed on the grounds that the Convention was exceeding its mandate; the Federalist, Alexander Hamilton, departed because the withdrawal of his fellow delegates deprived him of his vote. Had Hamilton remained in Philadelphia, there is little doubt that he would have backed Madison's proposal giving Congress the power to contravene sectional intransigency by creating corporations, and there is little doubt that he would have lost.

As for the importance of canals, Hamilton devoted only two paragraphs to internal improvements in his famous *Report on Manufactures,* the longest of which was a quotation from Adam Smith, the other a revision of several paragraphs in a draft report submitted by Tench Coxe as Assistant Secretary of the Treasury. Hamilton thoroughly recast Coxe's original, and where the latter's opinion that "the most useful assistance perhaps, which it is in the power of the legislature to give to manufactures and which at the same time will equally benefit the landed and commercial interests, is the improvement of inland navigation" was preserved by Hamilton in his own first draft, he soon thereafter changed it to the following, more accurate opinion: "It were to be wished, that there was no doubt of the power of the national Government to lend its direct aid, on a comprehensive plan . . . to the improvement of inland Navigation, which . . . must fill with pleasure every breast warmed with a true Zeal for the prosperity of the Country" (*Paper*: X: 21, 310). No subsequent executive champion of internal improvements aided by the national government, from Jefferson to John Quincy Adams, could avoid the fact testified to in the record of the Convention debates and here alluded to by Hamilton: not only did the Constitution not contain a provision for government assistance in carrying out internal improvements, but its makers had voted such a provision down.

Jefferson, in his first annual message to Congress, observed that "the four pillars of our prosperity," which he enumerated as agriculture, manufactures, commerce, and navigation, thrive most "when left most free to individual enterprise," but then with characteristic ambiguity he added a vague but (literally) inviting qualifier: "Protection from casual embarrassments . . . may sometimes be seasonably interposed. If in the course of your observations or inquiries they should appear to need any aid within the limits of our constitutional powers, your sense of their importance is a sufficient assurance they will occupy your attention" (*Messages*: I: 330–31). By the time of his fourth annual message, Jefferson's hints had become somewhat stronger, though they were still couched in tentative terms: "Whether the great interests of agriculture, manufactures, commerce, or navigation can within the pale of your constitutional powers be aided in any of their relations; whether laws are provided in all cases where they are wanting; whether those provided are exactly what they should be; whether . . . in fine . . . anything can be done to advance the general good, are questions within the limits of your functions which will necessarily occupy your attention" (373). In his second inaugural address, Jefferson was even more pointed: anticipating the day when the national debt would be retired, he proposed that "the revenue thereby liberated may, by a just repartition of it among the States and a corresponding amendment of the Constitution, be applied *in time of peace* to rivers, canals, roads, arts, manufactures, education and other great objects within each State" (379).

Although Jefferson, having by then had Gallatin's report published, returned in his sixth annual message to a call for a constitutional amendment that would provide government support for public improvements, the threat of impending war with, variously, Spain, Great Britain, France, and Algiers, took up considerable time and national treasure during the succeeding months. And when, in his eighth and final message, Jefferson once again brought up the (by then) highly theoretical use of future surplus revenues, "whenever the freedom and safety of our commerce shall be restored," his repeated use of interrogatives suggests either fatigue or a weakening resolve: "Shall it lie unproductive in the public vaults? Shall the revenue be reduced? Or shall it not rather be appropriated to the improvements of roads, canals, rivers, education, and other great foundations of prosperity and union under the powers which Congress may already possess or such amendment of the Constitution as may be approved by the States?" (456). These were all leading questions, however, gauged perhaps to lure Congress into action, for Jefferson ended by observing that though "the course of things" was at present uncertain, "The time may be advantageously employed in obtaining the powers necessary for a system of improvement, should that be thought best." Though undoubtedly sincere, Jefferson's suggestion was couched in syntax and phrasing far different from that used by him in declaring independence from Great Britain.

His successor inherited a "course of things" that rapidly headed toward the certainty of war with Great Britain, and though James Madison, in his own inaugural address, promised to "promote by authorized means improvements friendly to agriculture, to manufactures, and to external as well as internal commerce," he soon enough experienced the truth of Jefferson's observation that "useful works" are suspended "in time of war" (*Messages*: I: 468). Still, Madison was a more fervent advocate of internal improvements than was his fellow Virginian: despite the failure of his attempts to write them into the

Constitution, in the fourteenth paper of *The Federalist* Madison had argued that "the intercourse throughout the union will be daily facilitated by new improvements," thereby assuaging fears that the republic would be threatened by "the great extent of country which the union embraces" (86–87, 83). He then went on to point out how "the communication between the western and Atlantic districts, and between different parts of each, will be rendered more and more easy by those numerous canals with which the beneficence of nature has intersected our country, and which art finds it so little difficult to connect and complete" (87). By conceiving of rivers as natural "canals," Madison evoked the idea so dear to late eighteenth-century imperialists that the North American landscape was given its shape by Providence, a plan intended to facilitate westward expansion: Man, by completing the grand design, is thereby fulfilling a divine mandate.

In December 1811, as war with Great Britain seemed certain, Madison communicated to the Congress "an act of the legislature of New York relating to a canal from the Great Lakes to Hudson River," with a brief message in which he stressed the utility of canals in general, and in particular "the signal advantages to be derived to the United States from a general system of internal communication and conveyance" in times of war as well as peace: "As some of those advantages have an intimate connection with the arrangements and exertions for the general security, it is at a period calling for those that the merits of such a system will be seen in the strongest lights" (*Messages*: I: 497). This was a significant revision of Jefferson's emphasis, being a strategy borrowed from *The Federalist* (and Gallatin's report), for Madison apparently hoped to use the present crisis to move Congress in the direction necessary for the "introduction and accomplishment" of "a general system" of internal improvements, not excluding, presumably, the constitutional amendment proposed by Jefferson. The difficulty of military transport along the Canadian frontier during the ensuing war helped prove Madison's point, and at hostilities' end he returned to his theme, in his seventh annual message urging Congress to turn its attention to "the great importance of establishing throughout our country the roads and canals which can best be executed under the national authority" (567).

Madison took advantage of the auspicious occasion to enlarge upon his observations concerning roads and canals as stated in the fourteenth paper of *The Federalist,* providing a concise summary of the argument for internal improvements and for the need of government support at the national level:

> *No objects within the circle of political economy so richly repay the expense bestowed on them; there are none the utility of*

which is more universally ascertained and acknowledged; none that do more honor to the governments whose wise and enlarged patriotism duly appreciates them. Nor is there any country which presents a field where nature invites more the art of man to complete her own work for his accommodation and benefit. These considerations are strengthened, moreover, by the political effect of these facilities for intercommunication in bringing and binding more closely together the various parts of our extended confederacy. Whilst the States individually, with a laudable enterprise and emulation, avail themselves of their local advantages by new roads, by navigable canals, and by improving the streams susceptible of navigation, the General Government is the more urged to similar undertakings, requiring a national jurisdiction and national means, by the prospect of thus systematically completing so inestimable a work; and it is a happy reflection that any defect of constitutional authority which may be encountered can be supplied in a mode which the Constitution itself has providentially pointed out. (567–68)

No extended analysis is necessary to point out the difference in rhetorical strategies between Madison's argument for internal improvements and Jefferson's. Madison's onward rolling periods and strong, positive diction end by once again linking the benefices of nature to the providential design of the Constitution—which permitted amendments much as the accommodating landscape cried out for improvements. This time his appeal found an eager listener in John C. Calhoun, who turned his eagle gaze inland and proposed that in time of peace, war should be declared on Space, espousing the kind of continental thinking that Hamilton had earlier championed.

"The first great object," said Calhoun, in support of a bill setting aside National Bank dividends in a fund to be used for the construction of roads and canals, "is to perfect the communication from Maine to Louisiana. This may be fairly considered as the principle artery of the whole system" (*Works*: II: 195). Next would be a "connection" between the Hudson River and the Great Lakes ("In a political, commercial, and military point of view, few objects can be more important"); followed by connections between "all the great commercial points on the Atlantic . . . with the Western States"; and, finally, the perfection of "the intercourse between the West and New Orleans" (195). Refusing to stoop to details, Calhoun rose to broad generalities, and he was equally latitudinarian when it came to the problem of constitutional warrant: "If we are restricted in the use of money to the enumerated powers, on what principle can the Purchase of Louisiana be justified?" (194). When inspired to think continentally, Calhoun's thoughts ran to generous dimensions: "To legislate for our country requires . . . the most enlarged views" (191).

The Congress, convinced by Calhoun's arguments or by their own personal and political interests, sent up his bill to the White House, and Madison vetoed it, giving in his last Message to Congress the not surprising explanation that it was unconstitutional: "I have no option but to withhold my signature from it, and to cherish the hope that its beneficial objects may be attained by a resort for the necessary powers to the same wisdom and virtue in the nation which established the Constitution in its actual form and providentially marked out in the instrument itself a safe and practicable mode of improving it as experience might suggest" (*Messages*: I: 585). To the end of his long life, Madison served as a living source of commentary on the Constitution, and he never wavered in his belief that without the necessary amendment, national funding of internal improvements was unconstitutional, however desirable.

James Monroe inherited this dual belief and burden, and in his first annual message observed that "a difference of opinion has existed from the first formation of our Constitution to the present time among our most enlightened and virtuous citizens respecting the right of Congress to establish such a system of [internal] improvement," a right that, it was his own "settled conviction," it did not have, nor could it have "the right in question" without the necessary amendment (*Messages*: II: 18). With Madisonian consistency, Monroe in 1822 vetoed an act providing for the upkeep of the Cumberland Road on constitutional grounds, calling once again for the needed amendment. He accompanied his veto with a curious and unprecedented document (actually written four years earlier) entitled "Views of the President of the United States on the Subject of Internal Improvements"; in effect, a "doctrine" that has long since been overshadowed by Monroe's much more famous pronouncement concerning the interests of the United States in preserving the integrity and stability of the entire Western Hemisphere.

Moreover, Monroe's "View" is a document that delineates the complex situation that underlay the "Era of Good Feelings" his administration initiated (and Adams' terminated), a period of apparent prosperity and optimism that has been seen by recent commentators as a time of national anxiety. A new group of statesmen was emerging who had inherited the legacy of the Founding Fathers, and who were in a sense operating in their shadow, yet were faced with problems for which the Founding Fathers had left no clearcut solutions. Thus the figure of George Washington still stood on the threshold of a century now two decades old, still pointing "the way to prosperity by new roads and canals," but none of the Fathers who succeeded him as President had managed to remedy the constitutional deficiency that had been largely of their own making, and though "the way" still stood open, it as yet led nowhere.

Clearly, the need that Madison had expressed in the Constitutional debates, *The Federalist,* and in his own presidential messages, had become increasingly evident, even urgent. If the new generation of statesmen, like Calhoun, felt that urgency, then it was up to them to implement it, but first the deficiency in the Constitution must be amended—although great care had to be taken not to harm the delicate balance between state and national powers that was one of the Constitution's most admirable qualities. Monroe's "View" essayed to accommodate the fact that the republic for which the original Constitution had been designed now needed an additional instrument that would, without endangering the national stability guaranteed by the marvelous clockworks, assist in accelerating the spread of the empire. And Monroe was certainly the best possible person to suggest that the Constitution was in need of strategic repairs.

James Monroe affected the costume of the earlier generation of statesmen, a preference that may have been quaintly personal but certainly had a political implication well within his capacity to comprehend. For he was the last of the Presidents who had served in the American Revolution, yet he was the first President who had neither signed the Declaration of Independence nor had a hand in framing the Constitution. He had taken part in the Virginia debate over ratification, but had opposed the Constitution as an instrument of increased central powers, and in his subsequent senatorial career he tended to side with Jefferson against the policies of Washington. Still, over his thirty years of public service, as senator, diplomat, and Secretary of State, Monroe's early sectionalism had been considerably modified, and the structure and logic of his "View" reflected what was, by 1818, essentially an equilibrist position: a man of both centuries, Monroe was a President whose long-lived fear of centralized national power was counterbalanced by a recognition that the freedom of navigation on the Mississippi with which his fledgling diplomatic career was identified had ramifications only a national system of roads and waterways could accommodate.

In proposing that the Congress recommend to the States "an amendment to the Constitution to vest the necessary power in the General Government" to fund internal improvements, Monroe felt it was necessary to review the circumstances from which the Constitution emerged, a review that began not with the Convention of 1787, but with the situation immediately prior to the Revolution. As if reading the younger members of Congress a history lesson, Monroe then went on to rehearse the background of the Articles of Confederation, their failings, and the consequent framing of the Constitution, which in Monroe's accounting established "two separate and independent governments . . . one for local purposes . . . the other for national," both taking their power directly from the people (*Messages*: II: 148). The former opponent of the Constitution was now able to appreciate its neoclassical beauties: "It is owing to the simplicity of the elements of which our system is composed that the attraction of all the parts has been to a common center, that every change has tended to cement the union, and, in short, that we have been blessed with such glorious and happy success" (149).

Monroe then proceeded to examine, point by point, the powers granted the central government, a pedagogical exercise reinforcing in detail his contention that the Articles were designed to establish separate zones of force—not only the three branches of the government but the national government and the governments of the states: "The great office of the Constitution of the United States is to unite the States together under a Government endowed with powers adequate to the purposes of its institution, relating, directly or indirectly, to foreign concerns, to the discharge of which a National Government thus formed alone could be competent" (153). In short, the designated functions of the central government were those relating to foreign affairs. Given this thesis, there could be little doubt concerning Monroe's conclusion to his own examination "whether the power to adopt and execute a system of internal improvement by roads and canals has been vested in the United States" (155).

Having devoted more than seven thousand words to the history and structure of the Constitution, Monroe then went on to devote another fourteen thousand words to the consideration (and refutation) of the arguments previously advanced by such as Calhoun, which held that this or that clause of the Constitution could be construed as giving the national government the power to direct and execute internal improvements. While conceding that the central government was empowered to make "appropriations" for "important national purposes," among which could be included roads and canals, Monroe maintained that "for every act requiring legislative sanction or support the State authority must be relied on" (168). No other of the powers granted the central government applied to the construction and control of public works, and Monroe therefore concluded that the Constitution bestowed no right on Congress to "adopt and execute a system of internal improvement" throughout the Union.

At this point, Monroe's "View" seemed a massive defense of the strictest possible interpretation of the limitations placed by the Constitution on the national government. He then went on to stress "the advantages which would attend the exercise of such a power by the General Government," rehearsing what had by then become familiar truths: the "incalculable advantages that would be derived from such improvements," which ranged from commercial considerations to the lofty ideal of strengthening "the bond of union" (175–77). In these passages he sounds like Madison or Calhoun, but not for long: where Madison thirty years earlier had linked roads and canals to the consolidation and stability of the Union—which was most recently the theme of the nationalistic Calhoun—Monroe now added a strategically new element, associating internal improvements with territorial expansion. "It is believed that the greater the expansion within practicable limits—and it is not easy to say what are not so—the greater the advantage which the States individually will derive from it. With governments separate, vigorous, and efficient for all local purposes, their distance from each other can have no injurious effect upon their respective interests" (177). The advocate of "two governments" saw space no longer as a danger but as a blessing, "especially with the aid of these [internal] improvements," for "the expansion of our system must operate favorably for every state in proportion as it operates favorably for the Union" (177–78). In short, a system of internal improvements would work to the benefit of *both* "governments."

Declaring that the power of a nation derives from its "extent of territory," Monroe maintained that "at whatever point we may stop, whether it be at a single range of States beyond the Mississippi or by taking a greater scope, the advantages of [internal] improvements is deemed of the highest importance. . . . The further we go the greater will be the necessity for them" (178). Conceding that individual states "in making improvements should look to their particular and local interests," Monroe concluded that only "Congress would look to the whole and make improvements to promote the welfare of the whole" (179). None of this, of course, cancelled out his earlier conclusions regarding the original design of the Constitution: "If it is thought proper to vest this power in the United States, the only mode in which it can be done is by an amendment of the Constitution."

The remaining pages of Monroe's "View" are devoted to a brief history of the "revolutionary struggle," and conclude by elevating the work of the Founding Fathers so as to give a charge to the succeeding generation: "The establishment of our institutions forms the most important epoch that history had recorded. They extend unexampled felicity to the whole body of our fellow citizens, and are the admiration of other nations. To preserve and hand them down in their utmost purity to the remotest ages will require the existence and practice of virtues and talents equal to those which were displayed in acquiring them. It is ardently hoped and confidently believed that these will not be wanted" (183). Thus the first and last part of Monroe's treatise concerns "intervening space" in terms of time—American history—while the central section, the lengthy discussion of constitutional powers and the need for an amendment, is concerned with "space" in terms of governmental arrangements in the first instance, and in the second, expansion across a vast continent. The "space" that Calhoun had proposed conquering was that which separated the established states and territories, but, in Monroe's view, the most meaningful space was that which lay west of the Mississippi River. If the first part of Monroe's "View" expresses a strict constructionist view of the Constitution, what he proposes at the last is a massive revision, an amendment that would provide a progressive third part for the stable dichotomy of "two governments" by adding a mechanism accommodating national expansion. While continuing to admire the handiwork of the Founding Fathers (much as he preserved their dress), Monroe was prepared to undertake, by his own definition, a very large step in an imperial direction.

Compared to the forceful rhetoric of Jefferson's *Declaration of Independence,* to the smooth periods of Gouverneur Morris' *Constitution,* or even to the urgent yet cogent arguments of *The Federalist,* Monroe's "View" has all the excitement of a lawyer's brief. Yet the style somehow suits the times, and if the "View" is weak as a work of literature it lends a massive force of sanction to the expansionist phase of American history. Like Washington's farewell address, though hardly with such admirable brevity, Monroe's extended veto was a prolegomenon to future action, creating in effect the necessary legal basis for a congressional mandate that would permit a national system of roads and canals. Washington, once again, may have pointed the way, but future generations who benefited from that system would necessarily look to Monroe as the source of their blessings—or so Monroe may have hoped.

Ironically, the chief purpose of Monroe's "View" was never accomplished, for no amendment ever came forth from the Congress that would grant the central government the necessary powers to design and construct a national system of roads and canals. Content to work (as Calhoun had suggested) within the existing terms of the Constitution, Congress chose to work the power of appropriations for all it was worth, for the rise of political parties and factions demonstrated the beauties of sectional loyalty where longevity in office was concerned. Out of the mighty lumber of Monroe's magisterial veto, the once and future Congress created the staves, hoops, and butt ends of that enormous yet invisible structure that became a second "government," in fact, the Barrel known as "Pork." Internal improvements continued to be a symbol of progress but remained independent of national control, even while benefiting from national sponsorship engineered indirectly.

The future course of internal improvements in the United States only in part fulfilled the vision of Gallatin's great book; while facilitating the westward spread of empire, the building of roads and canals also promoted sectional divisions as much as national unity. That line of demarcation observed by early travelers along the Ohio River, like Thomas Ashe, was projected into a farther range, as states were divided between slave and free, a division that came to actual violence during the debate over the fate of Kansas. We can see from here that the compromise regarding slavery that was one of the expedient steps making ratification of the Constitution possible was also responsible for that irrepressible and unavoidable conflict between the North and the South, which increasingly threatened to split Washington's beloved Union forever.

But counterpart to that prolonged and imperfect compromise was the whole matter of internal improvements, which the Founders in their wisdom left to individual states, but which, over the first century of the republic, was increasingly and illegally funded by the central government, which condignly broke its own laws in the name of progress and prosperity. If the use of the sacredness of property as a sanction for chattel slavery may be considered as the statist equivalent of original sin, then the fostering of internal improvements by the United States Congress without legislating the amendment that would make such actions legal can be seen as fathering the kinds of lies that in a divine scheme of things ensure destruction. Their proponents saw improved rivers and canals as ties more closely binding the separate states. Today we may view them as threatening the dissolution of the very Union they were designed to solidify.

XI

Through Wood and Dale

i

Among the many documents stitched together by Jedidiah Morse in the first edition of his *American Geography* were two pamphlets by Tench Coxe that argued for the development of manufactures in the United States, arguments Morse paraphrased at length, thereby giving the ideas of Hamilton's protégé even further circulation. In one sentence, Morse summed up the geodynamics that lay at the heart of Coxe's attempt to amalgamate the agricultural (Republican) and manufacturing (Federalist) interests: "The produce of the southern states might be exchanged for such manufactures as can be made by the northern, to mutual advantage" (89). When Morse expanded his *Geography* in 1793, he took advantage of Alexander Hamilton's recent *Report on Manufactures* (as well as of Coxe's influential *Brief Examination of Lord Sheffield's Observations on the Commerce of the United States*) to expand his argument, which, again in terms of geo-economics, turned on the same point: "If the northern and middle states should be the principal scenes of [manufacturing] establishments, they would immediately benefit the southern, by creating a demand for productions [of agriculture]" (245).

Similar views were given space in Morse's *Geography* of 1789, in the shape of an extensive quotation from the manuscript journal of Elkanah Watson, "a gentleman who has travelled extensively both in Europe and America":

> *The northern and southern states differ widely in their customs, climate, produce, and in the general face of the country. The middle states preserve a medium in all these respects; they are neither so level and hot as the states south; not so hilly and cold as those north and east. The inhabitants of the north are hardy, industrious, frugal, and in general well informed; those of the*

> *south are more effeminate, indolent and imperious. The fisheries and commerce are the sinews of the north; tobacco, rice and indigo, of the south. The northern states are commodiously situated for trade and manufactures; the southern, to furnish provisions and raw materials; and the probability is, that the southern states will one day be supplied with northern manufactures instead of European, and make their remittances in provisions and raw materials.* (533)

The author was the same peregrinating Yankee who visited Mount Vernon in 1785, and Watson's viewpoint is detectably regional in its bias, while his rhetoric tends to be more expansive than the language used by either Coxe or Hamilton, who, while acknowledging the value of western lands (as speculative property), focus on the balance of trade to be carried on east of the Appalachians.

Watson's European travels during and immediately following the War for Independence qualified him to make comparisons between the Old World and the New, but, as is his regional bias, Watson's chauvinism is clearly evident:

> *When the extent of America is considered, boldly fronting the old world—blessed with every climate—capable of every production—abounding with the best harbours and rivers on the globe, and already overspread with three millions of souls, mostly descendents of Englishmen—inheriting all their ancient enthusiasm for liberty, and enterprizing almost to a fault—what may be expected from such a people in such a country?—the partial hand of nature has laid off America upon a much larger scale than any other part of the world. Hills in America are mountains in Europe—brooks are rivers, and ponds are swelled into lakes. In short, the map of the world cannot exhibit a country uniting so many natural advantages, so pleasingly diversified, and that offers such abundant and easy resources to agriculture and commerce.*
>
> *In contemplating* future America, *the mind is lost in the din of cities—in harbours and rivers clouded with sails—and in the immensity of population. . . . And when we consider the probable acquisition of people, by foreign immmigrations, and that the interior and unsettled parts of America are amply sufficient to provide for this number, the presumption is strong, that . . . at the end of one hundred years there will be ninety-*

> *six millions of souls in United America; which is two thirds as many as there are at present in all Europe.* (532–33)

"United America" was no slip of the pen, for Watson's continental thinking far exceeded Hamilton's range: "Europe is already aware of the rising importance of America, and begins to look forward with anxiety to her West Indian Islands, which are the natural legacy of this continent, and will doubtless be claimed as such when America shall have arrived at an age which will enable her to maintain her right." Throughout much of his long life (he did not die until 1842), Watson maintained a carefully neutral politics, but his stance in 1789 seems an initial stretching of what will become the expansionist wings of Jacksonian democracy.

Yet there is much about Elkanah Watson that reflects the Federalist cast of mind. Certainly, Watson remained loyal to his region: by his own claim, he declined Washington's invitation to settle in the Potomac Valley, and though he did spend a short period as a plantation owner in North Carolina, Watson eventually followed the pathway of Yankee migration to Albany, New York, where he busied himself with a series of Franklinesque improvements, not always to the gratification of the Albanians. The most important of these was the Western Inland Lock Navigation Company, incorporated in 1792, in the hope of bypassing falls and rapids along the Mohawk River, thereby opening a channel from the Hudson to Lake Ontario and beyond. Ever afterward, and most particularly as the Erie Canal neared completion, Watson was strenuous in promoting not only canals but his own contribution to their promotion.

It needs to be said that Watson's role in the building of the Erie Canal was at best tangential; nor was he particularly innovative in his scheme for improving navigation along the existing waterways of New York. "The channel," as Jefferson observed in 1784, "is already known to practice," for as early as 1724, Cadwallader Colden, friend to Franklin and John Bartram and the historian of the Iroquois Nations, submitted his *Memorial Concerning the Furr Trade* to the Royal Governor of New York, in which he pointed out the importance of the northern lakes and rivers to successful competition with the French of Canada. In 1784, even as Watson was visiting Washington at Mt. Vernon, Christopher Colles prepared a memorial to the New York State Assembly "for the improvement of the inland navigation between Albany and Oswego," a project that was extended by Colles, the next year, to Lake Erie. Of this there came little but an appropriation of $125 for an initial survey and a pamphlet by Colles, published in the following year, which contained the results of his survey and several memorable phrases:

> *Providence indeed appears to favor this design, for the Allegheny mountains which pass thro' all the States seem to die away as they approach the Mohawk River, and the ground between the upper parts of this river and Wood Creek is perfectly level, as if designedly to permit us to pass thro' this channel into this extensive inland country. . . . It appears highly expedient at this time to promote this undertaking, as the British may, and now are, endeavoring to draw the trade of the inland country towards Quebec; and it is an indubitable fact, that trade is like water, when it once passes in any particular channel, it is not easily diverted or drawn away into another.* (11, 13)

Thus Jefferson's fears about the primacy of the Mohawk–Erie "Channel" voiced in his *Notes on Virginia* were transferred by Colles to New York's rivalry with Canada, once (in Cadwallader Colden's day) a French economic field of force threatening the British fur trade, but now, in the hands of the British, a threat to the United States in general and New York in particular. Imperialism remained a constant factor; only the national (and regional) labels had changed.

"Commerce," wrote Elkanah Watson in a pamphlet written (but not published) in 1791, "like water, will seek its natural level, depending on natural or artificial causes; but once the current takes a settled direction, it will not be easy to divert its course" (*History:* 57). Watson used Colles' metaphor in a somewhat different context, alluding to "the great plan of Washington" that had been revealed to him in 1784 and that had, by Watson's account, so influenced his own thinking. Where Washington had hoped "to divert the commerce of this immense region in the west . . . to his beloved Alexandria," Watson now aligned the "great plan" in an easterly direction, past his own "beloved" Albany. But if Washington's Potomac scheme was transported to New York by Watson, Elkanah's "waters" (as commerce) were appropriated from Colles. From the very start, the Erie Canal was a highly eclectic idea, gathering its force from a number of sources, of which Elkanah Watson was merely the most prolific and insistent.

By 1807, Watson's Inland Lock Navigation Company was only one of several regional corporations that were attempting to fund internal improvements, led by men who were long on vision but short of cash. These included the Chesapeake and Delaware Canal, the Dismal Swamp Canal, and a canal connecting Buzzard's Bay with Massachusetts Bay by cutting through the length of Cape Cod; while on the other side of the mountains, Henry Clay, then a freshman senator from Kentucky, was pushing for the long-deferred canal around the Falls at

Louisville. All of these projects were heralded by their proponents as beneficial to the further cementing of the Union, and Senator Clay, already the emerging champion of internal improvements, argued for the government's funding of both the Ohio and Delaware projects by trading public lands for stock in the two companies. John Quincy Adams, still a Federalist, opposed Clay's measure on the familiar ground that the coalition of interested states would provide a disastrous precedent for similar combinations of vested and regional interests.

Adams was being protective of eastern concerns, but the Father of Standardized Weights and Measures was perhaps less against internal improvement *per se* than for an orderly and fair funding of them, and asked that the Secretary of the Treasury undertake a survey and produce a general plan that would give system and uniform direction to the improvement of roads, canals, harbors, and rivers. His resolution was voted down, but a subsequent and similiar request, introduced by Senator Thomas Worthington of Ohio, a long-time proponent of internal improvements, was successful, and Albert Gallatin spent the following year assembling his epochal report. Even as Gallatin was gathering his anthology of documents, a more obscure but no less dedicated citizen, Jesse Hawley of Geneva, New York, inspired by Jefferson's Sixth Annual Message, took advantage of a period in jail (the result of financial embarrassment) to write (under the name "Hercules") a series of articles proposing an overland canal from Lake Erie ("at Buffalo Village") to Utica, and from there down the Mohawk to the Hudson River (Hosack: 307).

New Yorkers soon discovered, however, that Jefferson's enthusiasm for internal improvements did not quite match that of his Secretary of State. In January 1809, a representative from the New York Assembly, Joshua Forman, took advantage of a business trip to Washington, D.C., to "converse with Mr. Jefferson" concerning the digging of a canal "between the tide-waters of the Hudson River and Lake Erie" (Hosack: 347). "In as laconic a manner as I could," Forman recalled, he recapitulated "some of the most important advantages it offered to the nation as inducements to undertake it," most of which had already appeared in Gallatin's report, but Jefferson's response was, if anything, more laconic. He told Hawley that his was "a very fine project, and might be executed a century hence. 'Why, sir,' said he, 'here is a canal of a few miles, projected by General Washington, which, if completed, would render this [Washington, D.C.] a fine commercial city, which has languished for many years because the small sum of 200.00 dollars necessary to complete it, cannot be obtained from the general government, the state government, or from individuals—and you talk of making a canal of 350 miles through the wilderness—it is little short of madness to think of it at this day'" (347). In 1822, as the Erie Canal was under construction, Jefferson could not "recollect" this conversation, but he was willing to concede its likelihood, nor was he alone in doubting the practicality of such a project. It took dreamers like Jesse Hawley and Elkanah Watson to foresee the great New York canal, while Jefferson saved his own visionary powers for improvements along the Potomac.

ii

Born in 1758, Elkanah Watson was James Monroe's exact contemporary, being of the generation that first saw the light of day during the French and Indian War. Though too young to participate in the founding of the republic, both men were old enough to fight in the war that made independence possible. Watson's military record, however, was not particularly notable, being limited to several alarums and excursions as a member of a "liberty company" of volunteers in Rhode Island. Born in Plymouth, claiming descent from Edward Winslow, Watson left his "native place" at the age of fifteen for Providence, Rhode Island, where he was apprenticed to John Brown, a leading merchant and the founder of the university that bears his name. Though Watson was kept from active military service by the refusal of both his father and Brown to release him from his articles of apprenticeship, he experienced the war from what turned out to be a unique and extended perspective.

Like many New Englanders, Watson was an inveterate diarist, keeping a series of daybooks that formed the basis of his memoirs, *Men and Times of the Revolution,* not published in full until 1856. Large sections, however, appeared as early as 1820 in his self-promotional *History of the Rise, Progress, and Existing Condition of the Western Canals in the State of New York . . . Together with the Rise, Progress, and Existing State of Modern Agricultural Societies, on the Berkshire System.* Watson was eager to beg comparison with George Washington for having fathered canals and improved agricultural methods, and the seal he adopted for his Berkshire Agricultural Society displays Washington in uniform standing before his plow—Cincinnatus Himself. But the portrait of Watson (by Ezra Ames) that graces his autobiography shows a face blending the features of Ben Franklin and P. T. Barnum, for the Plymouth phenomenon was a quintessential Yankee who combined the qualities of practical philosopher and promoter.

Though Watson's move from Plymouth to Providence was a decisive step, equivalent to Franklin's move to Philadelphia, his first important travels sent him on a much longer journey, on an errand that was essentially a business trip. At nineteen, as he recalled in his memoirs, Watson was sent by his employers overland to South Carolina and Georgia, carrying fifty thousand dollars "to be placed in the hands of their agents in the Southern States, and to be invested in cargoes for the European markets" (*Men*: 33). Watson's journey, undertaken during a period when the war was not going well for the patriot forces, was a singular exploit whose heroism was somewhat undercut by the fact that Watson rode most of the way in a sulky—the same practical, two-wheeled Yankee vehicle that would carry Manasseh Cutler over the mountains to Ohio ten years later.

En route, Watson passed through the Trenton battleground, but his own crossing of the Delaware was a detour necessitated by the presence of British troops in Philadelphia, forcing him north through Bethlehem, where he found room in an inn sheltering Lafayette "and other officers wounded at the Brandywine battle" (38). Watson had earlier (in 1775) delivered a wagonload of gunpowder to George Washington's headquarters in Cambridge, and had had the thrill of handing the manifest to the great man "in person," and as his sulky carried him south through Pennsylvania, he passed "within sound of the thunder of Washington's artillery at Germantown" (26, 39). It would be Watson's subsequent fortune to come temporarily into the orbit of a number of famous Forefathers, and the first part of his memoirs reads like a synopsis for an historical romance by Cooper. At the far end of his first trip to the South, he brought the "joyous news" of Burgoyne's defeat to Charleston, just in time to witness the burning of that historic town (51).

Watson's first southern tour brought him home in time to participate in the otherwise not notable Battle of Quaker Hill, which concluded with a mutual retreat of British and patriot forces. This mild baptism of fire took Watson into his majority, but, continuing in the employment of John Brown, he was sent to France, an errand that carried him into the circle of Benjamin Franklin. His experiences in Europe provided yet another rite of passage, for if he was amused when the French regarded him as something of a wild Indian, he was "enraptured with the ease and freedom exhibited in the table intercourse in France," which contrasted with the "cold ceremony and formal compliments" of New England social affairs (103). Setting up in business with a French partner, Watson prospered, and in 1782 he crossed the Channel to England, bearing letters of introduction from Franklin to such diverse notables as Dr. Priestly and Edmund Burke. Watson arrived in time to witness the formal recognition by King George of American independence, and had his portrait painted by Copley, in the background of which was "a ship, bearing to America the intelligence of the acknowledgement of Independence, with a sun just rising upon the stripes of the union, streaming from her gaff. . . . *This was, I imagine, the first American flag hoisted in old England*" (23).

Watson suffered financial reverses in England, and from "moving in the first circles of London," he was "humbled to the dust," but (at least in his memoirs) consoled himself by considering the uses of adversity, the usual Puritan antidote to depression: "Adversity tests our virtues, and tries sincerity . . . teaching us to look deeply into the treacherous volume of the human heart" (220). The next leg of his European tour brought him to that place sacred to Pilgrim chronicles, Holland. While traveling about England, Watson had interested himself in "general improvements" in manufactures, agriculture, roads, and canals, and his Yankee eye was arrested by what he found in the Netherlands: "All the embellishment and verdure of Holland is . . . the creation of the industry and energy of man. . . . This is a region of art, moulded by industry and labor into beauty and productiveness" (255–28).

Watson was particularly taken by the scenery along the Dutch canals, a veritable "fairyland" of gardens and snug villas, "picturesque and beautiful, not unlike splendid and extended paintings" (231). Yet Watson's heart, as when he witnessed the proclamation by King George that acknowledged an independent United States, always "swelled with my proud American blood":

> *Holland presents the aspect of an extensive cultivated garden; but it wants that variety of scenery, so essential to engage the imagination. England and France are more diversified and romantic, and are generally under almost as high improvement. In each of these countries, we meet, here and there, with an artificial forest; we admire their spacious and extended canals, their venerable castles, splendid country mansions, large and magnificent edifices, delightful roads, and numerous other objects of interest and attraction, which allure and fix the attention of an American. When we abandon the contemplation of these exhibitions, the results of art, and enter upon the broad domain of nature, we find her works on this side of the Atlantic but in miniature, contrasted with the vast lakes, the immeasurable rivers, bold harbors, giant trees, and lofty mountains of America.* (205, 230)

As in the passage from his journal quoted by Jedidiah Morse, Watson here sounds a theme that will be characteristic of American writers for the succeeding generations: Europe may have its artificial beauties, but for nature in its grandest manifestations, America stands supreme.

Having returned to America and paid his visit to Mount Vernon ("No pilgrim ever approached Mecca with deeper enthusiasm"), Watson eventually returned to Massachusetts, married, and after a period of indecision, removed to Albany, where he devoted himself, in the words of his son, to a period of "projection and execution," having spent his youth in "an attitude of observation" (304). Like Franklin in Philadelphia a half-century before, Watson busied himself with civic improvements, including streetlamps, more effective downspouts, and street pavements. But, once again, his most ambitious project was the opening of navigation between the Hudson and Lake Ontario, for if nature in America was superior to that in the Old World, it still needed some improvement to be accommodated to civilized use.

Before moving to Albany, Watson had taken a business trip to western Massachusetts in 1788, and hearing that an Indian treaty was to be signed at Fort Stanwix, he kept traveling west up the Mohawk, where he admired "the beauty of the country, the majestic appearance of the adjacent mountains, the state of advanced agriculture, exhibited in a long succession of excellent farms, and the rich fragrancy of the air, redolent with the perfume of the clover" (308–09). Leaving the Mohawk, he struck off overland through a "wilderness bordering upon the Indian territory" along a nearly impassable road, and from scenes of beauty he passed into a much drearier territory, including battlefields of the Revolution, still strewn with human bones (312). Frightened by a drunken party of Indians, and suffering from miserable accommodations, Watson at last reached Fort Stanwix, where he witnessed yet another historical moment, the signing of a treaty with the Iroquois that would open for settlement a "vast territory . . . without any impediments, to the flood of emigration which will pour into it from the East" (314). Indeed, Watson encountered the first settlers as he returned down the Mohawk, "numerous bateaux coming up the river, freighted with whole families, emigrating to the 'land of promise'" (316).

It was Watson's original intention to continue on down Wood Creek to Lake Ontario, but the bad weather "and the obstacles I found to exist in the creek" forced him to cancel his plans. They did not, however, dampen his conviction that "*a canal communication will be opened, sooner or later, between the great lakes and the Hudson*" (315). As Watson's italics suggest, the point of this passage is to establish the priority of his vision, but whether he did indeed entertain such thoughts in 1788 does not make much difference: George Washington had a similar vision six years earlier, as Watson knew, and Christopher Colles had already submitted his memorial to the state legislature. In making his pilgrimage, Watson evinced that uniquely American need to bear witness, to participate, as it were, in the future as he had borne testimony concerning so many events already a part of the past.

In 1791, Watson returned up the Mohawk to Fort Stanwix with a number of other canal enthusiasts and found that "emigrants are swarming into these fertile regions, in shoals, like the ancient Israelites, seeing the land of promise" (339). This time he did descend Wood Creek, which he described as "a natural canal, from ten to twenty feet wide," and entered Oneida Lake: "What a glorious acquisition to agriculture and commerce do these fertile and extensive regions in the west present in anticipations! And what a pity, since the partial hand of Nature had nearly completed the water communication from our utmost borders to the Atlantic ocean, that Art should not be made subservient to her, and complete the great work" (342). These remarks were supposed to have been uttered (or at least entered into his journal) on the spot, and the quality of Watson's vision continued to expand as he confronted "the Ontario ocean," which inspired the

> *delightful anticipation . . . of a free and open water communication from thence to the Atlantic, via Albany and New York. Looking into futurity, I saw those fertile regions, bounded west by the Mississippi, north by the great lakes, east by the Allegheny mountains, and south by the placid Ohio, overspread with millions of freemen; blessed with various climates, enjoying every variety of soil, and commanding the boldest inland navigation on this globe; clouded with sails, directing their course toward canals, alive with boats passing and repassing, giving and receiving reciprocal benefits, from this wonderful country prolific in such great resources.*
>
> *In taking this bold flight of imagination, it was impossible to repress a settled conviction, that a great effort will be made to realize all my dreams. . . . When the mighty canals shall be formed, and locks erected, they will add vastly to the facility of an extended diffusion, and the increase of its intrinsic worth. . . . In a word, I almost deplored the short span of human life, that I cannot witness the happiness of those blessed generations of Americans, yet unborn, who are destined to inherit these delightful regions.* (342–43, 347, 357)

This is the familiar optative mode of propagandists and republican visionary poets, but here keyed to a single factor, the Canal, which will give purpose and meaning to a fertile but at present untenanted region: "The further we explored these western waters, the more we were impressed with the vast importance of assisting nature, in the whole extent of the contemplated improvements, so that loaded boats, coming from the Hudson river, can reach our utmost borders without interruption" (358).

Anyone regarding the area on a map, observed Watson, cannot help but notice the way nature begs for the improvements of man, but once the same person "explores these waters in person . . . the first impression will not fail to be heightened by an enthusiasm bordering on infatuation" (358). There is, that is to say, a religious dimension to Watson's vision; and when he first surveyed the region bordering Lake Seneca known as Appletown, a region of "extensive orchards" once planted by Indians, the landscape arouses him to a rhetorical ecstasy: "The sun was just setting as we entered the lake, which opened upon us like a new creation . . . rising to our view in picturesque and romantic beauty . . . and my mind involuntarily expanded, in anticipating the period, when the borders of this lake will be stripped of nature's livery, and in its place will be rich enclosures, pleasant villages, numerous flocks, herds, etc.; and it will be inhabited by a happy race of people, enjoying the rich fruits of their own labors, and the luxury of sweet liberty and independence, approaching to a millennial state" (353–54). That Watson happened to own Appletown helps to explain the heights of his rapture, but his belief in progress finally is part and parcel of his faith in the rising value of his real estate: all that is needed for this place to become "the paradise of America" is "the vigorous arm of freemen" (356).

Everything turned on "the sublime plan of opening an interrupted water communication from the Hudson to Lake Ontario," a plan that in 1791 was largely relegated to paper. Christopher Colles in 1785 had written that "it is universally allow'd, that if the interior country were settled, Inland Navigation would follow apace; it is also generally agreed, that if Inland Navigation were first accomplished, the settlement of the interior country could thereby be promoted" (3). But as the difference between "universally" and "generally" indicates, the second proposition was not the common wisdom of the day. Enemies of canal construction, whatever their real motives, customarily pointed out that the situation warranting the canals in England—the need to connect areas of dense population—was lacking in America, where projects like Watson's led off from primitive towns into an untenanted wilderness.

Samuel Blodgett, another champion of public works, put it succinctly: "*England has made canals, let canals make America,*" but Watson tended to expand:

> *On this momentous subject* [*of western canals*] *a single question arises. Are we advanced to a sufficient state of maturity, to justify an undertaking of this magnitude? If we proceed on the European mode of calculation, waiting in the first instance to find the country, through which canals are to pass, in a state of maturity and improvement, the answer is at hand—No! But, calculating on the more enlarged American scale, and considering the physical circumstances of the country in question, should canals precede the settlements, it will be justified on the principles of sound policy. In return it will inevitably follow, that a vast wilderness will, as it were by magic, rise into instant cultivation.* (Bodget: 54; Watson: 359–60)

Watson was right, essentially, although the Erie Canal took a somewhat different route from the one he lined out, and its engineers, following Brindley's formula, avoided the clumsy system of river *cum* canal that Watson and his associates promoted. Still, he lived to see the day when his vision was realized—in 1818, at the age of sixty, when he took passage on the as yet unfinished Erie Canal. This was yet another historic moment, his being "the first packet-boat on her preliminary trip" (473). By 1818, Watson's enthusiasm was no longer idiosyncratic, but was common wisdom among a considerable number of influential men, and the completion of the Erie canal along with its sister waterway in Ohio did, as he had predicted, draw as tribute the rich resources of the midlands eastward, much to the benefit of Albany and New York, making the last the "Empire City" of an "Empire State."

When the fortunes of the Inland Lock Navigation Company declined, Watson returned to his native state, settling in Pittsfield, Massachusetts, where he became a founder and prime mover of the Berkshire Agricultural Society. Along with Chancellor Livingston, better known for the Louisiana Purchase and his association with Robert Fulton, he interested himself in improving the quality of wool production in New England, and, having purchased two prize merino sheep, he exhibited them in the *annus mirabilis* 1807 "under the lofty Elm Tree, on the public square, in Pittsfield" (420). "Many farmers, and even females, were attracted to this first novel and humble exhibition," which inspired Watson to mount "a display, on a larger scale, of different animals . . . and from that moment to the present hour, Agricultural Fairs and Cattle Shows, with all their connections, have predominated in my mind, greatly to the prejudice of my private affairs" (420–21). Where his canal scheme failed, Watson's promotion of agricultural fairs and societies succeeded, and he is chiefly remembered as not the "Father of Canals" but as the "Father of the Country Fair."

In this role, Watson made a major contribution to the improvement of agriculture in America, thereby furthering the aims of both Washington and Jefferson, being, as his son insisted, "a Republican in the highest and most emphatic acceptation of the term," which is to say, a Federalist (433). Having assisted in improving the breed of sheep in New England, thereby contributing to the "sheep mania" that soon followed the introduction of merinos to America, Watson addressed himself to improving the related art of weaving. He became a champion of "domestic manufactures," with the stated view of "arresting our colonial degradation and dependence on foreign countries, especially for articles of clothing. Perhaps the net gain to the nation may equal the benefit which agriculture will derive from these institutions" (*History*: 126–27). Still, in 1813, Watson's idea of "domestic manufactures" gave a literal meaning to the adjective, associating "domestic" with the home and "manufactures" with a lady and her loom. He recalled with pride the stratagems devised to lure farm women into displaying their wares, "such was their timidity, and dread of being laughed at" (127).

From such modest beginnings grew the factory system (and the Lowell Maidens) of New England, which would eventually displace agriculture in that region, thanks in part to the protective tariff—of which Watson was an inveterate champion. For a time, likewise, sheep production flourished in the Northeast, but with the opening of the Erie and Ohio canals, and the consequent access to better, cheaper, and vaster grazing grounds, the raising of sheep moved west, as did many Yankee farmers. The factory system in New England increasingly was associated with cotton fabrics, and the remaining farmers abandoned sheep for cows, particularly in the region around Watson's Pittsfield. In sum, the canal of which he was an early champion became the chief instrument in transforming New England from an agricultural to a manufacturing region, suggesting that those who attempt to read the future are condemned to make the mistakes that become the past we call "history."

With undiminished zeal for instituting reform, Watson eventually turned to the cause of temperance; but his enthusiasm for improved transportation continued, also. In his old age he interested himself in a railroad connecting Boston with the St. Lawrence River, a project that may have inspired his last (and shortest) speech: "exclaimed in the delirium of approaching death, with the strongest emphasis and most earnest gesticulation: 'Yonder is the track of the road, and at this point it must terminate'" (*Men*: 526). By 1842, the year of Watson's death, it was possible to travel by rails from Boston to Buffalo, a route that would in time continue westward, taking its impulse and trajectory from the idea and direction of the Erie Canal. Prophetic as always, Watson's last words placed him in some imaginary point in a future that would soon enough become the past.

iii

In 1844 the American artist William H. Brown executed in silhouette the figures of a number of prominent Americans, living and dead. Among the latter was De Witt Clinton, depicted as a portly presence whose decidedly unheroic shape was given symbolic meaning by a wall map of upper New York State, its many lakes and rivers bisected by the line of the Erie Canal, which terminated at the Hudson River a short distance from the back of Clinton's head. Clearly, the canal is to be seen as Clinton's brainchild, yet as contemporary witnesses and Clinton's first biographer testified, the canal as an idea was like the canal itself, a phenomenon fed by a number of streams. Still, everyone agreed, by the time the canal was completed, that it *was* Clinton's canal (or "ditch," depending on your viewpoint or politics). For Clinton was no Elkanah Watson, no outspoken visionary, but was instead that very necessary factor involved in the realization of any large scheme, a version of corporate man; for the digging of the Erie Canal, as a political and financial event, was as much a matter of management as of engineering. He was, with necessary qualifiers, a Fulton to Watson's Fitch.

Thus, where Watson tended to depict himself as a lone visionary, Clinton did not, and we may compare Watson's accounts of his several trips into the western country with the journal kept by Clinton when, in 1810, he made the requisite and by then ritualistic passage up the Mohawk and on to the Great Lakes. Clinton was motivated not by individual initiative but by a sense of civic responsibility, for he was one member of a commission appointed by the State Assembly, which had been charged to investigate the possibilities of "inland navigation from Hudson's River to Lake Ontario and Lake Erie." His journal in all respects differs from that kept by Elkanah Watson. Never revised by himself for publication, it renders a much more balanced account of the territory through which the commission passed, and where Watson's is high on rhapsodic vision, Clinton's is heavy with facts, recording

William H. Brown (1808–1883) and Edmund Burke Kellogg (1809–1872), after Brown. *De Witt Clinton* (1844). (Private collection)

Despite the given names of the lithographer, there is nothing beautiful or sublime about Clinton's silhouette, any more than there was about his canal. There is, however, an interesting sphericity to the man, which contrasts amusingly with the linear diagram on the wall.

everything of possible relevance, from the wildlife encountered to the going prices of various commodities—raw and manufactured, wholesale and retail—a cross between Bartram's *Travels* and the first U.S. Census.

Where Watson, like many another visionary, is humorless before his sublime sense of the future, Clinton, like William Byrd of Westover, maintains a dry wit as he confronts the present moment. Near Oneida Lake, on a section of Watson's Inland Locks Canal, the party lodged at the home of a family whose "obliging and simple" manners were clearly self-interested:

> *They had been forewarned of our approach, and their attention was turned towards the contemplated canal. As they are the proprietors of the soil . . . they were apprehensive that the canal would be diverted from them . . . and the old lady said she would charge us nothing* [*for our lodgings*] *if we straitened the creek and lowered the lake . . . The old lady, on being interrogated as to the religion she professed, said that she belonged to the church, but what church she could not tell. The oracle of the family was a deformed, hump-backed young man, called John. On all occasions his opinions were as decisive as the responses of the sybil; and he reminded us of the Arabian Night's Entertainment, which represents persons hump-backed as possessed of great shrewdness. John told us the story of Irish Peggy, a girl whom he described as going down* [*the creek*] *in a batteaux* [sic], *so handsome and well-dressed that she attracted him and all the young men in the neighborhood, who visited the charming creature; that on her return some weeks afterwards, she looked as ugly as she had been before beautiful, and was addicted to swearing and drunkenness; that she had been indirectly the cause of the death of three men; that one of them, a negro, was drowned in a lock, who had gone to sleep on the deck of the boat, in order to accommodate her and her paramour; that another fell overboard, when she had retired with her gallant, and prevented by it assistance that might have saved him; and that the third one experienced a similar fate. The Commodore* [*the Quaker and philanthropist Thomas Eddy, one of the commissioners*] *did not fail to extract a moral from John's story, favorable to the cause of good morals; and admonished him to beware of lewd women, "whose house is the way to hell, going down to the chambers of death."* (*Life*: 60–61)

In years to come, after the canal defined this region, the stories it generated were of a similar kind; nor would it ever shake its reputation for attracting wild life of the two-legged kind. The crews of Clinton's boat were alternately incompetent, barbarous, and inclined to theft, which did not bode well for future enterprise.

Yet his narrative is on the whole optimistic, and though Clinton and his companions found fault with the lock system built by Elkanah Watson's company (made of wood, it had quickly deteriorated), they acknowledged the improvements to navigation it had effected, and noted also that the canal had the additional advantage of keeping the cellars of houses dry along its route. If the Mohawk was too difficult to navigate, thereby discounting the chief element in Watson's scheme, still, the commission was greatly aided in its upstream survey by a map prepared for the Inland Lock Navigation Company; and if the river was too incommodious for their purposes, being "good only as a feeder," it had formed a valley "admirably calculated for a canal" (55, 54). Like Elkanah Watson, Clinton had an imperial eye—even if it lacked a visionary gleam—which like a surveyor's transit was invariably drawn along linear corridors heading west.

Like Watson, also, Clinton had already invested in the lands through which the canal would pass, and he notes, not only his own holdings along the route, but those of his fellow commissioners, one of whom—his cousin, the Surveyor General of New York, Simeon De Witt—was the proprietor of the flourishing village of Ithaca. Though wary of Yankees and their works, in abundant evidence along the Mohawk, Clinton praised evidences of industry and progress wherever he found them. From the pages of his journal there emerges overwhelming evidence of great activity, the corridor of settlement along upper New York State bustling with people as Elkanah Watson's vision of the future had foreseen.

And yet, because of the Embargo Act then in effect, Clinton's journal also records evidences of divisiveness; political, yet with a regional coloration. As a long arm of New England, the Mohawk Valley had become a stronghold of what Clinton—then a Jeffersonian Republican—described as "violent" Federalism; and as Jefferson had already discovered, the northern lakes provided convenient avenues for contraband traffic—a New York tradition dating back to the French and Indian Wars. Even the makeup of the commission reveals the characteristic party-split along the Mohawk: Clinton and Colonel Peter B. Porter were Republicans while Gouverneur Morris and Stephen Van Rensselaer were Federalists.

Whatever his political affiliation, Gouverneur Morris was a man of very large geopolitical views, who contributed a visionary dimension to the commission's survey. Where the majority of the commissioners agreed with the American-born civil engineer hired for the survey, James Geddes, that the new canal to be dug from the Hudson to Lake Erie should involve a traditional system of feeders and locks, Gouverneur Morris had a much more novel, even spectacular, plan. Morris, Clinton noted tersely in his journal, was "for breaking down the mound of Lake Erie, and letting out the waters to follow the level of the country, so as to form a sloop navigation with the Hudson, and without any aid from any other water" (54). It was not that Morris was unfamiliar with Brindley's adage concerning the proper use of rivers; quite the reverse. As Geddes later testified, "he seemed to have caught much of that spirit of the celebrated Brindley, who would make tunnels, high embankments, almost anything to avoid lockage," and like Brindley he stuck to his hobbyhorse with such "pertinacity . . . that it was almost impossible to call his attention to the impracticability of such a thing" (Hosack: 266).

In 1800, having recently returned from Europe, Morris had earlier made the ritual passage to Lake Erie, and was inspired to expansiveness by the sight of so large a pond: "Here," he wrote to a friend in Germany, "the boundless waste of waters fills the mind with . . . astonishment," but the lake's size was best measured by the sight of nine vessels "riding at anchor . . . the least of them 100 tons. Can you bring your imagination to realize this scene? Does it not seem like magic? Yet this magic is but the early effort of victorious industry. Hundreds of large ships will in no distant period bound on the billows of those inland seas. Shall I lead your astonishment to the verge of incredulity? I will: know then, that one tenth of the expense borne by Britain in the last campaign, would enable ships to sail from London through Hudson's River into Lake Erie. As yet, my friend, we only crawl along the outer shell of our country" (Hosack: 257). Morris was silent in his letter as to the means by which ships would be conveyed from the river to the lake, but in 1803 he confided his scheme to Simeon De Witt, who "very naturally opposed the intermediate hills and valleys, as insuperable obstacles." Morris' answer was, in substance, "that the object would justify the labour and expense, whatever that might be," and the seven intervening years had not changed his mind (261).

Morris' Herculean scheme served chiefly to impede the work of the other commissioners, and though he was overruled by them, such was his "pertinacity" that, as the author of the commission's first report to the legislature, in 1810, the stylist of the Constitution managed to smuggle in his pet scheme once again. As a contemporary witness put it, "the rare project of a canal on so extensive a scale . . . caused much opposition to the undertaking" (271–72). The following year, the other commissioners gave the revised report a more careful reading, and Morris' grand plan thenceforth disappeared from view, much as the man himself withdrew from politics, disillusioned by the direction the country was taking under Republican rule. Yet, in a metaphorical sense, the completion of the canal that was commenced in the year of Morris' death (1816) realized his vision, for though Geddes' system of locks was preserved, the Grand Canal tapped Lake Erie, drawing the flow of commerce and lake water to the Hudson. Whatever their reservations concerning the democratic impulse in America, visionary Federalists like Morris and Elkanah Watson responded to what can only be called the geopolitical imperative, most particularly when it manifests itself as a natural phenomenon.

iv

"On the 22nd of October, 1819," recorded Elkanah Watson, "the first boat sailed on the Erie canal, from Rome to Utica"; and on the following day, carrying the president and the board of canal commissioners "with a band of music," the boat ("called the *Chief Engineer*") returned to Rome, sent off by "the ringing of bells, the roaring of cannon,—and the loud acclamations of thousands of exhilarated spectators . . . who lined the banks of the new created river. The scene was truly sublime" (*History:* 79). As each new section of the Grand Canal opened, the magnitude of the celebrations would expand, but nothing perhaps would match the wonder of that first day, as progress itself seemed to take visible shape, even as (in Watson's words) "new agricultural societies were exhibiting in every direction," nor could cannons or crowds match the creation of the "new created river" for "electrification."

Watson quotes entire a letter that appeared in the *Albany Daily Advertiser* by "a gentleman in Utica" (perhaps himself) that attempts to capture the great moment:

> *On Friday afternoon I walked to the head of the grand canal, the eastern extremity of which reaches within a very short distance of the village, and from one of the slight and airy bridges which crossed it, I had a sight that could not but exhilirate* [sic] *and elevate the mind. The waters were rushing in from the westward, and coming down their untried channel towards the sea. The*

> *course, owing to the absorption of the new banks of the canal, and the distance they had to run from where the stream entered it, was much slower than I had anticipated; they continued gradually to steal along from bridge to bridge, and at first only spreading over the bed of the canal, imperceptibly rose and washed its sides with a gentle wave. It was dark before they reached the eastern extremity; but at sunrise next morning, they were on a level, two feet and a half deep throughout the whole distance of thirteen miles. The interest manifested by the whole country, as this new internal river rolled its first waves through the state, cannot be described. You might see the people running across the fields, climbing on trees and fences, and crowding the bank of the canal to gaze upon the welcome sight. A boat had been prepared at Rome, and as the waters came down the canal, you might mark their progress by that of this new Argo, which floated triumphantly along the Hellespont of the west, accompanied by the shouts of the peasantry, and having on her deck a military band.* (80–81)

There is a marked contrast here, an internal inconsistency, between the relatively sluggish progress of the "rushing" waters and the excitement of the crowd as it followed the advancing tide, nor does the phenomenon as described seem to warrant the rhetorical comparison to the "Hellespont." Yet for Watson and Gouverneur Morris (to whose "gigantic, and expanded mind" Elkanah generously credits the original idea for the Erie Canal), the mere appearance of the advancing water was a shallow token of the great force that was being released. As in the contingency of the *Chief Engineer* and the *Argo*, a new kind of epic was being enacted, to which the future destiny of not only New York but an emerging nation was linked. It was in 1819 also that the sternwheeler *Western Engineer* set off up the Missouri, breathing smoke and fire, westward-moving steamboat and eastward-moving canal boat providing a fit emblem for the expanding energies of a dynamic republic, for whom the Civil Engineer was emerging as a regnant symbol, a logical extension of the Surveyor as represented by George Washington.

XII

With Music Loud and Long

i

The long views (and lengthy) expressed in Monroe's message accompanying his veto of the bill proposing a national system of internal improvements were part of the expansive mood that followed the otherwise not very satisfactory conclusion of the War of 1812. Her boundaries now assured, her integrity validated, her sanctity twice blessed, the United States began to feel truly national and to think nationalistic thoughts. Thus the years immediately following war's end were characterized by an intense feeling of geopolitical euphoria, inspiring expressions of trans-sectional unity, sentiments approximating Jefferson's phrase in his first inaugural address: "We are all Republicans. We are all Federalists." This period has been called the Era of Good Feelings, but as with Jefferson's first administration, the mood was like that which attends the consumption of champagne, celebratory but necessarily brief; for Washington's dreaded, two-faced specter of section and faction was not dead, only sleeping. As a period, therefore, the years of James Monroe's administration are perhaps most interesting because of what followed after, being a demonstration that good will among men lasts only so long as the general welfare and individual interests remain in balance.

Still, while the euphoria lasted, Americans seemed to be having a good time, and enjoyed an extended bout of self-gratulation that provided a comic interlude between the war and the party wrangles that soon followed after. The mood was literally celebratory, for the Era of Good Feelings witnessed four notable public festivities: the tour of President Monroe in 1817; Lafayette's tour in 1824–1825; the opening of the Erie Canal in 1825; and the Jubilee of Independence, in 1826. Each of these celebrations had its own particular meaning, but all four were vehicles for the expression of a general exhilaration, and taken together they may be seen as manifestations of an art form best calculated to memorialize the national optimism. Celebrations are a kind of public theater, being the "spontaneous" outpouring of emotions that by necessity must be carefully arranged beforehand, lest the enthusiasm of crowds result in disaster—as during the "Fifth Celebration" that put a decided end to the Era—Jackson's Inaugural Brawl. The celebration is an art form particularly (though hardly peculiarly) American, being a popular manifestation carefully controlled by an elite. It is a tradition of oligarchical orchestration that dates back to the time of the Revolution and is epitomized by that uniquely American secular festivity, the Fourth of July.

In the late eighteenth and early nineteenth centuries, such events were associated with the kind of artworks requiring planning and production—illuminations, panoramas, allegorical paintings, floats, triumphal arches, fireworks—often created or engineered by professional artists. The celebration also inspired (usually through the Muse of Commission) original works of literature—poems, speeches, even dramas—yet these works were by nature ephemeral (like the fireworks they accompanied) and seldom outlived the occasions for which they were created. Moreover, if they somehow managed to transcend the momentous event, they tended afterwards to lose meaning and impact, there being no luster to a diamond once the light is removed. But we miss an important aspect of the culture of the day if we ignore or dismiss as merely transient these celebrations and the works they inspired. Not only were they signal manifestations of the good feelings assigned to the period, but we can detect by a careful examination of extant records numerous clues to the inner complexity of what might appear to be a simple outpouring of joy.

For one thing, though each of these four celebrations was at least in part inspired by past events—immediate or of some duration—they all projected the euphoria of completion into the future, and are characterized by that same enthusiasm for progress found in the visionary poems of an earlier generation: the "Rising Glory" tradition of Freneau, Barlow, and Timothy Dwight. They are attempts to harness Good Feelings as a kind of public power, urging on the sorts of improvements that will produce even Better Feelings, the kind of ultimate joy associated with the Enlightenment version of the millennium. As for the details of the demonstrations and displays that characterized all four celebrations, these were coeval with George Washington, whose inaugural journey from Philadelphia to New York in 1789 was accompanied by a massive mounting of tableaux and transparencies, salutes by muskets and cannon, triumphal arches and chauvinistic mottoes. And yet, by 1789 such demonstrations had become a national reflex dating from pre-Revolutionary days when similar devices were used to express proto-nationalistic fervor and anti-British sentiments, and the celebrations attending Washington's progress were but a repetition of the public festivities attending the adoption of the Constitution.

Very little of anything produced during the Augustan Age of Dr. Johnson (and Dr. Franklin) did not have some kind of utilitarian basis, and James Wilson, who had argued so persuasively for the adoption of the Constitution in Pennsylvania, took advantage of the celebration of its adoption in Philadelphia to point out the didactic value of such public demonstrations: "They may *instruct* and *improve,* while they *entertain* and *please.* They may point out the elegance or usefulness of the sciences and the arts. They may preserve the memory, and engrave the importance of great *political events.* They may represent, with peculiar felicity and force, the operation and effects of great *political truths"* (Silverman: 578–79). Here again we have the Horatian principle operating in aesthetic matters central to the business of a republic, and there is no more apt instance of the detailed operation of that principle during the first half-century of the republic than the activities of the Horace of New England, Elkanah Watson.

After his first impulsive display of merino sheep in Pittsfield in 1807, Watson came quickly to understand the value of public demonstrations for achieving his desired end of agricultural improvements; and from sheep he went to pigs, and from pigs to pickerel, private experiments that culminated in his first "Berkshire Cattle Show" in 1810. To promote this event, Watson organized a "procession"—in effect a primitive parade—which he later described as "splendid, novel, and imposing, beyond anything of the kind, ever exhibited in America. It cost me an infinity of trouble, and some cash, but it resulted in exciting general attention in the Northern States" (*History:* 123). A pioneer in so many activities, Watson must be given his due as America's first promoter—in the modern sense of that word—seeking always for novel methods by which "to give a powerful impulse on the public mind," a phrase that links him to the mechanical psychology of the day (to which Locke had provided the key), as well as to the Pavlovian theories of more recent times (124). "The grand secret," as Watson put it, "in all our operations, was to trace the windings of the human heart, and to produce *effects* from every step" (126).

Watson's cockades of wheatstraw, his "Farmer's Holidays" and "pastoral balls," his "stages [floats] drawn by oxen, having a broadcloth loom and a spinning jenny, both in operation by English artists, and another stage filled with American manufactures," his "pens handsomely filled with many excellent animals," his prizes of "silver plate," all were designed "not only to excite an ardent spirit of emulation" (the only ardent spirit of which he approved) "but to impress the minds of the audience, and produce some tincture of envy, so as to call forth more extended efforts" (123, 142).

True, this was pastoralism with an infernal mechanism inside it, envy being second cousin to original sin, and Watson's Jeffersonian ox cart carried a Hamiltonian loom; but (as Yankees are wont to boast) it worked:

> *Few [had] felt much interest in the proceedings [of the Agriculture Improvement Society] since little confidence was produced, as to any salutary or permanent utility. Hence, a different organization, to seize upon the human heart, to animate, and excite a lively spirit of competition, giving a direction to measure of general utility, became indispensable. To do this,—some éclat was necessary—music, dancing, and singing, intermixt with religious exercises, and measures of solidity, so as to meet the feelings of every class of the community, and keeping a fixt eye on the main object, all tend to the same great end, promoting agriculture, and domestic manufactures.* (160)

Watson often used a church as an auditorium for the ceremony of bestowing awards, a combination of necessity and design, for if the only available building of any size was generally ecclesiastical, he was enabled thereby to borrow a measure of grace. Still, the movement he organized was in essence quasi-religious, a marriage of theatrics and temple designed to effect "a political Millennium," a secular religiosity that may be witnessed also in the four celebrations staged during the Era of Good Feelings. All contain a subtle combination of utilitarian ends and idealistic sentiments, resulting in things of beauty with a definable practical base, once again testifying to the spirit of Horace moving over the face of American waters.

ii

Thus the declared purpose of President Monroe's tour was to inspect the fortifications along the Atlantic seaboard and the Great Lakes so that he might the better recommend their repair and maintenance, thereby assuring the nation that the commander in chief was vigilant to ensure that the depredations of the recent conflict would never be repeated. But, as with similar presidential inspections of years to come, the event was as political as logistical in nature, and though Monroe's purpose may have been as modest as his equipage when he set out, the eventual outcome of the journey was grandiloquent. Most authorities agree that in making his tour, Monroe was imitating Washington's presidential visit to the Northeast in 1789; both were undertaken at least in part to promote good will and cement a more perfect union. Moreover, though the costume was customary with him, Monroe could not have been completely unconscious of the figure he was cutting with his anachronistic knee-breeches and buckles. That dress has often been regarded as a quaint sign of his old-fashioned republicanism, but it can also be seen as an acute political gesture, identifying him with the older generation of patriots.

As Monroe passed through town after town he was met by delegations of Revolutionary War veterans and committees from local Societies of the Cincinnati, to whose formal statements of greeting he responded with equally formal (if improvised) declarations of comradeship and brotherhood. Not until Lafayette's much more elaborate and considerably enlarged tour (which was undertaken at Monroe's invitation) was there such a widespread indulgence of patriotic fervor imbued with nostalgia for a war that (unlike the most recent conflict) had ended in a decisive American victory. But the Virginian's sortie into New England was also undertaken as a placating gesture to the Federalists, and the word "Union" fell so often from Monroe's lips that he soon approximated the spread-eagle on his presidential seal.

Whatever his intention when he left the nation's capital late in June 1817, "traveling in the most private manner," President Monroe returned nearly two months later with a deep conviction that his chief accomplishment had been a public healing. In his second inaugural address, Monroe said he considered himself "rather as the instrument than the cause of the union which has prevailed in the election"; and the same may be said of the view he increasingly held as he made his way into the heart of Federalist territory, where, he reported to Thomas Jefferson en route, he saw demonstrated everywhere "respect for our Union & republican institutions" (*Messages*: II: 86; *Writings*: VI: 27).

Something of Monroe's self-revised purpose is suggested in a speech he gave at the far end of his eastern tour, in Kennebunk, Maine, an *extempore* address that began by explaining how the popular will had turned him from his original intention of attending only to "public and national objects" to a much more transcendent errand, the nature of which was expressed by what followed next in his speech: a series of phrases that placed him in a central position amongst a newly united people. Monroe said that he shared the "confidential hope, that a spirit of mutual conciliation may be one of the blessings, which may result from my administration," and he prayed that the hope would be realized:

> *The United States are certainly the most enlightened people on earth. We are certainly rapidly advancing in the road of national pre-eminence. Nothing but union is wanting to make us a great people. The present time affords the happiest presages that this union is fast consummating. It cannot be otherwise. I daily see greater proofs of it: the further I advance in my progress, in the country, the more I perceive that we are all Americans—that we compose but one family—that our republican institutions will be supported and perpetuated by the united zeal and patriotism of all. Nothing could give me greater satisfaction than to behold a perfect union among ourselves—an union, which I before observed, is all we can ever want to make us powerful and respected—an union too, which is necessary to restore to social intercourse its former charms, and to render our happiness, as a nation, unmixed and complete. To promote this desirable result requires no compromise of principle, and I promise to give it my continued attention, and my best endeavours. For the good of our common country, I feel that I am bound constantly to act.* (*Narrative*: 149–50)

But if Monroe went more than halfway to deliver sentiments welcome in New England, its citizens amply returned the favor, thereby compensating for their threat to secede at the start of the recent war. It was a suitable irony, therefore, that the steamboat carrying Monroe and his party into the Northeast was named *Connecticut* (Captain Bunker commanding); and it was in Hartford, seat of the notorious Federalist Convention, that Monroe's tour began to take on the look of a triumphal progress. It was there also that Monroe's increasingly augmented party encountered for the first time the decorations that would likewise be augmented as he continued on through New England: "The whole of the immense assemblage of soldiers and citizens," wrote S. Putnam Waldo, author of the "official" account of the tour, "escorted the President over the city bridge, which was elegantly ornamented with three lofty arches, thrown over it, composed of evergreen and laurel, imitation of the triumphal arches of Rome, under which the benefactors of the Commonwealth passed" (92).

The quality of Connecticut's enthusism was attested to by the recent election of Governor Oliver Wolcott, the once-fervent Federalist who had connived with Hamilton against John Adams, but who had more recently come to see the Republican light. This latter event lends additional meaning to "the elegant flag" waving "over the arches," which bore the "letter M; it being the same letter reversed, which on election day meant WOLCOTT" (93). Yankee ingenuity in the service of thrift is a constant nineteenth-century phenomenon, but in this case it suggested that a number of significant reversals were taking place in the former stronghold of Federalism; a decade hence, the "W" could stand for "Whig."

Of all the decorative symbols encountered during Monroe's tour, none was more ambitious than the elaborate effort that greeted the President as he crossed over the Presumpscot River into Westbrook, Maine, the bridge having been embellished with nineteen arches of evergreens and roses, each arch surmounted with a wreath, "one for each state, with the name of the state in large letters on the top of the arch. A twentieth was erected as symbolical of the union," which was surmounted by yet another arch, dressed also in evergreen,

> *and emblazoned in front with nineteen brilliant stars. A living Eagle, a native of our own forests, and the symbol of our martial prowess, perched on the summit of the twentieth arch, and under the canopy of stars, by which it was surmounted, apparently watching, with intense scrutiny and surprise, the concourse of people passing under him, heightened in the bosom of every beholder the interest of this lively spectacle. It was a delightful sight to behold this haughty monarch of the feathered tribe, the pride of the forest, encircled by the blaze of the stars he loves, stifling, for a moment, his untamed spirit of liberty; and gratefully spreading his pinions, as the chief of the nation passed, which had chosen him from the whole range of animated nature, as the emblem of its glory and strength.* (177–78)

S. Putnam Waldo writes with all the disinterestedness and constraint of a campaign biographer, and it is difficult to ascertain who has been "chosen," the eagle or the President, but in this case the confusion merely heightens Monroe's own sense of himself as "instrument" of a more perfect union.

Like the inverted "W" that greeted him in Hartford, the decorated bridges across which Monroe passed as he made his progress through New England may have been inspired largely by convenience; yet like the arches that made up the chief decoration, the bridges conveyed a symbolism in keeping with the occasion. From the earliest days of the Puritans, bridges were an important aspect of progress in New England, where the east-to-west movement of settlement and commerce necessitated numerous crossings of rivers and streams, most of which run north-to-south, a pattern entirely different from that which obtained along the accommodating rivers of Monroe's Virginia. From Edward Johnson's observation, in his *Wonder-Working Providence* (1654), that "the constant penetrating farther into this Wilderness, hath caused the wild and uncouth woods to be fil'd with frequent ways, and the large rivers to be over-laid with Bridges passeable, both for horse and foot," to Timothy Dwight's joining "vast bridges" to "long canals" in the visionary conclusion of *Greenfield Hill* (1794) as "patriot works," bridges in New England were regarded as essential to a well-ordered world (Johnson: 234; *Poems*: 517). In his multi-volume *Travels,* Dwight frequently delivered encomniums about the virtues of the bridges over which has passed (e.g., "this is undoubtedly the best bridge which has been erected in the United States"); and S. Putnam Waldo likewise notes that "the toll bridge across the Connecticut River will not suffer by comparison with the first works of this kind in the union" (*Travels*: III: 38; Waldo: 96). Finally, given Monroe's evolving mission—to unify the disparate sections of the Union—the decorated bridges of New England may be seen as something more than local manifestations of civic pride: each bridge that the President crossed in his eastward progress marked yet another symbolic passage cementing the parts of the nation.

In order that his account of Monroe's tour not be merely a scrapbook of excerpted speeches, Waldo resorted to a number of devices to fill out the volume, including his own assumptions about what must have been the workings of the presidential mind as Monroe passed through the land. He described in compact form every major town the President passed through, rendering a brief account of significant architectural features, the local manufacturers and points of interest, and a capsule history of significant local events of national importance—in effect approximating a tour guide of a more conventional kind. All of these digressions have a collective implication in that they constantly stress the element of growth and change, the national progress that Monroe himself evoked as a parallel to his own. This is particularly true of Waldo's occasional descriptions of the landscape through which the President passed, scenery rendered for the economic benefits it promised.

Thus the Hudson is not only "the noblest stream in the Middle, Northern, and Eastern States," but it has a special importance because of its navigation, which "has called the inventive faculties of man into operation upon the most important subject that ever exercised mechanical ingenuity; the navigation of rapid streams" (77). In his orotund way, Waldo is alluding to the steamboat, the invention of which "forms an era in the progress of useful arts." He then passes on to mention a parallel achievement by New Yorkers, "the canal now constructing to unite the waters of the western Lakes, with that of the Hudson" (77–78). While declaring the Erie Canal "one of the grandest schemes ever conceived, and, when accomplished, the greatest ever executed," Waldo ends his encomium lamely, departing from his usual willingness to view the passing scene through Monroe's eyes: "The President while in this place, from his limited tarry, could hardly estimate the importance of it to his extensive native country." Even as Monroe was staying over in Boston, binding the nation's wounds by enjoying the celebration of Independence Day, the first spadeful of dirt was turned that began the great project.

Yet the President, in his return swing, again failed to visit the site, an omission undoubtedly dictated by a number of practical considerations of space and time—which the canal would eventually remove—and hurried on westward, to Detroit and Pittsburgh, where he was warmly welcomed, not as a resurrected Washington, but as the first President of the United States to be seen "upon the western waters," thereby completing the intended circle of his healing errand by identifying himself with the future and the West. Despite his self-cast role as the national binder of sectional wounds, James Monroe the Virginian could not easily smile upon the merger of waters connecting Lake Erie to the Hudson River, but preferred to bask in his fame in Pittsburg as "the early, uniform, active friend of the western country, who . . . secured to us the invaluable right of free communication with the ocean through the Mississippi, an attainment second only to national independence itself, and inseparably connected with it" (256).

iii

Even to details of costume, President Monroe's tour into the northeastern states was a literal dress rehearsal for the much greater swing through the entire circle of states performed a few years later by Marie Joseph Paul, Marquis de Lafayette. Where Monroe by his dress invited comparison with the Founding Fathers and by his Virginian origins, with Washington, Lafayette was both a bona fide sire of the nation and an adoptive son of the greatest Father of all. Where Monroe's journey was by administrative necessity abbreviated in extent and duration, Lafayette, carried on by wave after wave of increasing public adulation, like many another actor before and after him kept extending his farewell tour until he had visited all twenty-four states, had spent more than a year in the process, and had at last received a full measure of *la gloire,* having been virtually apotheosized as the personification of America's rising republican glory, circa 1775.

Where Monroe had sought to bind the nation's wounds and unify Republicans and Federalists to provide a solid geopolitical base for future expansion and prosperity, Lafayette (whatever his intentions) was made over by his hosts into a figment of an idealized past. Styled "the people's Guest" (suggesting his passive role), he was a ghost also, a figure resurrected from a glorious past, and was carried from city to city like a precious relic or plaster saint, for although the general was very much alive, his life was chiefly associated in America with the nation's past. Lafayette borrowed glory from the events in which he had taken part, but he was valued also as a living witness to the marvels of the present, to which he obligingly and repeatedly bore testimony. In him the glorious past of the republic was joined to the progress evinced by a miraculous rate of change, both symbolized by Lafayette's passage through the land. If Monroe stood for "Union," a Virginian extending a friendly and forgiving hand to his New England–Federalist countrymen and a glad hand to future states beyond the Alleghenies, Lafayette was an embodiment of "Nation," that mystic combination of hallowed past and wonderful present essential to future movement of an imperialistic cast.

An actor in the War for Independence that was now a matter of history, Lafayette in 1825 was conceived of by Americans chiefly as an audience of one, and as he passed through a vast panorama of progress his attention was repeatedly called to the changes that had taken place since his last visit forty years earlier. Thomas Hart Benton, looking back from the vantage point of a quarter of a century, regarded Lafayette with the eyes of a young American in 1825. "He saw his own future place in history, passing down to the latest time as one of the most perfect and beautiful characters which one of the most eventful periods of the world had produced" (30). But then Benton went on to quote from Henry Clay's famous address as Speaker of the House to the Guest of the Nation, the rhetoric of which was alive with a sense of Lafayette's singular role as witness:

> *The vain wish has been sometimes indulged, that Providence would allow the patriot, after death, to return to his country, and to contemplate the intermediate changes which had taken place—to view the forests felled, the cities built, the mountains leveled, the canals cut, the highways opened, the progress of the*

> *arts, the advancement of learning, and the increase of population. General! your present visit to the United States is the realization of the consoling object of that wish, hitherto vain. You are in the midst of posterity!* (30)

Like Barlow's Columbus comforted in his cell, Clay's Lafayette is treated to a vision of spectacular progress, the result of the terrific energies of the American people.

Lafayette's tour was widely reported by the exuberant, chauvinistic press of the day, and a narrative of Lafayette's journey was written by Auguste Levasseur, who accompanied him on his travels; but the most significant literary use of Lafayette's Tour was by James Fenimore Cooper in his *Notions of the Americans* (1828), which is in some ways a complex version of Waldo's *Tour of James Monroe*. For where Waldo linked the various speeches and celebrations by a thin concourse of digressive commentary, Cooper subsumed the events of Lafayette's year in the United States in a dissertation on the character of the Americans and their institutions. Cooper, according to Lafayette's own declaration, "was one of the first New York friends I had the gratification to take by the hand," and he frequently entertained his American friend during Cooper's subsequent tour of France (*Corres:* I: 100). *Notions* was written in part as a response to Lafayette's request that Cooper put together an account of his tour, a contract that the American writer declined, preferring to use the general for his own purpose: he appears and reappears in Cooper's extended, semi-fictitious travelogue, like a famous personage in a historical romance, a stock figure connoting national glory.

Significantly, the only speeches recorded by Cooper out of the hundreds generated by the event were those delivered by Clay and Lafayette on the occasion of the general's visit to Washington, D.C. For Cooper's emphasis, like Clay's, is on the panorama of progress, the remarkable evidence of the accomplishment achieved by the unchecked energies of the republic, and like the tours of Monroe and Lafayette, Cooper's *Notions* is an expression of the Era of Good Feelings. It has been criticized for its uncritical view of America, for in attempting to counter the prejudiced reports of British travelers, Cooper described an impossibly virtuous nation. Yet, Cooper's *Notions* is best taken for what it is, not what it might have been. Cooper's book is a gauge of the nationalistic feelings engendered by Lafayette's visit, a boisterous spirit of republican optimism that has as its epicenter the moving point of the general's passage through America.

Cooper would eventually become disillusioned with the direction taken by Jacksonian democracy—and by the republic itself—but in

1828 his *Notions of the Americans* brims with the good feelings, on whose rising tide the author had set sail for Europe two years earlier. His European experience had given him a certain cosmopolitanism, but served, if anything, to intensify his republican ardor. Though *Notions* takes the form of letters from America written by a European, the "traveling Bachelor" is given an American companion and commentator, "John Cadwallader, of Cadwallader, in the State of New-York," who is something more than a thinly veiled "James Cooper, of Cooperstown," though the two share opinions as well as initials. For Cadwallader is both a composite and a consensus figure, a paragon of republican virtue: "Truly, there was something so naïf, and yet so instructed—so much that was intellectual, and yet so simple—a little that was proud, blended with something philosophical, in the temperament and manner of this western voyager, that he came over my fancy with the freshness of those evening breezes . . . on the shores of the Dardanelles. To be serious, he was an educated and a gifted man, with a simplicity of thought, as well as of deportment, that acted like a charm on my exhausted feelings." (I: 2). "Educated and gifted," Cadwallader is also an apostle of "common sense"—of the sort espoused by Franklin, not Tom Paine—and the phrase appears throughout the book like a refrain: according to Cadwallader, in America common sense has been established "as the sovereign guide of the public will" (170).

To this composite portrait, this confederated American, Cooper adds another, almost mystical element, making of John Cadwallader a virtual personification of republican virtue. The magical moment takes place at Mount Vernon, where the Bachelor and his American friend have gone in the company of Lafayette, traveling down the Potomac by steamboat. While Lafayette visits Washington's tomb, the Bachelor and Cadwallader are "permitted, by an especial favour" to visit the house and its grounds—not yet a national shrine but the home of Bushrod Washington (II: 186–87). The house is "entirely empty," yet it seems to the visitors to be filled with the presence of the dead hero: "More than once, as my hand touched a lock to open some door," records the Bachelor, "I felt the blood stealing up my arm, as the sudden conviction flashed on my mind that the member rested on a place where the hand of Washington had probably been lain a thousand times. That indescribable, but natural and deeply grateful, feeling beset me, which we are all made to know when the image of a fellow-mortal, who has left a mighty name on earth, is conjured before us by the imagination in the nearest approaches to reality that death, and time, and place, and the whisperings of an excited fancy will allow" (187).

Inspired by "a secret desire, rather than an expectation, of finding something more than what reason told me to expect," the Bachelor moves

> *from parlour to parlour, in my haste, until my companions were left behind, and I found myself alone in the upper office of the mansion. I shall never forget the sensation that I felt, as my eyes gazed on the first object encountered: It was an article of no more dignity than a leathern firebucket; but the words "Geo. Washington" were legibly written on it in white paint. I know not how it was, but the organ never altered its look until the name stood before my vision distinct, insulated, and almost endowed with the attributes of the human form. The deception was aided by all the accessories which the house could furnish. Just at that instant, my friend, who is a man of tall stature and grave air, appeared in the adjoining door, without speaking. I felt the blood creeping near my heart with awe, nor did the illusion vanish until Cadwallader passed before me, and laid a hand, with a melancholy smile, on the words, and then retired toward the grounds, with a face that I thought he would gladly conceal.* (188)

This moment occurs as Lafayette is paying his respects to his dead comrade-in-arms and adoptive father, a contingency given further meaning by the received sense in America that Lafayette's visit was a resurrection of sorts: "To most of those whom he passed," writes the Bachelor, "His form must have worn the air of some image drawn from the pages of history" (I: 45).

Lafayette is less a human presence than a monument, "an imposing column that had been reared to commemorate deeds and principles that a whole people had been taught to reverence," an idea graphically conveyed in a portrait painted during his visit by Samuel F. B. Morse, in which the hero is seen at the top of a flight of stairs, pausing before he enters a building of neoclassical design. His hand rests on a pedestal displaying the marble busts of Washington and Franklin, and we may assume that the great general is waiting, in old age, on the porch to the pantheon of American heroes. In Cooper's *Notions,* likewise, Lafayette appears generally at a remove, idealized, noble, radiating dignity and virtue but from a distance; while in John Cadwallader the American past is given new life, assuring us that the nation's glorious history has a continuity, that the spirit of Washington and Franklin will not pass from the land. Lafayette is a figment of the past, but Cadwallader is a living representative of the present moment who points forward to the future. His very intitials suggest a sacred, even a saviour-like, role, yet put forth also a secular, even imperial, second meaning, as a reincarnation of Republican-Federalist unanimity on the Roman model. If the

French General is the Guest, then the British Bachelor is his surrogate, the Visitor; while John Cadwallader is the Host, with all that the word implies.

iv

Counterpart to his river voyage down the Potomac to Mount Vernon is the trip taken by Lafayette up the Hudson River, the general and his party boarding the steamboat *James Kent* in New York City for Albany. Though they are, once again, not in Lafayette's party, the Bachelor and his American guide accompany him on the voyage, which gives Cooper an opportunity, through the eyes of his British narrator, to render a description of "the glorious scenery of this renowned river" (I: 203). In the Bachelor's informed opinion, "having seen the Rhine, the Rhone, the Loire, the Seine, the Danube, the Volga, the Dneiper, and a hundred others," the Hudson, or "the noble river of the north, as it is often called in this country," loses nothing but rather gains by the comparison, in that it "embraces a greater variety of more noble and more pleasing natural objects, than any one of them all" (202, 203). By means of his traveling Bachelor, clearly a European mouthpiece, Cooper imposes a series of aesthetic imperatives on the natural course of the Hudson, sectioning it off into "divisions" according to an arbitrary scale of landscape views, each of which may be detached from the whole and hung upon a wall.

The Bachelor's divisions approximate Pownall's description of the different stretches along the Hudson, though by 1824 the Palisades that earned the governor's admiration had been turned to practical uses, as witnessed by the "quarries which are wrought in its side," while, opposite, the pastoral scene still prevailed: "Orchards, cattle, fields of grain, and all the other signs of a high domestic condition, serve to heighten the contrast of the opposing banks" (203–4). And yet this "first division" is not particularly satisfying aesthetically, for though "the parapets and their rival banks form a peculiar scenery, the proportions of objects are not sufficiently preserved to give to the land, or to the water, the effect which they are capable of producing in conjunction. The river is too broad, or the hills are too low" (204). But once the river passes within "the Highlands, the objection is lost," for this, the "second division," is "beautifully romantic," being a region of "broken and irregular" mountains, "which nature had apparently once opposed to the passage of the water" (204). The "third division" is a "succession of beautiful lakes" above the Highlands, any one of which "may compete with any of Italy, and to one in particular there is a noble back-ground of mountains . . . which are thrown together in splendid confusion" (203). We have clearly (and literally) moved from an eighteenth-century aesthetic into high Romantic values, where worth is identified with wildness.

But it is in the fourth and last division, north from the town of Hudson to Albany, that the noble stream finally "acquires more of the character of a river according to our European notions," resembling the Seine "above Caudebec," with scenery that is "picturesque and exceedingly agreeable" (205–6). And yet what the Hudson still lacks, even here, is not a purely artistic but an historic perspective, for it does not boast the "inexhaustible recollections" of rivers like the Rhine, the historic associations prized by European aestheticians like Sir Archibald Alison. Only "time" can render it "unequalled" by Europe's ancient streams. And yet, the Hudson "is not without a character of peculiar moral beauty," for if ruins are lacking, "improvements" are not: "The view of all the improvements of high civilization in rapid, healthful, and unequalled progress, is cheering to philanthropy; while the countless villas, country-houses, and even seats of reasonable pretensions, are calculated to assure one, that, amid the general abundance of life, its numberless refinements are not neglected" (206).

This was likewise the opinion of that bona fide European traveler and letter-writer, Crèvecoeur, nor was Lafayette blind to the charms of progress as an aesthetic phenomenon. Moreover, after the Bachelor and his American friend leave the Hudson at Albany (at which point they also part company with Lafayette), the emphasis on the moral (and aesthetic) qualities of progress increases. "Our faces were turned west," notes the Bachelor; and something similar happens to the Englishman's sensibility, who turns from viewing America through foreign eyes to correcting European misconceptions of the New World. In the Old World, observes the Bachelor, they "picture their images" of America from what they hear, and since "they hear of churches, academies, wild beasts, savages, beautiful women, steam-boats, and ships," they bring together "wolf, beauty, churches, and sixty-gun frigates in strange and fantastic collision" (244–45). The truth is quite different, for even in a "thriving settlement" in "what is called the new country . . . the only difference between the aspect of things here and in Europe, is in the freshness of objects, the absence of ancient monuments, the ordinary national differences in usages and arrangement, an air of life and business, always in favour of America, and a few peculiarities which blend the conveniences of civilized life with the remains of the wilderness" (245). In sum, the American scene differs from the European in two particulars, the absence of ruins and the presence of energy, for though the landscape lacks visible evidence of man's former presence—anything of historic significance—it is teeming with signs of present activity, thereby auguring much for the future.

Tocqueville came to America in 1831 expecting to find a civilization that took its character from the frontier—much as it had been described by Crèvecoeur—but discovered that the frontier, at least along the line defined by the Mohawk Valley, took its character from civilization; and this is the Bachelor's observation, too: where the "tide of emigration has set steadily towards any favoured point for a reasonable time, it is absurd to seek for any vestige of a barbarous life among the people. The emigrants carry with them . . . the wants, the habits, and the institutions of an advanced state of society" (I: 245–46). In the "new country" the "shop of the artisan" is put up at the same time as "the rude dwelling of the farmer," and the log cabin is soon replaced by "dwellings of wood, in a taste, magnitude, and comfort, that are utterly unknown to men of similar means in any other quarter of the world" (246). "The little school-house is shortly erected at some convenient point, and a tavern, a store . . . with a few tenements occupied by mechanics, soon indicate the spot for a church, and the site of the future village. From fifty or a hundred of these centres of exertion, spread swarms that in a few years shall convert mazes of dark forests into populous, wealthy, and industrious counties" (246).

A paradigm of this emerging American scene is provided by a description of the Mohawk Valley as seen from "the very brow" of a mountain, a spot permitting "an uninterrupted prospect . . . that could not have been contained in a circumference of much less than two hundred miles" (248). The scenery is chiefly characterized by "an equal distribution of field and forest . . . in proportions which, without being exactly, were surprisingly equal," and the appearance, from a distance, is similiar to that of "a fertile French plain, over which agriculture has been conducted on a scale a little larger than common. You may remember the divisions formed by the hues of the grains, of the vineyards, and of the grasses, which give to the whole an air so chequered and remarkable. Now, by extending the view to the size I have named, and enlarging these chequered spots to a corresponding scale, you get a tolerably accurate idea of what I would describe" (248). As on the Hudson, the Bachelor's aesthetic is dictated by European norms: the Mohawk Valley resembles a French plain. It pleases because it is an orderly and balanced mixture, an Old World (neoclassical) ideal. Only its magnitude distinguishes it as American.

The dominant feature of the landscape is provided by signs of human industry: "Countless farmhouses, with their capacious outbuildings, dotted the fields, like indicated spots on a crowded map," symbols indicating a topography of plenty, the "fields alive with herds" while "pyramids of hay and of grain" can be seen as far as the eye can travel (249). Only one thing is missing from this scene "to give it the deepest interest," and that is the historical element, "the recollections and monuments of antiquity," the same dimension found lacking on the Hudson. But to this observation John Cadwallader objects: "'You complain of the absence of [historical] association to give its secret, and perhaps greatest charm which such a sight is capable of inspiring. You complain unjustly'" (250).

What follows is essential to understanding the mood of Cooper's fellow Americans as the republic approached the Jubilee of Independence, for John Cadwallader imposes upon the Bachelor's topography a grid of sorts revealing a progressive aesthetic: "'Examine this map. You see our position, and you know the space that lies between us and the sea. Now look westward, and observe how many degrees of longitude, what broad reaches of territory must be passed before you gain the limits of our establishment, and the consequent reign of abundance and civilization'" (250). Where the scene before them is European save for the signs of antiquity, the map held out by Cadwallader is uniquely American, a vast terrain that keys the essential difference between the Old and New Worlds: "'The moral feeling with which a man of sentiment and knowledge looks upon the plains of *your* hemisphere, is connected with its recollections, *here* it should be mingled with his hopes'" (250; italics added). Americans may have their

> *recollections, but they are the sort drawn from the memories of the living. I remember this country, Sir, as it existed in my childhood. . . . Draw a line from this spot, north and south, and all of civilization that you shall see for a thousand miles west, is what man has done since my infancy. . . . That view before you is but a facsimile of a thousand others. I know not what honest pleasure is to be found in recollection, that cannot be excited by a knowledge of these facts. These are retrospects of the past which . . . lead the mind insensibly to cheerful anticipations, which may penetrate into a futurity as dim and as fanciful as any fictions the warmest imaginations can conceive of the past."* (251)

This is the perspective essential to the visionary aesthetic in America, shared by Cadwallader with Crèvecoeur, Barlow, and Elkanah Watson.

The American guide on the mountaintop is playing the role of prophetic angel to the Bachelor's Columbus: "'The speculator on moral things can enjoy a satisfaction here, that he who wanders over the plains of Greece will seek in vain. The pleasure of the latter, if he be wise and good, is unavoidably tinged with melancholy regrets, while here all that reason allows may be hoped for in behalf of man'" (251). Where the European must derive moral feelings by beholding the ruins of the past, with their connotations of the fall of empires, in America moral excitement is generated by "'a rapid and constantly progressive condition,'" a reminder of Thomas Cooper's adage concerning the "falling" condition in Europe and the "rising" of fortunes in America (252; Imlay: *TD*: 186). Cadwallader recalls having stood where the two now stand and having seen nothing below but "'the shadows of the forest,'" yet "'he who comes a century hence'" will find, not farmland and a village, but "'the din of a city rising from that very plain, or find his faculties confused by the number and complexity of its works of art'" (252). Moral meaning in America is derived not so much from what has passed from the landscape as from what will be added to it. The American aesthetic is clearly one of change, of constant revision of the landscape toward one crowded with "works of art," not nature. It is an aesthetic of impermanence, also, for the scene before them is beautiful on the moral plane because it will not be there a century hence.

Much as the Hudson River as described by the Bachelor contains the scenic divisions detected by Pownall, so Cadwallader's American scene is consistent with that progressive schema that likewise filled the Royal Governor with admiration and awe. What is curiously missing from the Mohawk panorama, however, is the river that occupies a strategic place in Pownall's diagram of progress, and is so essential to the meaning and shape of the valley: "The natural river which divides this glorious panorama in nearly two parts, with its artificial rival, and the sweet meadows that border its banks, were concealed beneath the brow of the last precipitous descent" (249). In an earlier part of *Notions,* however, Cooper had already more than made up for this omission, at least concerning the Mohawk's "artificial rival," the "narrow, convenient river" that was the Erie Canal. Like the map of extended settlement, the canal is an index of the unfolding republican plan: to behold it is to "easily understand the perspective of American character" (128). By reference to a map of the United States, the Bachelor demonstrates the instrumentality of the canal in drawing down the Hudson to New York City the produce of the interior regions. "The circumstances to which this town is to be indebted for most of its future greatness, is the immense and unprecedented range of interior which, by a bold and noble effort of policy, has recently been made tributary of its interests"

(126). But it is in the closing pages of *Notions* that the implication of the Erie Canal is most clearly demonstrated—as a paradigm of American enterprise and the national character.

Throughout the book, Cooper, by means both of the Bachelor and Cadwallader, expatiates on the practical and utilitarian bent of Americans, part and parcel of their native common sense. Although "that despot—common sense" works against the success in America of Romantic literature—"the smallest departure from the incidents of ordinary life would do violence to every man's experience"—it has produced much in the way of "useful arts": "Though there is scarce such a thing as a capital picture in this whole country, I have seen more beautiful, graceful, and convenient ploughs in positive use here, than are probably to be found in the whole of Europe united. In this single fact may be traced the history of the character of the people, and the germ of their future greatness" (II: 114–15). These same inventive Americans have perfected an axe that is "admirable for form, for neatness, and precision of weight," and likewise have given "form and useful existence to the greatest improvement of our age," the steamboat, and it is to this class of artworks that the canal also belongs (116).

For Americans, rehearses the Bachelor, "the actual necessities of society supply an incentive to ingenuity and talent" (320). The earliest attempts to build the Erie Canal were made by European engineers, who worked from "rules of science," adapting principles to results that were "calculated to meet the contingencies of European men." Unable "to allow for the difficulties of a novel situation . . . their estimates were always wild, uncertain, and fatal, in a country that was still experimenting." Equally important, the European engineers arrived too early on the scene, and tried to force their opinions and plans on a people not yet ready for canals: "It was wise to wait for a political symptom in a country where a natural impulse will always indicate the hour for action" (320).

When that impulse finally came, it motivated men "of practical knowledge, who understood the people with whom they had to deal, and who had tutored their faculties in the thousand collisions of active life"; not scientific theorists, but men with "the utmost practical knowledge of men and of things" (321). This, then, is the complex yet simple key to understanding the way things are in America, the secret "cause of the prodigious advance of this nation." The Erie Canal was just such a work as was "necessary to demonstrate to the world, that the qualities which are exclusively the fruits of liberty and of a diffused intelligence, have an existence elsewhere than in the desires of the good" (322). The canal proved to the world that the American was in perfect tune with his own condition, that he could "point to his ships, to his

canals, to his bridges, and, in short, to everything that is useful in his particular state of society, and demand, where a better or cheaper has been produced, under anything like circumstances of equality" (322).

With earlier champions of internal improvements, Cooper declares through his traveling Bachelor that "in a country so united in interests, but so separated by distance, a system of extended and easy internal communication is of vital importance. Without it, neither commerce, nor political harmony, nor intelligence, could exist to the degree that is necessary to the objects of the confederation" (324). In a sense, Cooper is living in the future envisioned by those such as Barlow, for in the world of 1826, canals *do* career round every hill, and the landscape *is* filling up with evidence of freedom's productive aegis, at a rate of progress unexampled in human history: "A nation that lives as fast as this, does not compute time by ordinary calculation. . . . The construction of canals, on a practical scale, the mining for coal, the exportation of cotton goods, and numberless other improvements, which argue an advancing state of society, have all sprung into existence within the last dozen years" (328, 332). But it is a future in which poetry finds little place, in which the important arts are those that serve utilitarian ends, for all creation is measured "on a practical scale."

Lafayette last appears in *Notions* in Boston, where he has gone to officiate at the laying of the cornerstone for his granitic counterpart, the Bunker Hill Monument, and though the Bachelor reports the event, he gives equal space in the same letter to an account of Lafayette's review of the firemen of New York City, who "are more expert and adventurous than those of any other town in the world" (II: 307). The intended monument may be "a column of granite . . . worthy of the occasion," but equally worthy and perhaps more meaningful are the fire engines of New York:

(*Facing*) John W. Hill (1812–1878). *Erie Canal Aqueduct Crossing the Mohawk River* (ca. 1831). Preliminary sketch and finished watercolor. (Courtesy of the New-York Historical Society)

Executed for a series of aquatints, Hill's depiction of the aqueduct stresses the triumph of the Line over Nature. The waterwheel on the left-hand side of the picture likewise signifies technological supremacy, as well as emphasizing the Horatian criterion of usefulness. "Picturesque" additions to the composition, like the fishermen on the bank and greenery growing from the stonework (augmented to tower over the wheel), are interesting acknowledgements of contemporary taste, but the overwhelming impression is of linear—hence neoclassical, not romantic—elements. Even the Mohawk, a once-wild river with an Indian name, seems flattened out and subdued by the heavy architectonic presence of the aqueduct.

> *No nobleman's carriage is more glossy, neater, or considering their respective objects, of more graceful form. They are also a little larger than those we see on our side of the Atlantic, though not in the least clumsy. When La Fayette had passed in front of these beautiful and exquisitely neat machines, they formed themselves in a circle. At a signal the engines were played, and forty limpid streams shot upward, toward an imaginary point in the air. It appeared to me that they all reached that point at the same instant, and their water uniting, they formed a* jet d'eau *that was as remarkable for its conceit as for its beauty.* (309)

The concert of firehose fountains was yet another architectonic display for the benefit of the nation's guest, "waters uniting" in a symbol of union. In reading Cooper's description of the modern fire engine, however, it is hard not to think of Washington's humble leather fire-bucket, mute testimony to the progress accelerated by necessity in America. "No rifleman could have sent his deadly messenger with surer aim, than the water fell upon the little torch-like flame" (308–9). The Minuteman of 1776 has become the Fireman of 1826.

It would be Cooper's fate to return in 1833 to an America that had undergone an accelerated leap into the future, and he would become something of a permanent alien-in-residence, the rapid pace of "American time" having left him behind. He became in effect his own Lafayette, a monument of the past still living in the present; but unlike the French hero, he would be a reluctant witness to the panorama of change, a Columbus beyond consolation. His *Notions of the Americans* is therefore of all the more value as a document, freezing as it does a moment of perfect balance and preserving a Cooper who shared his countrymen's mood during the Era of Good Feelings and gave it permanent expression even as it passed away. It is a Cooper, moreover, who imposes on the landscape both an imported, European Romantic aesthetic and a uniquely American ideal, neoclassical in its celebration of balance and proportion but thoroughly modern in its love of utility. If, as he maintained, the Erie Canal was a key to the secret of American life, then Cooper and his book provide an insight into the meaning of the canal, like the fire engine of New York City, an "exquisitely neat machine."

Where Monroe, the Virginian, bypassed the canal and hastened to the West to bask in the warm regard of settlers who remembered his activities on their behalf, Cooper does not follow Lafayette during the general's own western tour, but remains in the east. Monroe ended his trip back in Washington, D.C., on the banks of the Potomac, while Cooper's *Notions* seldom leaves the Hudson Valley, a regional emphasis that the Erie Canal would, at least in terms of geopolitical alignment, help promote. The dominant symbol greeting Monroe during his tour was the triumphal arch, connoting wholeness and unity, but the canal, as Cooper demonstrates, is a linear image: like the individual jets of water making up the firemen's fountain in Manhattan, it moves straight as a rifleman's bullet from the East, and was destined to rival the Mississippi and displace the Potomac as a symbol of "free communication with the ocean" by the West. The Hudson might contain a full panorama of aesthetic possibilities on the European scale, but it was the Erie Canal, so latent with promise for the future, that gave the lordly river its American dimension, a linear diagonal pointing into the West.

XIII

A Miracle of Rare Device

i

The Erie Canal, as John Quincy Adams declared, was made possible by "the persevering and enlightened enterprise" of a "single member of our Confederation," but it was also made possible by a geopolitical freak (*Messages*: II: 316). For if Providence, as Christopher Colles observed, had lowered the Appalachian mountains between the Mohawk River and the Great Lakes, history had given New York State a fortunate shape, a generous wedge of territory that not only encompassed the Hudson River and Lake Champlain, but stretched in a corridor all the way to the shores of Lake Erie. Where other states with transmontane aspirations had to deal with neighboring states, New Yorkers had to deal only with themselves, and though considerable political and economic objections were made to the Erie Canal by residents of the lower (and most populated) regions of the state, statesmen like De Witt Clinton were eventually able to convince a majority that the canal would shower benefits on the entire populace. By 1825, politicians were scrambling to claim credit for the achievement, and Clinton had been unfairly shoved aside in the rush; but the mood attending the official opening of the canal was one of universal jubilation and self-congratulation, and the governor got full credit for his contribution. For the Erie Canal had undeniably been the work of New Yorkers, and its completion had the same effect on the people of that diverse and often divided state that the tour of Lafayette had had on the nation as a whole. Though a regional event, it spoke in large terms of Union.

Much as the excitement engendered by Lafayette's tour considerably overmatched that generated by Monroe's, the opening of the completed canal resulted in a celebration that more than rivalled New York's welcome for the Nation's Guest. It marked the climax of a mood that had been growing ever since Governor Clinton had announced to the legislature in 1820 that ninety-six miles along the western part of the canal were open and doing business. The accomplishment served to silence regional and political opposition, a hush that turned to awe as the canal progressed at a remarkable rate; and then, like unbelievers caught up in a religious revival, former opponents of the project, such as Martin Van Buren, began vying for preeminence in championing it. By October 1823, the entire eastern section had been completed, and the western end had been carried as far as the town of Erie, allowing boats to pass from the waters of the Genesee to the Hudson, inspiring an outburst of joy that was closely followed by a committee of arrangements.

With that penchant for celebrations that was becoming an American instinct, dignitaries and functionaries gathered at Albany in 1823 to observe, literally and figuratively, the opening of the lock that would mingle the waters of the canal with the Hudson; ceremonies that gave a ritualistic formality to the relatively impromptu expressions of joy observed by Watson in 1820:

> *The pencil could not do justice to the scene presented on the fine autumnal morning when the Albany lock was first opened. Numerous steam boats and river vessels, splendidly dressed, decorated the beautiful amphitheatre formed by the hills which border the valley of the Hudson, at this place; the river winding its bright stream far from the north, and losing itself in the distance to the south; —the islands it embraced; —the woods, variegated by the approach of winter, a beauty peculiar to our climate; —the wreathed arches, and other embellishments, which had been erected for the occasion, were all objects of admiration. A line of Canal boats, with colours flying, bands of music, and crowded with people, were seen coming from the north, and seemed to glide over the level grounds, which hid the waters of the Canal for some distance, as if they were moved by enchantment.*
>
> *The first boat which entered the lock was the De Witt Clinton, having on board Governor Yates* [Clinton's successor], *the Mayor and Corporation of Albany, the canal commissioners and engineers, the committees, and other citizens. Several other boats succeeded. One (not the least interesting object in the scene), was filled with ladies. The cap-stone of the lock was laid with masonic ceremonies, by the fraternity who appeared in great numbers and in grand costume.*
>
> *The waters of the west, and of the ocean, were then mingled by*

> *Doctor* [*Samuel L.*] *Mitchell* [sic], *who pronounced an epithalamium upon the union of the River and the Lakes, after which the lock gates were opened, and the De Witt Clinton majestically sunk upon the bosom of the Hudson.* (Colden: *Memoirs*: 60–61)

And this was but a modest demonstration compared to the ceremonies that marked the final opening of the completed canal, which involved a 350-mile-long occasion that was unmatched in its time and may yet be unrivaled for duration of display. Starting out in Buffalo on the twenty-sixth of October, a flotilla of canal boats made its way eastward in a cruise lasting a week, to Albany; whence, on the third of November, towed by steamboats, the little fleet continued down the Hudson to Manhattan. On November fourth two events took place simultaneously, an "Aquatic Procession" and a parade, followed by an evening of feasting and dancing, the last in a series of similar festivities that marked each stop along the route, along with the firing of cannon, the making of speeches, the giving of prayers, the burning of bonfires, and the mounting of countless decorated arches, eagles, illuminations, banners, and a hot-air balloon or two.

As with the tours of Monroe and Lafayette, the grand celebration inspired by the opening of the canal resulted in a book, a production gauged to match the accomplishment it was designed to commemorate. A thick quarto, with large type and generous margins, and packed with pictures and maps, the *Memoir*, printed in observance of the occasion, resembles the gilt-edged edition of Barlow's *Columbiad*, the both got up in epic size to accommodate a heroic subject. Like the canal, the book was a pioneering achievement of American technology, being the first published in America using Senefelder's new process of lithography, which the editor saw as one "of the most useful and beneficial improvements in this remarkable *Age of Discoveries*," comparable to "the immortal Fulton's application of steam to the purposes of navigation" and to "Browere's discovery of the Art of making Facsimiles of living subjects from nature. Among these three discoverers, we are proud to find two of them native Americans, *viz.* —Fulton and Browere: others might be mentioned but we forbear the details" (Colden: 397a). A final addendum to the book suggested, however, that the printerly millennium had not yet arrived: "There are some typographical errors . . . scattered throughout this volume," among which, ironically, was the misnumbering of the page on which the tribute to American technology is made.

Still, the *Memoir* was no mean achievement in the history of American book-making, gathering together both written and graphic materials by a number of authors and artists that bore witness to the grandiloquence of the occasion: "The chief object has been to adopt the mode of expression, verbal or pictorial, by which, with the utmost precision might be communicated, a correct idea of our Grand Canal Celebration—its attendant circumstances and history, in the most concise, but comprehensive manner practicable. But when thus both systems are united, we have the most perfect mode of conveying ideas to distant ages and nations, that mankind are as yet possessed of" (396). The word "communication" in 1825 was beginning to gather the complex meanings it has today: the canal itself was a communication between eastern and western regions, a conduit not only of commerce but culture—as its earliest promoters had promised—and the *Memoir* was an attempt to communicate, by combining the written word with images drawn on stone, the feelings of those who actually witnessed the celebration of the opening of that line of communication. It was not a matter of mere record, therefore, but of capturing the magnitude and full meaning of the event, thereby conveying it, not only to contemporaries absent from the event, but to posterity. If Barlow's *Columbiad* of 1807 expressed the expectations for American progress inspired by the Enlightenment, the *Memoir* of 1825 recorded the realization of one of those expectations, and by celebrating a massive internal improvement by means of improved printing technology, the book as well as the canal were works of art that illustrated the kinds of progress Enlightenment savants had promoted.

Like Cooper's *Notions,* the *Memoir* not only conveyed the euphoria of the Era of Good Feelings, it captured a moment in American history when, at least to a number of influential people, no clear distinction could be made between canals, steamboats, lithography, and "the Art of making Facsimiles of living subjects from nature" as triumphs of creativity: "It has been the universal practice, even with the learned," observed the editor of the *Memoir* in his conclusion, "to use the terms 'Arts and Sciences,' as indicative of the rank which those two modes of expression take, in their real value to mankind," and it was made quite clear by the editor that the hierarchy still held. The mechanic "Arts" had their triumph in designing and building the canal, and now it was the turn of the graphic and literary "Arts" to record the event—not the canal itself, which was its own best monument—but the celebration that attended its opening.

The author of the main text of the *Memoir* was Cadwallader D. Colden, the grandson and namesake of the Cadwallader Colden who was the friend of Franklin and John Bartram, and the man who first suggested the route that became the Erie Canal. Colden had already made his literary mark as the biographer of Robert Fulton, and by 1825 he had served as mayor of New York and as a member of Congress. He was a lawyer and a Federalist, and that he was a friend to internal improvements, should proof be needed, is borne out by the tenor of the *Memoir* itself. "The Oceans and the Mediterranean Seas of our Continent are united," Colden announced in his opening sentence. "Canals, extending more than four hundred miles, have been completed in little more than eight years, by the energies and resources of a single State, within the territories of which no white man had set his foot at the beginning of the seventeenth century" (3).

There follows salvo after salvo of short, mostly single-sentence paragraphs comparing the sluggishness of internal improvements under royal rule (which Colden's grandfather and namesake, as a loyalist, had supported) with their rapid pace after the Revolution: "There is as much difference between man, the subject of a despotic government, and the citizens of a free representative republic, as there is between waters diverted to some artificial channel, and the deep current of Niagara, pouring through its natural course, irresistible, but by the hand of the Almighty" (9). Then, at a critical moment, Colden pauses in his chronicle to record the sound of a cannon, which he hears as he writes, the last of many placed along "a line of five hundred and thirteen miles," which "repeats the signal, that the first boat from Lake Erie has entered the Western Canal, on her way to this city" (11–12). Interrupting his historical narrative to insert the record of this historic moment, the present merging with the past to become part of that past, Colden asks his reader a rhetorical question: "Who that has American blood in his veins can hear this sound without emotion? Who that has the privilege to do it, can refrain from exclaiming, I too, am an American citizen; and feel as much pride in being able to make the declaration, as ever an inhabitant of the eternal city felt, in proclaiming he was a Roman."

Clearly, in Colden's mind, the canal was a watery Appian Way, the aqueduct as road, a great and monumental public work, "so Herculean" an achievement as literally to dwarf the former achievements of the Inland Lock Navigation Company: "The old Locks, at the falls, now form part of a communication from the Erie Canal, into the Mohawk River. When we stand on the lofty and magnificent stone aqueduct which is thrown over the Falls . . . we look down on the old Canal, passing below the new structure, creeping at our feet, through its narrow channel and straightened [*sic*] locks" (25–26).

The Erie Canal is by contrast "a combination of the beauties of nature and art, seldom to be met with," a cross between the architectural wonders of the ancient world and Niagara Falls, as instance "the great embankment across the Irondiquot, over which the western section of the Canal passes," which is

> *one of the greatest works on the Canal. This aerial water-course extends more than a quarter of a mile, on a mound of earth, seventy feet in height, from a stream, flowing through a culvert, at its base. The passenger looks down from the narrow eminence, on the tops of aged forest trees, rooted in the bottom of the valley. There are works upon the Canals, which are, undoubtedly, of a more artificial character, and may appear to some more magnificent, but when the length, and height, and magnitude of this embankment is considered, and when, above the tops of trees, boats are seen passing on its summit, which is but little wider than is necessary for the Canal and towing path, it must excite great admiration.* (64–65)

Still, it should be said that Colden does not devote much space to the architectural beauties of the canal, nor is he content with merely describing its origin and progress. Like Elkanah Watson and Gouverneur Morris, Cadwallader Colden has the New Yorker penchant for futures. Having observed that "our former predictions, however extravagant they may have appeared, have been more than realized," and having declared that the present "occasion does not require accurate measurements and precise calculations of distance," Colden is off, following a route much like that pointed out by Cooper's John Cadwallader on his map, an equivalent to which is included in the illustrations of the *Memoir* (77).

"Supposing," proposes Colden, that no other artificial water communication "be completed than that with the Mississippi, which will be formed by the Canal that Ohio is now executing, we shall then have a perfect line of internal navigation, from the city of New York, by the Hudson, the Erie Canal, Lake Erie, the Ohio Canal, the Ohio, Mississippi, Missouri, and Jefferson rivers, to the foot of the Rocky Mountains, a distance of nearly five thousand miles" (74). Having in one breathless sentence brought his reader to the foot of the Rockies, Colden surmounts that impediment to western progress with a conjunctive bound: if a canal over the mountains should prove "impracticable" "we shall, in time . . . have a turnpike" (75). This carries Colden to the "Lewis River and the Oregon," a line of "internal navigation" to the Pacific Ocean, and since, "from Astoria at the mouth of the Oregon, to China, would, in a steamboat . . . be a passage of some fifteen or twenty days," there would thus "be formed a northwest passage to India, for which Hudson was searching when he discovered the river which bears his name" (74).

Visionaries like Cadwallader D. Colden provide such pleasures as only hindsight can bring; yet, if we are tempted to smile at his easy way with huge geological barriers, rapids-torn rivers, and immense oceans, we should also remember that this kind of euphoric optimism would be necessary for the building of the Union Pacific Railroad, much as it had earlier sped the progress of the canal. Moreover, Colden was in the long view absolutely right, though it would be rails, not canals, turnpikes, and rivers, that would make Hudson's dream possible. Even in the relatively short view he was right, as when he observes that "the Canals of New York and Ohio will make a change in the course of waters on the American continent, which it could hardly have been believed the power of man could have effected" (75).

But in thinking continentally, Colden cannot resist reminding his audience—his *Memoir* is directly addressed to the mayor of New York City, but clearly a much larger audience was anticipated—that the Erie Canal is the triumph of a single state, a state whose appeals for aid were turned down or ignored by the national government and even by the western states whose own well-being would be most directly involved. The *Memoir* is therefore an intensely sectional and even parochial document, as Colden's distorted perspective of western geography suggests. Small wonder that Colden stressed the benefits the canal would bring to New York—as both state and city: "New York will stand alone at the entrance of this extensive channel, and must be a greater emporium than ever called herself the mistress of commerce" (52). Colden also sounds the larger theme, pointing out that the Erie Canal will serve as a good example to other states in the Union, as evidenced by the Ohio Canal already under way; but he always returns to New York. "No Canal can be opened in the United States which will not be a benefit to us. However remote, it will be a channel through which commerce will be attracted by our great emporium, the local situation of which precludes a rival" (40). For if the canal has made a "seaport" of every town on the inland waters and Great Lakes, so "the trade of all of them must centre in this metropolis. . . . There is already another New York grown up from that which existed before the Canals were commenced" (64, 91).

Toward the end of his *Memoir,* Colden gives a renewed emphasis to "the prosperity of the whole," evoking the disinterested spirit of George Washington and the other "great and wise men who have been concerned in projecting and executing these works," observing that "the establishment of steam navigation, and the opening the Canals, have not only consolidated the interests of our own State, but indissolubly united every part of the Union" (93–94). This tension between the interests of individual states and the good of the nation is a constant factor in the pamphlets and periodical writings preceding or inspired by the success of the Erie Canal: whether the town was Boston, Baltimore, or Philadelphia, strengthening the Union through canals and, ultimately, rail lines was an ideal inseparable from promoting the welfare of Boston, Baltimore, or Philadelphia. The proportion of regional to national concerns is best demonstrated by the words of the aldermen of New York on the occasion of the canal's opening, which they regarded as "an event calculated to exalt the character of the State, and to promote its best interests, not only as it affords the means of extending its commerce, but that it will bring together the citizens from the remotest corners of the State, and thus enable them to cultivate that friendship and mutual good feeling which the Citizens of New York are so desirous of promoting, and which is so necessary for cementing our Union" (165).

A more balanced perspective was provided the *Memoir* by letters sent to the Canal Committee (of the Corporation of New York) by such honorables as living ex-Presidents, surviving signers of the Declaration of Independence, and General Lafayette, who all wrote expressions of thanks for the commemorative gold medals sent to them in boxes of cedar and bird's-eye maple (made by "Mr. Duncan Phyfe"), the wood for which had been "brought from Lake Erie in the first Canal Boat" (197, 348). James Monroe declared that "the accomplishment of the great work undertaken by the State of New York, by which the western lakes are united with the Atlantic Ocean, through the Hudson river, forms a very important epoch in the history of our great republic. By facilitating the intercourse, and promoting the prosperity and welfare of the whole, it will bind us more closely together, and thereby give a new and powerful support, to our free and most excellent system of government." John Adams, with barely a month left of life, affixed his unsteady signature to the statement that "the great Canal" in New York was "the pride and wonder of the age," and Thomas Jefferson, in his own still firm but nonetheless fated hand, observed (as he had earlier feared) that "this great work will . . . bless" the descendants of its builders "with wealth and prosperity, and [will] prove to mankind the superior wisdom of employing the resources of industry in works of improvement rather than of destruction."

But the most interesting letter came from James Madison, who had remained steadfastly loyal to the idea of internal improvements, so long as they were made consistent with the Constitution: "As a monument of public spirit conducted by enlightened Councils," he wrote, "as an example to other States worthy of emulating enterprize, and as itself a precious contribution to the happy result to our country of facilitated communications and intermingled interests, bringing nearer and binding faster the multiplying parts of the expanding whole, the Canal which unites the great Western lakes with the Atlantic ocean, is an achievement of which the State of New York may at all times be proud, and which well merited the homage so aptly paid to it by her great commercial Metropolis." Moving from state enterprise out to the welfare of the nation at large, Madison in a single sentence brings everything back to, as all things will subsequently flow toward, the city of New York.

As if to seal that certification, Richard Riker, the recorder of the city and the chairman of the committee responsible for the ceremonies recorded in the *Memoir,* saw to it that these letters, reproduced in facsimile (another triumph of lithography), were placed (unpaged) at the end of the book, in effect closing the text with a glorious endorsement, the canal and its celebration becoming one with the Constitution and the country. By such a device were the most disinterested expressions of patriotic goodwill made to serve a sectional end. New York was well on the way to becoming the Empire State.

ii

The building of the Erie Canal was a triumph of logistics as well as of inventiveness and engineering. Cadwallader Colden enumerated the "several labor-saving machines" devised for the project, including one for "prostrating the forest trees that grew on the Canal line," a mechanism as theatrical as it was powerful: "To see a forest tree, which had withstood the elements till it attained maturity, torn up by its roots, and bending itself to earth, in obedience to the command of man, is a spectacle that must awaken feelings of gratitude to that Being, who has bestowed on his creatures so much power and wisdom" (55). But Colden also gave credit to less spectacular but perhaps more important conceptions, like the plan, devised by the canal commissioners, of "dividing the . . . Canal line into small sections, and procuring the work to be done on these by contract. . . . [N]othing has contributed more to the successful accomplishment of the work" (53). This strategy of independent yet coordinated and articulated parts (devised to mitigate sectional jealousies by spreading the profits to be earned along the entire length of the route) was not unlike the geopolitical plan that made possible the union the canal was designed to strengthen.

As a symbol of union, moreover, the canal stood also for power, conveying not only canal boats and produce, but water, in sufficient quantity to "afford . . . a surplus, which . . . can be applied to move machinery," thereby furnishing "agriculturalists and manufacturers . . . those means of moving their mills and machinery" (90). Harnessing that great eastward flow of western force that Gouverneur Morris saw rushing toward the Hudson, the canal was a vast conduit of natural power to which an endless series of machines might be hitched, much as it provided innumerable "connections . . . with boundless fields producing raw materials, and with markets" (91). Confronting "the extent to which the canal will promote agriculture, manufactures, and commerce," as when contemplating the spectacle of Niagara Falls, "the human mind" was rendered "hardly capable of comprehending" what it held in view.

The promotion of "agriculture, manufactures, and commerce" was the role of an enlightened central government, in the opinion of President John Quincy Adams, and the reciprocal relationship between Canal and Constitution is worth consideration. The Constitution was the creation of the Founding Fathers and the Canal was the work of the Sons, the latter a mechanism designed to improve the workings of the former, conceived as a great Republican machine. If John Adams wrote *A Defence of the Constitutions of the United States of America,* his son wrote *Report on Weights and Measures,* documents evincing a considerable reduction of focus with no loss either of an Enlightenment faith in mechanisms of order or a Puritan skepticism regarding the workings of human nature: "When law comes to establish its principles of permanency, uniformity, and universality," wrote the son, "it has to contend not only with the diversities arising from the nature of things and of man; but with those infinitely more numerous which proceed from existing usages, and delusive language" (13).

In New York in 1825, however, such somber musings were out of place, and what the Erie Canal was in long—a marvel of planning—the celebration of its opening was in brief, the broad outlines of which are suggested by the many documents gathered together in the lengthy appendix to the *Memoir*. In them, we are admitted backstage and are allowed to view the inner workings of the pageant—the correspondence of committees, which from the time of the Boston Tea Party helps explain the strategy of "spontaneous" events. In those documents having to do with the grand flotilla and the great procession, phrases such as "obeying with utmost exactness previously arranged signals," "uniformity and effect," "conform to regulations," and "great regularity," appear with considerable regularity themselves (119, 134, 153, 210). Those professional and tradesmen's societies "who intend[ed] to unite in the celebration" were requested to send to the recorder's office "an estimate of their respective numbers, to the end that the Grand Marshal may apportion a sufficient space, in which each society may conveniently form and move in the Procession" (155). The grand marshal for the procession, General August Fleming, was responsible for assigning "to the Societies and Bodies that had signified their intention of uniting in the Celebration, their stations in the Procession," which were "designated by the cross-streets, in which they would respectively form" (209).

The procession was formed in the lower part of Manhattan, below the gridiron system of streets devised in 1811, yet the organization and formation of the parade suggests the extent to which that regular and uniform plan had pervaded the thinking of New York's leaders. Devised to serve convenience, the grid system provided a preordained order readily available for civic purposes of whatever sort, and though the streets of lower Manhattan are decidedly more organic, the great procession was formed up in a paramilitary fashion, even to the four "divisions" into which the cross streets were organized, three of which were commanded by militia officers; a "regular" plan further enforced by the "Regulations" by which the parade was conducted:

> *1. The procession will be formed (six in front) under the direction of Major General Fleming. . . . It will form on the west side of Greenwich Street—its right on Marketfield-street. The line of procession will begin to form at nine o'clock, A.M., and be ready to move at eleven o'clock, A.M.*
>
> *2. Its right will wheel and pass, the whole line moving at the same time, so that all may see each other.*
>
> *3. The procession will pass up Greenwich Street to Canal Street and to Broadway—up Broadway to Broom Street and to the Bowery, down the Bowery to Pearl Street, down Pearl Street to the Battery. The procession will reach the Battery by three o'clock, P.M., when the whole Aquatic Procession will have returned from the ocean, and be stationed off the Battery.*
>
> *4. The whole aquatic party being ready and the boats duly arranged, the procession will pass the boats at the margin of the Battery, upon the broad circular walk. This close approach of the boats to the Battery is intended to give to the City Procession, and the Aquatic Party, a view of each other, and to enable the Corporation to unite the two together in the Grand Procession.*
>
> *5. The Corporation, with their guests, preceded by the aborigines from Lake Erie, with their canoes, will fall in the rear of the*

> *City Procession, following it under the direction of the Grand Marshal, to the City Hall, where all will disperse.* (154–55).

As the regulations emphasize, the line of march was defined by the Battery, which served as a pivot and focus, the line becoming at that critical point two lines, the City Procession and the Aquatic Party becoming in one transformational moment both actors and audience, performing and admiring at the same time. Thus the "good feeling that animated the Procession" was the result of very careful orchestration, ensuring that the spirit of American enterprise, like the waters of the canal itself, was conducted through the proper channel (266).

Coordination between the activities of the flotilla and the procession was facilitated by "means of the Telegraph under the superintendence of Captain John Greene, who promptly complied with the request to transmit and answer our dispatches"; the apparatus in question presumably being the semaphore erected by Christopher Colles on Castle Clinton, which saw service during the War of 1812 (265). Rapid communication was essential to the orderliness of the festivities, and a number of pages in the appendix to the *Memoir* are devoted to the arrangements by which the signal cannon were obtained and set up along the Hudson and the Erie Canal, whose sequential firing (interrupting Cadwallader Colden's narrative history) would carry the word from Buffalo to Sandy Hook that the canal-boat flotilla was under way. Like the telegraph, many of the cannon had served in the War of 1812, and having "lately sounded our country's glory on the Northern Lakes, [now] proclaimed the accomplishment of a work which . . . stands unrivalled in the annals of the world" (164–65). One of the arguments used in promoting the canal was the cost of carrying those very cannon to the scene of war; for, as Colden reminded his readers, "the expense of transporting cannon from Albany to the Lakes was at one time, more than double what the pieces cost" (42).

The contingency of telegraph and cannon has a further significance, for the fiery information that interrupted Colden's *Memoir* was undoubtedly heard by Samuel F. B. Morse, perhaps as he worked in his studio on the great full-length portrait of Lafayette. In 1827, Morse would attend a course of lectures on electricity in New York, the first in a series of steps that resulted in yet another artist's entering the field of mechanical invention. Like Fulton and his steamboat, Morse and his telegraph greatly accelerated the speed of communications—and progress—in the United States, helping to fill out the great design. Thus the son of the pioneering geographer Jedidiah Morse would become a pioneer of a different sort but toward the same end as that proposed by his father, further binding the parts of the Union.

Still, in 1825, the news of the departure of the flotilla seemed to travel marvelously fast to witnesses of the event, as the official narrator of the festivities testified. "In the short space of one hour and twenty minutes, the joyful intelligence was proclaimed to our citizens"; while the progress of Governor Clinton's aquatic cohort took more than a week, the ordinarily slow pace of a canal boat further impeded by the ceremonies that greeted the increasingly augmented flotilla along the route (295). Like the festivities that were launched once the flotilla reached Manhattan, these various tributes and demonstrations along the canal route were planned with careful attention to parts. Thus it was Jesse Hawley, the "Hercules" who first committed to paper the idea of the Erie Canal (and who was by 1820, according to Elkanah Watson, "Collector of the Port of Rochester"), who delivered the send-off address in Buffalo, and Clinton's flagship, the *Seneca Chief,* was accompanied at the start by boats with carefully chosen names, including the *Superior,* the *Commodore Perry* (a "freight boat"), and the *Buffalo* ("of Erie"). Perhaps the most novel of the canal boats to leave from Buffalo was *Noah's Ark,* the work of a prominent citizen of New York City, Mordecai Noah, who presumably hoped to promote his projected Zion on Grand Island in the Niagara River.

The *Ark,* according to the "Narrative of the Festivities," was "literally stored with birds, beasts, and 'creeping things.' She was a small boat, fitted for the occasion, and had on board, a bear, two eagles, two fawns, with a variety of other animals, and birds, together with several fish—not forgetting two Indian boys, in the dress of their nation—*all products of the West"* (296). This covenantal craft apparently did not find the waters of the canal in sufficient flood to carry its cargo, because it dropped behind the rest of the boats and never did arrive in New York City. Perhaps more in keeping with the occasion was the craft that joined the procession at Rochester, the *Young Lion of the West,* "loaded with flour in half-barrels, butter, apples, &c., and . . . a quantity of cedar tubs and pails, of very elegant workmanship . . . and some brooms of a superior quality, and had on deck a collection of wolves, foxes, raccoons, and other living animals of the forest," including the inevitable eagles (198).

The *Young Lion* was also equipped "with a pair of brazen lungs," and was given frequent occasion to "mingle his roar" with the sound of cannon "with which he was saluted on his passage down" (322). Not only cannonades greeted the flotilla as it moved slowly from town to town along the canal, but a great variety of ingenious and traditional displays: banners, festooned bridges and aqueducts, illuminated buildings, and signs proclaiming patriotic slogans—as at Montezuma, where "De Witt Clinton and Internal Improvements" met the boats and

"Union of East and West" signaled their departure (302). At the many ports of call, the travelers were met by banquets, toasts, and addresses to the divinity and to the assembled guests, in effect an extended prelude to the even greater display got up in Manhattan by Richard Riker.

Like the ceremonies in the city, those along the canal route involved considerable symbolism, and where possible, local touches were provided; as at Lockport, the first stop after Buffalo and the place where the waters from the West descended by a veritable Niagara of double locks, a singular technological marvel "whose workmanship will vie with the most splendid monuments of antiquity" (298). There, "the cannon used on the occasion were those with which Perry conquered upon Erie—the gunner was a Lieutenant who had belonged to the army of Napoleon—and the leader of the band was the cabin-boy of Captain Riley, who suffered with him in his Arabian captivity. During the passage [through the locks] the company were introduced to the venerable ENOS BOUGHTON, of Lockport, the pioneer of the Western District—the man who planted the first orchard, and built the first framed barn West of Utica!" Along with Governor Clinton and his keg and murals, the *Seneca Chief* carried a token cargo, including potash, an Indian canoe, and the bird's-eye maple and cedar to be made into boxes "for the medals to be struck on the occasion" (198). It was, in short, the kind of occasion that provides its own poetry, in the form of self-consciously contrived event.

Still, despite careful planning, things went occasionally awry—as at Weedport, when, during the firing of salutes, "a twenty-four pounder was accidentally discharged, and . . . two valuable young men lost their lives" (303). Nor did everyone along the way share the feelings of unqualified joy. At Rome, the inhabitants had lost a long fight to keep the new as well as the old canal within the town limits, and the flotilla was met there by a procession carrying *"a black barrel"* filled with water from the old canal, which was slow-marched by muffled drums to the edge of the new canal and emptied. And yet, having staged their demonstration, the people of Rome joined the rest of the world along the canal and welcomed the visitors, thus demonstrating that "howevermuch they have been disappointed, [they] are not behind any of their fellow-citizens, in appreciating the value of the Canal as a state and national work" (304). In sum, as the rest of New York went, so (finally) went Rome.

At Schenectady, likewise, the civic mood was sullen, for the town anticipated a great loss of revenue now that the carrying trade from the old canal to the Mohawk River had ended; a gloominess matched by the weather, for a day of "drizzling rain" was followed by a "dark and dreary night" (307). But then, with the arrival of the flotilla at Albany, the sun broke through and "a more beautiful day never dawned upon our land." As with the divine disposal of the landscape west of the Mohawk, "it seemed as though a benignant Providence, smiling upon the labors and triumphs of human genius and enterprise, had purposely chained the storms in their caverns" (312). The addition of steamboats to the fleet increased the spectacle, all the vessels "gorgeously decorated" and arranged in line:

> [A] *brisk north-west wind caused the gay banners and streamers to flutter in the air, so as to be seen to the best possible advantage. And the beauty of the scene was still further heightened by the large columns of steam rushing from the fleet, rising majestically upwards, and curling and rolling into a thousand fantastic and beautiful forms, until mingled and lost in surrounding vapors. Every boat was filled with passengers, and each was supplied with a band of music. The delight, nay, enthusiasm, of the people, was at its height. Such an animating, bright, beauteous, and glorious spectacle had never been seen at that place; nor, at that time, excelled in New York.* (313)

The contemporary notion of beauty expressed here derives largely from marvels produced by modern technology, the whole embued with the kind of magic only steam power can produce: "At times," as the fleet passed "among the group of islands between Albany and Coeymans . . . a boat, at some distance astern, appeared to be swiftly darting across the river; and again, at another point could only by discovered [by] the variegated flags and streamers [showing] through the intervening though scattered shrubbery, whose verdure had lost its freshness, and had been speckled with pale red and yellow by the early autumnal frosts. And now again, when the broad bosom of the Hudson was unbroken from bank to bank, the whole squadron appeared in line, like a fleet from the dominion of the fairies" (314).

As in this description from the official "Narrative" of the festivities, much of the pageantry on the Hudson and in New York Harbor was connected not with the canal boats but with vessels powered by steam. The latest marvel of technology was represented by the steamboat *Washington* (Captain E. S. Bunker, commanding), which met the fleet in the harbor, "heaving up the foaming billows as though she spurned the dominion of Neptune," an "entirely new boat . . . uniting all the improvements in steam-boat architecture" and incorporating most of the decorative motifs associated with the celebration:

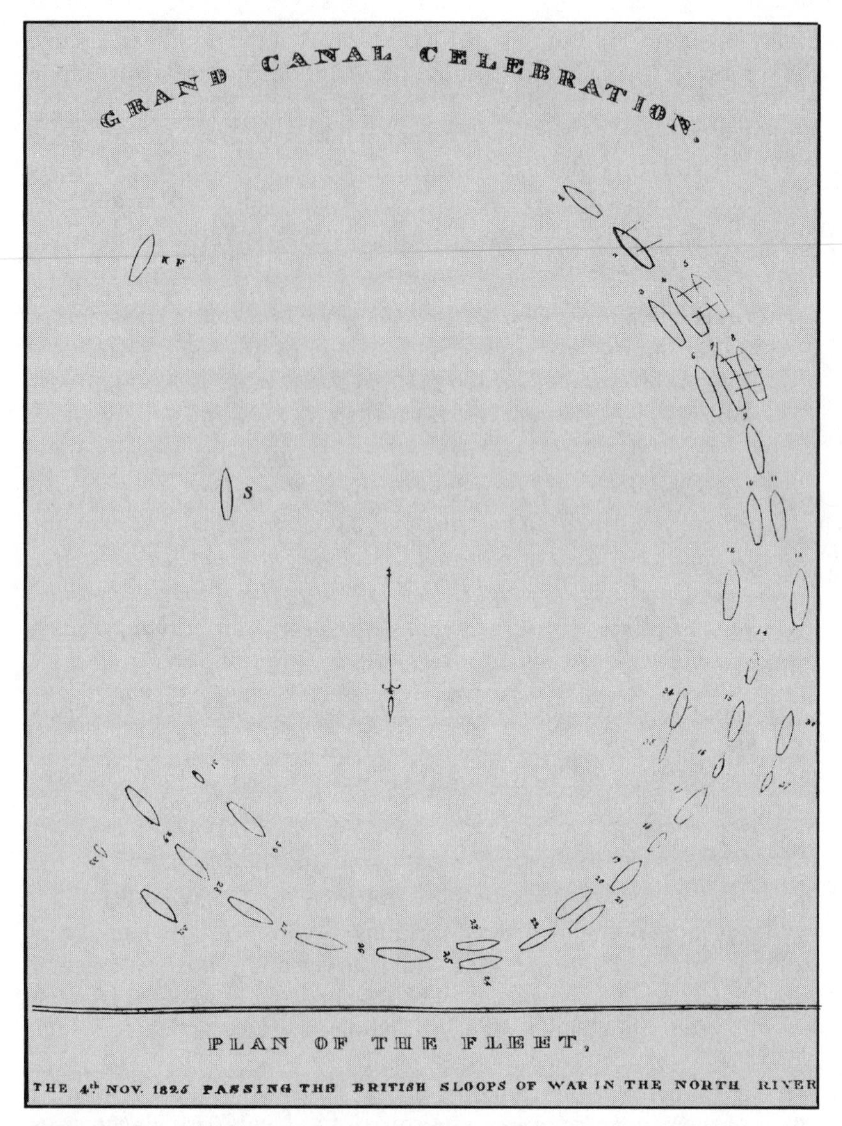

Anthony Imbert (active in America from 1825–1838). *Plan of the Fleet*. From the *Memoir* and appended documents celebrating the completion of the Erie Canal (1825).

This trial effort by Imbert, a French lithographer, was the first of the plates executed for the *Memoir*. Here again, as in the Savage engraving of Washington and his family, we find technology and improvements in fine arts inextricably interwoven.

Samuel Valentine Hunt (1803–1893). ***Bombardment of Port Royal.*** **From John Stephens Cabot Abbott,** ***A History of the Civil War in America,*** **2 vols. (1862–1866).**

A diagram of a highly destructive tactic that certifies the extent to which the celebration of geopolitical union in 1825 provides a meteoric prefiguration of the war between the states thirty years later.

The design of the tafrail represented the renown of Washington and Lafayette. The centre was a trophy of various emblems—the laurel and the olive—standards—swords—the balance—the caduceus of Mercury, &c. The trophy was surmounted with a bald eagle. Each side of it was decorated with a bust—on the right, that of Washington—on the left, the bust of Lafayette. The former was crowned with the civic wreath and the laurel—the latter with the laurel only. The Genius of America was crowning her hero, and the spirit of Independence, waving the flaming torch, binding the brow of Lafayette. Each of these figures was attended with emblematic medallions of Agriculture and Commerce. The whole was based on a section of the globe, and the background was a glory from the trophy. The corners of the tafrail were each filled with a cornucopia, which gracefully completed the design on which neither painting nor gilding had been spared to enchance the effect. (317)

Of the fourteen toasts (one for each state) given aboard the flotilla in New York Harbor, the last was "given standing" to "the memory of ROBERT FULTON, whose mighty genius has enabled us to commemorate this day in a style of unparalled [*sic*] magnificence and grandeur" (288). Cadwallader Colden in his contribution to the *Memoir* paid tribute to the great inventor for his treatise on canals and his contribution in that regard to Gallatin's report, but it was finally with steamboats that Fulton would be identified.

The magical power with which navigation was invested by steam was most clearly demonstrated by the maneuvers of the flotilla during the aquatic procession in the harbor, the entire fleet of boats passing "round" the two British sloops of war anchored there: "While performing this circular maneuver, the British bands struck up 'Yankee Doodle'; in return for which act of courtesy, the American bands, as they passed the other side, successively played 'God Save the King'" (322). Clearly, the wounds of the most recent conflict between the British and their American cousins had already healed—a newfound wholeness certified by the great rotating circuit of the flotilla. And, as in the straight line of advance down the Hudson, the precision of this nautical maneuver, preserved in the *Memoir* by a lithographic plate (the first executed for the book), was made possible "by means of steam navigation," which overcame the vagaries of wind and tide: "It was impossible to behold this wonderful movement upon the waters, unaided by a breath of air, without feeling how vast a debt of gratitude we owe, and the whole world owes, to the mighty genius of our immortal countryman, Robert Fulton" (122).

That Fulton's creation could be associated with a millennium of sorts was born out by the decorations mounted on board H.M.S. *Swallow,* "representing Britannia, Columbia, the Eagle, the Lion, and an English and American Sailor, Neptune, Liberty, and the flags and shields of both nations, all classically arranged, denoting good feeling, fellowship, and union of sentiment" (322). So the Lion lay down with the Eagle; but that the British warships were powered by sail alone adds a sinister dimension to both the circling maneuver and the escutcheon of universal peace. As the author of the narrative of the festivities observed, the navigational versatility made possible by steam, used in 1825 in a demonstration of good will toward a former adversary, could be adapted in the future to other, less pacific, ends.

Such an armada, he pointed out, would provide New Yorkers a formidable weapon: "They could choose their own time, position, and points of attack; and tremendous must be the power that could successfully oppose, and superhuman the skill that could baffle, an expedition of this kind, directed by the hand of valor, and sustained by the unconquerable spirit of freemen!" (323). In 1825, Samuel Francis Dupont was serving as a young midshipman aboard the *Constitution,* but by 1861 he would have risen to the rank of commodore, and, having become a foremost theorist in the use of steam-powered navigation, Dupont would draw up his fleet for the battle of Port Royal in South Carolina in just such a formation, a great ellipse that, slowly rotating in position, reduced two Confederate forts at once and inspired Herman Melville's poem, "Dupont's Round fight," a blend of martial and aesthetic terminology celebrating the "victory of Law" (*Battle Pieces*: 49).

Yet Dupont's "round Fight" was planetary in motion by appearance only, for like the great procession up and down Greenwich Street in 1825, the ships actually formed a double, rotating column to present parallel ranks of fire: the "geometric beauty" celebrated by Melville was essentially the terrible beauty of the Line—the soldier's, the sailor's, and the surveyor's ideal. Such was the beauty also of the Erie Canal as a technological marvel, and given the effect that that linear diagram would have on the delicate balance between state and nation, between the interests of section and union, it might be said that the slowly moving columns that churning paddles made into a circle at the mouth of the Hudson River—a vast engine turning in the outpouring current that had floated the *Seneca Chief* from Buffalo—was a fatal machine, a penultimate expression of changing forces that would in time disrupt the very Union whose preservation the canal was supposed to guarantee.

The man chiefly responsible for the canal, De Witt Clinton, barely outlived his triumph, and died suddenly in 1828. Ten years later, Henry A. S. Dearborn (son of the general who nearly lost the War of 1812), a native of New England and an ardent supporter of internal improvements, proposed that there be placed at the entrance to the Erie Canal at Buffalo "a colossal bronze statue of the illustrious DE WITT CLINTON, a hundred feet high,—holding aloft in one hand, a flambeau, as a beacon light, to designate, in the night, the entrance, and pointing with the other, in the direction of the route of the Erie Canal" (*Letters:* 22–23). But if, in the words of the narrative of the celebration, the canal was Clinton's "perennial monument," it remained the only one he would have of similar proportions.

Moreover, like the celebration itself, the canal was of relatively short duration. It was barely finished before changes became necessary, a process of revision that continued until as a marvel of communication it was surpassed by the railroad, of which Henry A. S. Dearborn was a perfervid champion. Indeed, as Dearborn bears loud witness, it was the very success of the canal that spurred on the enthusiasm for rail lines that (literally) followed. "IT IS DONE!" was the message signalled by the chain of cannon planted along the Erie Canal and the Hudson River, but in a certain sense the work was never completed, and the events it set in motion had just begun: "The longest canal in the world" would in time act as a geopolitical lever, aligning the states of the Northeast with the territories of the West and further isolating—and alienating—the South. As the maneuvers of the steam flotilla suggested, the canal was a counterpart to Fulton's other inventions, being a mechanism intended to promote universal harmony that was in truth a war machine, much as the celebration of its opening was an expression of the Good Feelings, the chief occasion of which was a euphoric reaction to a war lately won.

XIV

Mingled Measures

i

When the canal-boat flotilla reached Albany, among the speakers greeting the voyaging dignitaries was "Wm. James, Esq.," of that city, whose speech was given a generous paraphrase in the narrative of the festivities included in Colden's *Memoir:* "A glowing picture of the future greatness and happiness of our western and northwestern territories was presented; and the glory of the nation, its territory, its institutions, its wealth, its liberty, and its spirit in local and general improvement, successively furnished themes for the imagination on the orator" (310). James' speech is representative of many such delivered during the preceding week, and if it epitomizes the aura of Good Feelings the event was embued with, it is chiefly notable for being the work of a man best known to posterity as the grandfather of Henry James, Junior, who would not be famous for celebrations of American progress in the West.

Even with the considerable literary talent then available, the documents engendered by the celebration are marked by the absence of well-known New Yorkers. Irving was living abroad, and though Cooper would give an enthusiastic account of the canal in *Notions of the Americans,* in 1825 he was tied up in litigation, French lessons, and *The Last of the Mohicans.* Of the major voices of the earlier generation, only Phillip Freneau was still living, but though the completion of the canal was a fulfillment of his prophecy in *The Rising Glory* that "hoarse Niagara's stream" would be "taught a better course," he was not included among the official celebrants. Freneau nonetheless picked up his tireless pen to celebrate the event, citing the canal as a monument of enterprise that owed its grandeur to no "slave" (perhaps a reference to the use of slave labor on Washington's pet project), being the work of freeborn artists, "who with skillful hand / Conduct . . . rivers through the land" (*Last Poems*: 21–23). Freneau also regarded the canal as having articulated "the great idea" of the federal period, which was *"to lead the veins the system through,"* an apt enough figure, but perhaps too lame an inversion to keep up with the pace of modern times. For whatever reason, the great idea was left to lesser (and younger) talents to commemorate.

Given the singular occasion of the celebration, it is fitting that the author of "Narrative of the Festivities," appended to the *Memoir,* was a journalist, Colonel (an honorific title) William Leete Stone. A former Federalist turned Republican, Stone was an important contributor to the rich and turbulent world of Knickerbocker letters; and if he did not represent the highest level of literary genius associated with Gotham at that time, he, like William James of Albany, was very much in tune with his times. The same may be said of the poet officially associated with the event, Samuel Woodworth, famous even in 1825 for "The Hunters of Kentucky," inspired by General Jackson's victory in New Orleans, as well as for his first attempt at water poetry, "The Old Oaken Bucket." Woodworth was himself an editor of note, and as such had, like many in his day and later, been first trained as a printer; circumstances that gave particular relevance to the way his poem was made part of the parade that wound its way through lower Manhattan. Each of the tradesmen's associations in the city had decorated "cars of gigantic structure" (Colonel Stone's phrase) with appropriate murals and symbols, "upon which their respective artisans were busily engaged in their several occupations" (324). The float constructed by the Printer's Society (to which Woodworth belonged) displayed "the venerable JAMES ORAM . . . seated in the Library Chair of Dr. Franklin attended by four Boys costumed as Heralds and Mercuries, who distributed copies of the Ode, as they were printed, to the assembled multitude" (253).

The union of the poetic muse with technology is intrinsic to Woodworth's contribution, the which, though like his bucket wooden in execution, is like his better-known work in perfect harmony with the age: "Art" as celebrated in New York in 1825, as in Barlow's big poem, includes the voyage of Columbus, "the path of the pilgrim band," Franklin's lightning rod, and "an unshackled Press," for

> *'Tis this which call'd forth the immortal decree,*
> *And gave the great work its first motion:*
> *'Tis done! by the hands of the brave and the free,*
> *And* ERIE *is link'd to the Ocean.* (252)

Thus, the chief union celebrated by Woodworth is (as the last line above suggests) the "marriage" of Lake Erie to the Ocean, for it is to that end

Charles P. Harrison (1783–1854) and John Hill (1770–1850), after Harrison. *Sacred to the Immortal De Witt Clinton* (ca. 1828). (Private collection)

Liberty crowns the bust of Clinton amidst symbolic trophies of his accomplishments. Foremost in the composition is a "History of the Grand Canal," behind which we catch a peep of the Erie itself winding around hills (the requisite serpentine line of beauty) toward a rectangular lock. The concept, composition, and symbols are neoclassical, dominated by the toga-draped Liberty and the massive bust of Clinton (here replacing Washington as the patron saint of civil engineering). The art is entirely in the spirit of the many decorations used during the 1825 celebration of the opening of the canal.

that "Proud ART o'er NATURE has prevailed!" and "GENIUS and PERSEVERANCE have succeeded!" The Lake is figured as a pagan goddess, "attended by a sparkling train" of Naiads from the West, who

> *meets the sceptered father of the main,*
> *And in his heaving bosom hides her virgin face.*
> *'Tis done! the monarch of the briny tide,*
> *Whose giant arm encircles earth,*
> *To virgin* ERIE *is allied,*
> *A bright-eyed nymph of mountain birth.* (251, 250)

Woodworth's friend and colleague, George P. Morris, editor of the *New-York Mirror* and later made famous by his "Woodman, Spare That Tree!" marched beside the Printer's Society float in 1825, and in 1842 he paid his fellow poet the tribute of imitation, by means of a broadside celebrating the opening of the Croton Aqueduct in similar terms:

> *Gushing from this living fountain,*
> *Music pours a falling strain,*
> *As the Goddess of the Mountain*
> *Comes with all her sparkling train.*

The chief importance of the similarity, however, has less to do with the state of the poetic art in New York during this period than with the frequent association of figures from classical antiquity with events decidedly modern in implication. Much as the builders of the canal asserted its linearism by architectonic features reminiscent of Roman aqueducts and viaducts, thereby eliciting from residents of the Empire State admiring comparisons with ancient imperial example, so poetry celebrating the event tended also to depict neoclassical figures in connection with the achievement, borrowing status from Rome.

The dominant conceit in Woodworth's poem, the marriage of the lakes to the ocean, was the creation of Dr. Samuel L. Mitchill, who in 1823 had presided over the union of the canal and the river, as we have seen. No better representative of the Enlightenment in America in 1825 could have been found than Dr. Mitchill, who was Jeffersonian in both his Republicanism and his wide range of "philosophic" interests. One of the most prominent scientists of his day, a polymath and a politician, Mitchill was something of a belles lettrist and neologist as well, not averse to writing a medical treatise on "septon, the principle of dissolution" in heroic couplets, and coining the word "Fredonia" as a substitute for the awkward "United States." Having been used as a title for a bathetic epic celebrating the War of 1812 and as a place name for a town along the route of the Erie Canal, "Fredonia," happily, disappeared.

As a friend to improvements of every kind, Dr. Mitchill was the ideal person to preside over the wedding of waters—an Enlightenment version of an ordained minister—and the ceremony he devised, like the imagery (and architecture) engendered by the Enlightenment spirit in America, was heavily dependent on neoclassical forms. "Man delights in types and symbols," intoned Dr. Mitchill as he officiated at the marriage of Lake Erie and the Atlantic, and then went on to expatiate upon the origin of river-gods (generally male) and their attendants (usually female). "It was a modern and happy improvement in these matters to ascribe a female form and attributes, to the chief allegorical personage of the fountains, and to distinguish her by the becoming title . . . 'LADY OF THE LAKES,'" a studious translation in which Sir Walter Scott's popular poem of 1910 had undoubtedly played a part (274, 273).

But if Dr. Mitchill's mythological lady was a mixture of ancient example and modern literary improvements, her passage eastward was made possible by the miraculous intervention, not of pagan gods and goddesses, but of American technology:

> *Her progress through an artificial river, more than three hundred miles long, was unexampled. At her annunciation obstacles of every kind disappeared. Was an excavation necessary for her accommodation?—the rocks disparted and made room. Was an embankment required? —the ground rose to its proper elevation. Were locks and reservoirs necessary to go up and down declivities? —they sprung into being and performed their functions. . . . Was water demanded to facilitate conveyance? —the ponds and brooks joyfully furnished their stores."* (275–76)

Mitchill's "Lady" was given a specific and sacramental form during the ceremonies accompanying his speech, for the first of the flotilla of canal boats to travel the entire length of the canal, the *Seneca Chief,* carried not only De Witt Clinton but a keg of Erie water, and the first emptied the second (after "preserving a portion of the water" to be sent to Lafayette in a bottle "of American fabrick") into New York Harbor, a mingling of fresh and salt water that was reciprocated by sending back aboard the *Seneca Chief* to Buffalo a keg of "briny fluid" to be emptied into Lake Erie. This artifice symbolizing the opening of intercourse between lake and ocean preceded Dr. Mitchell's address, which was

ALLIANCE OF NEPTUNE AND PAN
UNION OF ERIE WITH THE ATLANTIC

followed by a much more complex mingling of waters, an apocalyptic emptying of vials in keeping with the millennial mood of the occasion. Mitchill was an indefatigable correspondent and had managed to gather together for the ceremony bottles containing water from the great rivers of the world. These he emptied into the bay, describing in turn the unique qualities of each river, and then went on to announce that, having spilled his millennial waters, "'the virtue infused with them . . . has spread from this spot by a combination of mechanical impulse, chemical attraction, and diffuse propagation, through the whole mass of waters, with an electrical rapidity and a magnetical subtilty, that authorizes me to pronounce the *circumfluent Ocean republicanized!* IT IS DONE!'" (278).

Dr. Mitchill's aquamorphic metaphor has its geopolitical counterpart, expressed by himself and most of the major contributors to the *Memoir:* "'Greenbay and Chicago now consider Michillmackinac and Detroit, heretofore frontier posts, as but places of refreshment on the [eastward] voyage'" declared the Doctor; and Colonel Stone portrayed the canal as "a *new* and *additional river,* twice the navigable length of the Hudson, and traversing a region, whose population and agricultural wealth will soon rival, and even surpass, those of its banks" (331). As Cadwallader Colden claimed, "every place on the shores" of the Great Lakes had been "converted by the Canals into a sea-port" (64), a commercial advent that is given a symbolic reading also, this time by "a classic emblematical production of the pencil" that decorated the canal boat bearing Governor Clinton on his voyage from Buffalo to New York.

> *This piece, on the extreme left, exhibited a figure of Hercules in a sitting posture, leaning upon his favorite club, and resting from the severe labor just completed. The centre shows a full length figure of Gov. Clinton, in Roman costume; he is supposed to have just flung open the lock gate, and with the right hand extended* (the arm being bare), *seems in the act of inviting*

(*Facing, above*) Archibald Robertson (1765–1835). *Design for the Official Badge of the Erie Canal Celebration.* From *Memoir* (1825).

(*Facing, below*) Archibald Robertson. *Titlepage Decoration for "An Account of the Grand Canal Celebration at New York."* From *Memoir*.

In both of these designs by Robertson (who, incidentally, instructed Washington Irving in art), we find native American and technological symbols struggling with the decorums of neoclassical design.

> *Neptune, who appears upon the water, to pass through and take possession of the watery regions which the Canal has attached to his former dominions; the God of the Sea is upon the right of the piece, and stands erect in his chariot of shell, which is drawn by sea-horses, holding his trident, and is in the act of recoiling with his body, as if confounded by the fact disclosed at the opening of the lock; Naiades are sporting around the sea-horses in the water, who, as well as the horses themselves, seem hesitating, as if half afraid they were about to invade forbidden regions, not their own. The artist is Mr. Catlin, miniature portrait-painter.* (296)

Mr. Catlin would soon depart for regions even farther west, abandoning both the miniaturist and the allegorical modes for the graphic depiction of a people whose doom was hastened by the opening of the canal. But, for the moment, he was one of several artists who contributed, not only to the decorations ornamenting the opening of the canal, but to the book commemorating the work.

Chief (in prevalence) among these last was Archibald Robertson, a Scots-born miniaturist who, like Catlin, rose to the epic occasion, providing for the *Memoir* a panoramic "view of the Fleet" of the steamship and canal boats "Preparing to Form in Line" in New York Harbor. Robertson was in charge of the department of fine arts set up for the celebration, and designed the official badge as well as the medallion sent to the Signers and other notables, "classic productions" that testify to a certain inconsistency in the iconography inspired by the event. Where Dr. Mitchill and Samuel Woodworth employed an erotic conceit, Robertson, like Catlin, preferred a masculine set of symbols; but, surprisingly, it was the Scot who chose a figure indigenous to the landscape. Thus the design for the badge and medallion depicts "Pan seated in a canoe, loaded with the products of our fields, forests and Mediterranean Seas, from Lake Erie, being piloted by an aboriginal native of the western forests," his arm "interlocked" with that of Neptune, who is seated "in his naval car, attended by a Triton winding his conch" (344). This overloaded pairing was designed to convey a complex idea, a male marriage, in which Neptune accompanies "his brother Pan to his native woods on the shores of Lake Erie" and the Indian navigator seems to have been added as if in uneasy recognition that classical deities are somewhat out of place in "the native woods" of America—a mix similar to that of Bartram's Arcadian "Siminole."

Where Mitchill's nuptials celebrated the union of male and female principles, Robertson's voyage celebrated brotherhood (as with the

William Guy Wall (1792–1864?) and John Hill, after Wall. *The Palisades*. From *The Hudson River Portfolio* (1820–1825). (Courtesy of the New-York Historical Society)

Again, a pioneering art form, the aquatint, heralds progress in America, while the aesthetic impulse crops elements still deemed unpicturesque. Wall's composition should be compared with Pownall's depiction of the Palisades, for here the roughness formerly associated with the western side of the Hudson becomes a relatively smooth plane emphasizing the linearity of the composition.

earlier opening of the locks at Albany in 1823, Masonic symbolism seems here implied), and the emblematic frontispiece he designed for the *Memoir* features another fraternal crest, "a young eagle . . . ready to spring off to flight" supported by, "on the dexter side, an armed Indian native of the forest; and on the sinister, a mariner; both together united to convey the idea of the amphibious character of our State—and, moreover [as] . . . the modern representatives of the Heathen God Pan, and his brother Neptune, the God of the seas" (354). In this design Robertson is entirely American in his symbolism, and he adds to his translated iconography three more symbols at once literal and highly figurative to a contemporary eye: behind the Indian is seen "a Canal, with locks," while "on the left hand of the mariner is the steam-gallery Chancellor Livingston, under steam . . . the whole illuminated by a radiating rising Sun, giving relief to the thick volumes of rolling smoke, issuing from the flue pipes of the Chancellor Livingston" (354).

Like the decorations that greeted Monroe and Lafayette, the emblems devised to celebrate the completing of the Erie Canal—despite occasional iconographic confusion—indicate that the art of allegorical design had flourished in the half century since the Revolution. Yet there is a tension implicit in Robertson's design between inherited classical and native American symbols, as well as a proposed unity: the eagle may be ultimately Roman (or perhaps scriptural) in origin, but the species is North American, and though the rising glory by 1825 was traditional, it is given a much more modern implication by the realistic smoke from the *Chancellor Livingston*. The combination asserts a harmonic union between national aspirations and technological achievements, yet, like the substitution of an Indian for Pan and an American tar for Neptune, Robertson's glorified smoke is vernacular evidence that Old World symbols are inadequate to New World occasions.

Yet another sign of aesthetic conceptual inadequacy in confronting technological innovation appeared in a collection of aquatints published by the Anglo-American engraver, John Hill, in 1821 to 1825, which celebrate the beauties of the Hudson River from source to harbor. After watercolors by William Guy Wall, most of the engravings are conventionally European in their treatment of American vistas, following the example of Hill's earlier portfolio, *Picturesque Views of American Scenery* (1820). But several seem to struggle compositionally to come to accommodating terms with the changing landscape, most specifically Wall's depiction of a steamboat moving crisply along before the backdrop of the Palisades, a composition severely linear in conception, with an emphasis on flat planes unbroken by the curving forms preferred by artists of the period. What is noticeably missing from the picture is a cloud of smoke issuing from the steamboat's stack, suggesting a curiously divided attitude in the artist, who is true to the linearity implicit in both the steamboat and the cliffs (which had been used as quarries from time to time), yet whose training resists inclusion of the necessary cloud found in Robertson's emblematic composition.

George Catlin similarly yielded to the linear implications of the canal in several lithographs he executed for the *Memoir* before abandoning the modern scene entirely for landscapes populated by primitive men. But the most interesting attempt to come to aesthetic terms with the Erie Canal is a series of watercolors painted by John Hill's son, John William, depicting scenes along its route, a bustling celebration of the line and plane that also reflects an emerging (Jacksonian) interest in the labors of the common man, seldom mentioned in contemporary ac-

John W. Hill and John Hill, after Hill. *Unidentified Town and Canal Locks* (ca. 1831). (Courtesy of the New-York Historical Society)

An engraving prepared for aquatint, this picture is one in an intended series from watercolors by the younger Hill celebrating the Erie Canal that was never published. Hill is departing radically (like Wall) from the English picturesque tradition, accentuating linear elements of composition in order to emphasize the triumphs of American technology. Also emphasized, in the opening years of the Jacksonian era, is the American worker, whose activities nearly obliterate the canal itself.

counts of the digging process. Yet the series was never brought to press and soon thereafter the American aesthetic would be entirely dominated by European example, particularly in the work of Thomas Cole, who preferred wilderness scenery or distant prospects of settled terrain for subject matter, thereby removing the mechanical evidence of civilized man.

The impact of Romanticism is also observable in the works of writers who were emerging even as the Erie was being dug, most notably perhaps in the prose of one of its most outspoken champions. In *The Last of the Mohicans* (1826), Cooper paused in his melodrama long enough to pay tribute once again to Clinton and his canal, but his picture of "the tourist . . . float[ing] steadily toward his object on those artificial waters" is chiefly intended to emphasize the hardships endured by Americans attempting to traverse the same region in the mid-eighteenth century (186). Like his friend Cole, Cooper preferred his novelistic woods and waves to be undefiled by evidence of man's inventions. Judge Temple in *The Pioneers* (1823) is a Republican visionary, who, where "others saw nothing but a wilderness," visualized "towns, manufactories, bridges, canals, mines, and all the other resources of an old country" (353). But where the Judge has the "good sense" to suppress "in some degree, the exhibition of these expectations," his foolhardy cousin, Richard Jones, does not, and in a manner approximating that of Elkanah Watson—the regional gift to the spirit of progress—Jones is a champion of "beauties and improvements" and an advocate of "system," who intends to "run our streets by the compass . . . and disregard trees, hills, ponds, stumps, or, in fact, anything but posterity" (353; 189–99).

In 1813, ten years before Cooper's novel, there had appeared Daniel Bryan's Miltonic celebration of Daniel Boone as a veritable avatar of the "Spirit of Enterprise," urged on by a pantheon of neoclassical abstractions to prepare the way for Civilization by cutting the Wilderness Road; like the Erie Canal figured as "the first grand CONDUIT . . . / Through which, to the dark wastes of savage night, / From their bright Eastern seas, REFINEMENT'S FLOODS / Effulgent flow!" (153–54). The process by which refinement spreads its glory "Along Ohio's smooth majestic stream, / And Mississippi's mighty flood" is accelerated by the arrival of Boone's wife and daughters, his "darling Consort and sweet lovely Maids," who introduce in their persons "the Power of Virtuous and Refined Beauty" to the dark and bloody ground (153–54; 44; 154). This progressive scheme is in harmony with the Jacobin dialectic of Imlay and Ashe, and Bryan's poem is clearly a late exercise in epic engineering whose terms look forward to the poetics of the Erie Canal celebration. As such, it is a last (if lengthy) gasp of the visionary mode, much as the celebration itself was an aesthetic atavism by 1825.

James Fenimore Cooper also regarded the introduction of women into the western landscape as the definitive signal that civilization had arrived, with its pinafores and pianofortes, and the vision Judge Temple has of his personal promised land resembles that which Bryan's Boone has of Kentucky, when from his mountaintop he views the wilderness through "prescient FANCY's telescopic tube" (112). But in 1823, Cooper's version of Daniel Boone is an aging hunter in retreat from civilization, not a harbinger of "Refinement and Enterprise" but a champion of wilderness ways who delivers scathing homilies about the wastefulness of the "pioneers," who give Cooper's title an ironic dimension. Bryan's Barlovian allegory may evoke the spirit and machinery of the Erie Canal celebration, but Cooper's novel attests that even as New Yorkers gave emblematic shape to their enthusiasm for such public works as benefitted their state, there were those among them who were beginning to solve the aesthetic problem of reconciling machinery to Romantically conceived landscapes by removing the machine from the garden.

ii

No periodical published in America during the opening years of the nineteenth century was so unabashedly a friend to internal improvements and modern invention as was *Niles' Weekly Register,* a chronicle whose first six years, from September 1811 to September 1817, bracketed the War of 1812 and its euphoric aftermath. The editor, Hezekiah Niles, an industrious and calculating Quaker from Philadelphia, was a mixture of progressive idealism and opportunism who, during the period in question, was a Jeffersonian Republican but eventually became a Whig and a devout Unionist. A champion of protective tariffs and a national bank, Niles' rage for order extended even to his own periodical, the "great bulk" of which by 1817, as he observed, "called loudly for a GENERAL INDEX," so that "a collected view of all the facts, papers, and circumstances belonging to [the] many things inserted" could be obtained.

To the end, as Niles explained in the preface to the index, a "gentleman well fitted to the task was found, who with Joblike patience and unwearied industry, completed the index for the first ten volumes." The job-worker in question was a young poet named John Neal, who gave an acerbic account of the experience in his autobiography. Neal characterized the *Register* "as the accumulated rubbish of years," the which, without an index, was "comparatively worthless," while with it, was "a golden treasure for the future" (210–11). He described the publisher-editor as "a shrewd calculator, far-seeing, crafty, and sagacious; a truly honest man, a patriot, and a christian," who paid Neal for his four 'months' hard labor a mere two hundred dollars and ". . . a copy of his 'Register'" (214).

Having acquitted himself of a punishing task and quit a penurious employer, Neal returned to creative writing, and 1818 he finished his second book and last long poem, *Battle of Niagara*, a Byronic exercise in the obscure sublime inspired by the late war, the events of which he had been so recently recording in alphabetical order, along with the latest devices for improving the look of the landscape and the lot of mankind. The Battle of Lundy's Lane provided the specific occasion, although nothing in the poem except the site reveals that vital fact, for so successful was Neal in conveying the Burkean sublime that very little in the poem is seen clearly. "Some parts very powerful" noted Longfellow in his journal of Neal's best novel, *Rachel Dyer* (1828), and the same judgment may be made of *Niagara*, parts of which were anthologized in school readers for years afterwards (*Memorials*: 104). One part in particular, a lengthy apostrophe to Lake Ontario patterned after Byron's famous address to the ocean in *Childe Harold's Pilgrimage* is as powerful as anything written in rhyme in America before the emergence of Poe. Whether or not the poem was composed as a negative reaction to Neal's enforced exposure to *Niles' Register* is difficult to ascertain, but it is clear that *Battle of Niagara* by "Jeru O'Cataract" runs against the full tide of Enlightenment faith in internal improvements reflected in Niles's bulky compendium.

"This was written when I was a prisoner," announces the pseudonymous author in an address to the reader, "when I *felt* the victories of my countrymen," thereby creating a persona circumstantially acquitted from participation in the war (a period spent by Neal as a smuggler and a profiteering merchant) but perhaps alluding also to Neal's experience shut up with the *Register* for four long months. Whatever else the falls of Niagara may stand for in his poem (and it is not always clear), they are "the bounding of water that's free!" Like the American eagle celebrated in the poem, "the cataract hymn of an unfettered tide" is a symbol of "mountain-born" Liberty (ix–x). Lake Ontario, likewise, as the most immediate source of the falls, is presented as a virginal body of "free waters" untouched by civilized man:

> *Here sleeps ONTARIO. Dark blue water hail!*
> *Unawed by conquering prow, or pirate sail,*
> *Still heaving in thy freedom —still unchained!*
> *Still swelling to the skies —still unprofaned!*
> *The heaven's blue counterpart . . .*
> *Be ever thus Ontario! —and be free:*
> *The home of wild men, and of Liberty.* (27–28)

Associated by imagery with the American Indian—"a naked monarch—sullen, stern, and rude"—Ontario is also the natural asylum of the red man, who there is "sublime—he is a god!" He is also "Great Nature's master-piece" and an American Adam, "like him" who walked "the banks of paradise, and stood alone" (30).

But should the white man come with something more substantial than canoes, Neal warns, and "with mightier burthens . . . oppress . . . the smooth dark lustre of thy breast," then Lake Ontario is doomed: where Daniel Bryan celebrated the happy union of civilization and western waters, Neal sings no epithalamion but an anticipatory elegy:

Farewell to thee! and all thy loveliness!
Commerce will rear her arks—and Nature's dress
Be scattered to the winds: thy shores will bloom
Like dying flow'rets sprinkled o'er a tomb;
The feverish, fleeting lustre of the flowers
Burnt into life in Art's unnatural bowers. (32)

Instead of the "wild luxuriance" of natural "garlands," there will be hot-house flowers, and where fountains of freedom now "sing and sparkle to the skies," there will be artificial jets of "prisoned water . . . made to play / In one eternal glitter to the day" (35). Excoriating the "floppery of silly Taste, / That grieves to see wild Nature so unchaste." Neal bids Ontario to remain "grand—luxuriant—desolate," and heaps scorn on "wretched fools" who

Would have Creation work by lines and rules.
Their's [sic] *is the destiny—be theirs the curse,*
In their improvements still—to mount from bad to worse. (35–36)

From a line recalling Byron worshiping at the altar of Pope, Neal swings into a derivative yet powerful apostrophe, imitative of Byron at his most romantic pitch:

Be ever dark Ontario! and be wild
In thine own nakedness—young nature's child!
Still hand her festoons o'er thy glittering caves;
Still far from thee the pageantry of slaves! . . .
May Architecture never rear her spires
Or swell her domes to thy warm sunset fires;
Where now, o'er verdant pyramids and pines,
And dark green crowns, the crimson lustre shines!

Enough has now been done—thou art but free:
Art but a refuge now for Liberty. . . .

Roll not thy waves in light, Ontario!
Forever darkly may thy waters flow!—
Through thy tall shores and blooming solitudes
Sacred to loneliness—and caves—and woods—
Roll not thy waves in light—or thou will see
Their bosoms heave no longer darkly free:
But whitening into foam beneath their load,
While Commerce ploughs upon her flashing road;
And thou mayest stand, and hearken to the cry
Of thy young genii mounting to the sky:
And feel the fanning of the last free wing
That's shaken o'er thy brow, as it goes wandering. (36–37)

Neal is here worth quoting at length, for this is a remarkable exercise, if only in its rhetorical reversal of the catchwords associated with the visionary mode. "Light" is here again opposed by "Darkness," but with reversed emphasis: only so long as the lake remains "dark" can it remain free, otherwise it will bear the burden of Enlightenment's darling, Commerce. Many right-thinking folk in 1818 thought of Liberty in terms not of wild waters but of streams conducted in orderly fashion through artificial channels for the service of men. Not so John Neal, virtually an abolitionist in the service of lakes and streams, expressing sentiments that look forward to the radical attitude toward the material aspects of progress that we associate with Transcendentalism—this in a year when Emerson was a junior at Harvard and Thoreau was two.

A champion of the natural, the wild, the free—whether figured as dark waters or red man—Neal attacks those apostles of Reason who would have "Creation work by lines and rules," and he thereby reverses the thrust of so many poems and pamphlets published during the previous half-century. Like Cooper in *The Pioneers* he opposes the unquestioned law of the line, but where Cooper will admit that technology has its blessings, that the idea of Progress is essential to the American aesthetic, Neal takes an angrily reactionary stand, championing Nature and the Indian at the cost of civilized advance and anticipating thereby the themes and imagery with which subsequent American writers will oppose what they increasingly conceived to be a mistaken national direction. Neal in effect proposes a new aesthetic, in which utilitarian considerations, instead of enhancing the natural prospect, detract from and ultimately destroy it. But in 1818, Neal was clearly ahead of his time, and so therefore are we, and need to fix our attention by way of conclusion on a mediating figure, whose uneasiness in the face of technological change was much more subtly even equivocally stated than was Neal's.

XV

O, That Deep Romantic Chasm

i

John Neal might assault the law of level and line, but most Americans of his day could only marvel over the accomplishments of the civil engineers who designed and built the Erie Canal, wonders of technology aided to produce a grand effect by Palladian architectonics. Contemporary witnesses bore frequent testimony to the wonderful sight provided by a canal boat's easeful and silent passage through the opening land or its apparent flight through the air by elevated aqueduct, a phenomenon that inspired Cooper's paean to the canal and Clinton. Even in its comic manifestation, whether being ridiculed by comparing its actual size with the grandiloquence of the rhetoric boasting of its importance or by describing the crowded and thereby unpleasant conditions aboard the barges (where accommodation for passengers was minimal), the Erie Canal quickly found acceptance as part of an increasingly mobile American life. Unlike the railroad—to which it gave added impulse—which was difficult to place within the picturesque categories then available to artists, the canal could be easily reconciled with aesthetic norms, if occasionally by means of careful selection of natural backdrops, the more easily to accommodate it to the burgeoning Romantic spirit.

From the first, as we have seen, the steamboat was conceived of in aesthetic terms that were harmonic with contemporary modes of architecture, and by the time of the Era of Good Feelings, as Henry Clay's apostrophe suggests, it was regarded as an exceptionally beautiful machine, both because of its graceful appearance and because of the shapely future it seemed warranted to convey. Thus, in 1824, as the steamboat *James Kent* headed upstream from Manhattan with the Nation's Guest aboard, the residents of the town of Hudson awaited its arrival with a complex mood of pleasurable apprehension, rendered by a reporter for the *Northern Whig*, a witness to the event. "All eyes," he

George Catlin (1794–1872). *Deep Cutting Lockport*. From the *Memoir* (1825).

Catlin was the most accomplished artist, after Robertson, who contributed to the memoir. Here, as in his other depictions of the canal, the emphasis is on the linear element, celebrating such triumphs of technology as the deep excavation (enabled by Dupont's blasting powder) necessary to keep the channel level at Lockport (the highest point along the canal). This was quite a different note from that found in his much more famous delineations of American Indian life, though it is a mistake to view Catlin as a romantic because of his subject matter there.

wrote, "were bent in painful suspense toward the quarter where he was expected," the operative pronoun covering both General Lafayette and the *James Kent*, so that when the steamboat finally hove into view, the sight was gratifyingly "splendid," a fulfilling of expectations theatrical in effect (Brandon: I: 230). For the importance of the moment had been enhanced by the boat's decorations, converting it into a floating stage set framing the famous visitor in an appropriate manner, a marvel of artifice given added meaning by modern technology. "She seemed like an object of fairy creation, so silently, so proudly she stemmed the waves of the majestic Hudson," exulted the reporter. "It was one of those scenes which is seen but once"; not only the boat itself, but the grand setting provided by "the village of Athens and the Catskill mountains . . . filling up the background of the picture," while in the foreground "the whole extent of Warren Street [had been] cleared of almost every object that would impede the view, the triumphal arches that so triumphantly spring up for so triumphant an occasion," so that "the entire field of vision" was occupied by "a landscape which can safely be pronounced unrivalled."

Like William Wall, the reporter quietly left out the necessary cloud of smoke from his composition, and instead moved on from "the natural beauty and sublimity of the picture" to "its moral sublimity," a concomitant of the picturesque in the aesthetic scheme of William Gilpin: "The countless numbers of animated beings, in every direction; the glow of feeling excited by the circumstances which called them together; the distant rumbling of the artillery along the mountains, all combined to form a subject worthy of only a master's hand. It would require the talent of our own Irvine [*sic*] to depict it in its true colours."

William Henry Bartlett (1809–1854) and J. T. Willmore (1800–1863), after Bartlett. *View of the Erie Canal Near Little Falls.* (Courtesy of the American Antiquarian Society)

Bartlett, the English landscape artist who imposed a European aesthetic on the American scene, was at pains to seek out a view of the canal that would promote the element of the picturesque (here with a decidedly Gothic atmosphere). Both Bartlett and Catlin employed chiaroscuro, but each to different ends. Catlin's perspective lines run straight to the vanishing point, while Bartlett's composition employs a series of planes, emphasizing, not the accomplishments of engineering, but the towering cliffs and forests of the natural scene. Catlin celebrates technology; Bartlett obscures it.

At the time, Washington Irving's talent was being devoted to depicting the picturesque aspect of foreign lands, though he was already famous in New York and in most of the United States for two charming stories of life along the Hudson River. Yet it is doubtful that Irving would have shared the enthusiasm of the *Northern Whig* for the beauties of steamboats set against the Catskill Mountains. The namesake (and biographer) of George Washington might have worked up a sentimental tear on the occasion of Lafayette's return, but the racket and splash of the celebration itself would have been less than inspiring to a writer who stressed his retiring sensibility and love for the antique. For Irving, the steamboat was an unwelcome agent of change, "confounding town and country" and driving the picturesque element of ancient age from view.

A Federalist in his youth, Irving, in his *Knickerbocker History of New York* (published in the same year as Gallatin's report), provided a satiric antidote to that massive infusion of Enlightenment progressivism as displayed by Jeffersonian republicanism, finding in the ancient chronicles of Gotham situations and characters analogous to the contemporary political scene, from Jefferson to Washington's noisy shadow, General James Wilkinson. More recently, in his "Legend of Sleepy Hollow," he had used the ancient antagonism between Yankees and Yorkers to frame a parable of rustic virtue resisting valiantly (and violently) an agent of exploitive change, asserting the sanctity of rural stability against the destructive forces of westward expansion. Most important, in "Rip Van Winkle" Irving had poked gentle fun at the noisier aspects of democracy, but in so doing had in the idea of a bewildered old man, twenty years late in returning from a day of hunting, hit upon a figure perfectly expressive of uneasiness in the face of rapid change, yet spelled out in terms familiar to champions of western progress.

ii

Irving had been in Europe during the euphoric aftermath of the War of 1812, and therefore missed the universal rise of what even its exponents were willing to call the "mania" for internal improvements, which finds no place in the gentle satire of "Rip Van Winkle." After all, the future shock that Rip feels is the result not of accelerated but of merely normal progress, whose effects are exaggerated because of his long absence [illegible] The emphasis of the story, moreover, is on politics, not on changes wrought on the landscape by technology, and one of Irving's little jokes is that political revolutions have little real effect on the populace. As with the changing of the lettering on the village tavern sign, the transfer of allegiance from King George to President George is an easy matter, and Rip likewise, after his initial bewilderment, effortlessly finds a place in the village society.

Still, Irving's contemporaries were quick to adapt his fable to the 1820s as an expression of the uneasiness that can make the expression of Good Feelings occasionally shrill. Consider the example of William Leete Stone, author of the ebulliently optimistic narrative of the Erie Canal festivities, who shortly thereafter took a trip through northern New York State, the region of his boyhood, and whose diary of the journey contains a mixture of awe, self-satisfaction, and regret, providing a useful commentary on his public celebration of progress. Visiting Utica, "which, nine years ago, the period of my last visit to it, ranked only as a flourishing village," Colonel Stone found it had "now grown as if by magic, to the dimensions of a large city" (219). Stone had "heard much of the march of improvement in Utica since the completion of the Grand Canal. But I had no idea of the reality." As a journalist and celebrant of the up-to-date, Colonel Stone liked to work into the fabric of his writings references to the latest works of (preferably) American literature, and in beholding the wonders of Utica, he reached for the most obvious allusion: "Rip Van Winkle himself, after his thirty [*sic*] years' repose . . . was not more amazed than I was at the present aspect and magnitude of this beautiful place."

As in his narrative of the canal celebration, Stone's amazement is of the positive kind when confronting the evidence of progress in Utica, but as he neared home, his feelings about change took a different tack, and he was relieved to discover, in visiting Cooperstown, "the favorite spot of my boyhood," that very little transformation had occurred; in large part (though Stone does not dwell on the fact) because the village had been bypassed by the route of the Erie Canal. As a result, the scene was virtually untouched by progress, and Stone was able to concentrate on the contrast in his own personal circumstances, in which change had been very much for the better, resulting in a familiar formula in American life: "I had left [here] a poor young man without experience in the world, with but little knowledge, without means and without friends to aid or influence my destiny, or to push me forward in the great world. And through the blessing of a kind Providence, I now returned, accompanied by an intelligent wife, in prosperous if not affluent circumstances, and known for more than fifteen years in political life and ten years as the editor of one of the oldest and most respectable daily papers in our country" (212).

Proud of his own progress in life, amazed by the rapid advances of civilization effected by the Erie Canal in remote areas, relieved, however, to find that Cooperstown remained untouched by progress, Colonel Stone was saddened when, in visiting the old family farm, he found that "everything had changed" except for "the only pine which within my recollection had ever stood upon the farm." This alone survived, remaining "as in the bright and sunny hours of childhood. May the woodman's axe never be upraised against it!" (216). Out of this perilous balance between a public celebration of progress in America and private musings over the loss change can bring would come George P. Morris' celebration of the Croton Aqueduct and his plea to the woodman, or Samuel Woodworth's epithalamion for the union of lake and ocean and his nostalgic recollection of an old well bucket. If the 1820s were the Era of Good Feelings, the mood from time to time seems to have inspired "feelings" associated with sentimentality, likewise captured in many of Irving's pieces in his *Sketchbook,* which for more reasons than "Rip Van Winkle" recommended itself to American readers of the day.

"Never before was there such a fleet collected," observed Colonel Stone in describing the demonstration in New York Harbor which ended the Erie Canal celebration, "and it is very possible that a display so grand, so beautiful, and we may even add, sublime, will never be witnessed again" (321). Even nature had contributed to the display, providing "uncommonly fine" weather, so that the harbor waters lay still as a "natural mirror"; "indeed the elements seemed to repose, as if to gaze upon each other, and participate in the beauty and grandeur of the sublime spectacle. . . . It was one of those few bright visions whose evanescent glory is allowed to light up the path of human life—which, as they are passing, we feel can never return, and which, in diffusing a sensation of pleasing melancholy, consecrates, as it were, all surrounding objects, even to the atmosphere we inhale" (321–22). Confronted by a massive demonstration of technological accomplishment, Stone fell back on a stock *ubi sunt* posture, the assembled fleet in effect becoming the flower of a moment's blossoming, a memento of mutability. Stone's odd moment of melancholy is hardly equivalent to Neal's apostrophe to an unburdened Ontario, yet it does testify to a certain uneasiness in confronting the evidences of change, there being that in the Romantic spirit that resists the bright promise of canals and steamboats and seeks instead an embowered shade and a melancholy reflection of the transitoriness of life.

Moreover, much as Irving's parable of the sleeper awakened held more meanings for his contemporaries than the author perhaps intended, it was also sufficiently flexible to be adapted to even more radical changes in the American scene. Thus, when "Rip Van Winkle" gained renewed popularity after the Civil War as a play, the story was given a modern point (and a plot) by framing it as a melodrama in which the foreclosing of the mortgage on Rip's home place is threatened, a subtextual reference to the antagonism between banks and the agrarian ideal of independent freeholders. The writer thereby transformed Irving's tale into a populist allegory, reflecting the contemporary American anxiety over the threatened disappearance of the rural way of life and the stability it was associated with, a point that is latent in the reading given Irving's parable by Colonel Stone. Finally, that Irving's original situation was borrowed from a European source makes its application to the American scene all the more marvelous. For Irving was also working unawares within a long-established American tradition, dating back to the seventeenth century.

In 1699, Benjamin Tompson of Dedham, Massachusetts, the most indefatigable poetaster of Cotton Mather's day, wrote a broadside welcoming to New England the newly appointed royal governor, Lord Bellomont: it begins by conjuring up the ghost of a long dead Indian sachem, who is struck with amazement and terror by the evidence of civilized progress that has replaced his forested domain. And in 1740, in one of his earliest almanac efforts, Dr. Nathaniel Ames (also of Dedham) summoned the shade of William Blackston, one of the "Old Planters" who had settled on Shawmut before the arrival of Winthrop's fleet. Like Tompson's sachem in 1699, Blackston is dumfounded by the evidence of progress he finds in 1740: "If ever I liv'd here, / Trees were as men, now men as trees appear!" Bewildered and dismayed by change, Blackston "willingly . . . re-dy'd" (134). Like the sachem, Ames's Blackston is a figure from a past no longer relevant, and he serves thereby as testimony to the marvels of change.

In the visionary exercises that comfort Joshua and Columbus by means of angelic guides who point to the wonders of a future age, republican poets like Dwight and Barlow gave new (and somewhat different) life to the notion of a resurrected witness from the past, an idea virtually incarnated in General Lafayette, as we have seen. But unlike his earlier literary counterparts, Rip (who is no ghost, after all) easily slips into the modern scene by mostly ignoring it. He returns to his old ways, becoming a village fixture and the children's favorite. For Irving's is a profoundly conservative fable, stressing not so much the challenge of change but the persistence of ancient use, and thereby reversing the implications of the emerging tradition of dead men awakening to bear testimony to the marvels of American progress.

iii

If we return Rip to his place at the start of the story, we find that the nameless town "of great antiquity" with its houses built in the old Dutch manner completely shares Rip's proclivity for doing nothing, typified by the convivial group that shares his idle hours. Sitting before the inn "telling endless sleepy stories about nothing" or engaged in "profound discussion" "when by chance an old newspaper fell into their hands, from some passing traveller," they immerse themselves in events that are already irrelevant to the present moment (769, 772). In this scene, it is Rip's wife who is out of place, her rampaging tidiness most American in its emphasis, in contrast to the decided European character of the town, and it is her threat to her husband's happiness that drives Rip up into the mountains for his great evasion. There he encounters the strange men in antique garb playing at ninepins who remind him "of the figures in an old Flemish painting . . . which had been brought over from Holland at the time of the settlement," and who are drinking the liquor that puts him into his long sleep (775). This act, so essential to what plot there is, links Rip with the antique, European element, bonding him to the ninepin players up in the mountains much as he had before been bonded to the village idlers below.

The contrast and the continuity between the village "at the foot of those fairy mountains" and the hollow up in the hills is amplified by landscape description. As he wanders into the mountains that fateful day with gun and dog, Rip finds himself on a "green knoll, covered with mountain herbage, that crowned the brow of a precipice. From an opening between the trees, he could overlook all the lower country for many a mile of rich woodland. He saw at a distance the lordly Hudson, far, far below him, moving on its silent but majestic course, with the reflection of a purple cloud, or the sail of a lagging bark here and there sleeping on its glassy bosom, and at last losing itself in the blue highlands" (774). This is a height equivalent to that from which Daniel Boone traditionally catches his first glimpse of Kentucky; but instead of a land of promise, Rip (also a hunter) sees sleepy fulfillment. But the height provides an alternative view, and turning from the eastern picture of somnolent beauty, Rip looks west, "down into a deep mountain glen, wild, lonely and shagged, the bottom filled with fragments from the impending cliffs and scarcely lighted by the reflected rays of the setting sun" (774). Here, once again, is the familiar aesthetic division, for from Pownall, to Bartram, to Charles Brockden Brown, the western prospect is associated with aspects of wilderness sublimity.

It is the ravine that holds Rip's attention, and he lies "musing" on the scene until "the mountains began to throw their long blue shadows over the valleys." It is this delay, also, that leads Rip into his meeting with the strange ninepin players, and his last thought before the first of them calls to him is of "encountering the terrors of Dame Van Winkle" for having tarried too long; and though the potion he drinks while with the ninepin players may be a punishment for having penetrated their secret world (he actually drinks it on the sly), it is also the anodyne that relieves him of the torments of his termagant wife. Clearly something is going on here, the contrast between sleepy village and haunted hollow given further point by the scenic contrast of the "beautiful" river and the "sublime" mountains. Hunting provides Rip a temporary escape, but his penetration of the mountains' wildness proves a permanent asylum.

The world of the ninepin players is entered by a "narrow gully, apparently the dry bed of a mountain torrent," which leads to "a deep ravine, or rather cleft between lofty rocks," and is described as "a hollow like a small amphitheatre, surrounded by perpendicular precipices, over the brinks of which impending trees shot their branches, so that you only caught glimpses of the azure sky and the bright evening cloud" (775). The contrast between "azure sky" and dark hollow is like that between the river valley and the ravine, and the players are similarly a melancholy lot, who go about their "play" without conversing, in vivid contrast to Rip's good-natured companions outside the inn below. Rip's first encounter with the game comes as he climbs up the gully, for the balls striking the ninepins sound like thunder, providing a somber prelude to his entrance into the hollow itself. This is a place that shares the mountain gloom of the "alternative" view, the Gothic realm of shadows, the habitation for haunted men who play their game as if under a curse.

Having awakened from this long sleep, Rip returns to the village, to his once-familiar world, and though the changes there at first bewilder him, he soon adjusts to the new system. What is most important to him, however, is not that "the country had thrown off the yoke of old England" but that "he had got his neck out of the yoke of matrimony," for "the changes of states and empires made but little impression on him; but there was one species of despotism under which he had long groaned and that was petticoat government" (783). Through "old Peter

Vanderdonk . . . a descendant of the historian of that name," Irving explains that "the Kaatskill mountains had always been haunted by strange beings," reputed to be Hendrick Hudson and his crew, "the first discoverer[s] of the river and country," who "kept a kind of vigil there every twenty years . . . being permitted in this way to revisit the scenes of [their] enterprise, and keep a guardian eye upon the river and the great city called by his name" (782). To call the village of Hudson a "great city" in 1817 is clearly a facetious joke, but the main point of the "explanation" is that Irving's Henry Hudson, like Barlow's Columbus, is one of those symbols of enterprise who, when summoned up as a ghostly visitor, casts a wary (because "guardian") eye on the changes that have occurred.

In terms of the meaning of the tale, however, Hudson is associated, not with the river and progress, but with his secret hideaway, and he seems less approving of modern improvements than engaged in play. As opposed to conventional notions of progress, associated with the spread of cities, light, and female influence, the world of the ninepin men is not only inside the dark and wild aspect of nature, it is entirely masculine. The "guardian" function of Hudson and his fellow dwarfs is to serve as Rip's protective spirits, removing him from events until he can return to a world without his wife; a world, moreover, that will permit, even approve, his idleness, Rip having arrived "at that happy age when a man can be idle with impunity" (783). That this was the intention of the ninepin men is suggested by the news that Rip's wife had died "but a short time" before he awoke from this long sleep, her death resulting from a stroke suffered while arguing with a "New England pedlar," her counterpart as an interloping agent of industriousness (782).

Rip's salvation by means of wilderness encounter is paralleled by a later story by Irving, "Rolph Heyliger," in which another scapegrace journeys into the wild western side of the Hudson in order to discover (from the leader of a band of huntsmen) the location of a hidden treasure that will make him a wealthy man. And Brom Bones, who uses a legendary ghost to drive his Yankee rival from Sleepy Hollow—a rival associated with Yankee enterprise in the form of real estate speculation—likewise has a wild element in his makeup, despite the indolent prosperity of the Dutch farming life with which he is associated. This notion of a salvational wildness will gain ideological meaning in the hands of Thoreau, for whom a retreat into wilderness zones was a regular ritual, and who used it to combat champions of technological "improvements." But in the story of Rip Van Winkle, as in Cooper's *The Pioneers*, the proto-Romantic assertion of wilderness sanctity is set forth in terms familiar to Enlightenment discourse.

Again and again we have observed in American writing of the late eighteenth century a dichotomy established by emphasizing the opposing characteristics of two-sided rivers: One side is wild and unsettled, the other pastoral and cultivated, a division often channeled into the theme of two confluent rivers, the one turbulent and unnavigable, the other accommodating to man's uses. This opposition is essential to the Enlightenment rhetoric advocating western settlement in the propaganda engendered by Pownall, Jefferson, Bartram, Cutler, and Imlay, but it is here turned on its symbolic axis by Irving, who thereby enforces the dramatic reversal of values found in Neal's *Niagara*. Where another early Romantic writer, Brockden Brown, presented the wilderness zone as a dangerous maze at whose center stands a bearded wild man driven mad by isolation and guilt (Rip's gothic counterpart), Irving posits an equivalent arena as the place where his hero can find respite from civilized order. Crossing a river to a salvational zone is an early Puritan myth in America, conflating Exodus with the imperial impulse, as Manasseh Cutler's Ohio sermon suggests. But in Puritan terms, the western bank is converted to a pastoral area, a Canaan of productive plenty (much like the Van Tassel farm that Ichabod Crane hopes to possess), while in Irving's parable it must remain (like Neal's Ontario) forever wild in order to retain its salvational power.

What Irving's story is about, finally, is personal, not national, independence; equated, not with the kind of enterprise permitted by liberty, but the lack of it allowed by freedom from female tyranny. The story is an extended joke at the expense of the kinds of norms celebrated by Daniel Bryan, in which the domestic, female spirit is essential to the taming of the wild West. Penetrating a bastion of that wildness, Rip finds an antidote to the domestic spirit, and in Irving's version, though Liberty has a mountain birth, it is personified by dwarfed maleness, not some neoclassical nymph. It is with male "spirits" that the mountains are associated, and "thunder" is what the noise of the balls sounds like to Rip, not literal but highly figurative explosions. As Rip and his mysterious companion with the keg climb up through the gully, "mutually relieving one another," they are met by "long rolling peals . . . that seemed to issue out of a . . . cleft between lofty rocks," rumblings that can be compared to other thunderous sounds, which when made in male company can inspire rolling peals of laughter, the signal of masculine independence and defiance of social norms imposed by polite society. In 1820, as Dr. Samuel Mitchill bore eloquent witness, the spirit of Rabelais was alive and well in Gotham; and that, finally, is the "Wink" in "Winkle."

iv

A celebration of the male bond and liberty, "Rip Van Winkle" is not a Hudson River story but a tale of the Catskills: the opening paragraph evokes not the majestic river, but the regal mountains, portrayed in rampant male imagery as "swelling up to a noble height, and lording it over the surrounding country" (769). What we see of the "silver Hudson" is a distant prospect and that only briefly, for the river defines the civilized, somnolent valley. Where for earlier, Enlightenment writers, like Gilbert Imlay and Daniel Bryan, American rivers carry alternatively a heroic and an erotic burden (equivalent to that traditional opposition of sublime and beautiful banks), the both connoting progress; Irving tends to use the Hudson as symbolic of repressive social norms, to which he opposes the freedom of the western mountains beyond.

Likewise, where earlier writers like Imlay and Bryan regarded the passage of civilized avatars into the virgin land as a fruitful and positive movement—a "marriage" like that typified in the ceremonies attending the opening of the Erie Canal—Irving presents an alternative version to that ritual passage in his burlesque *History of New York*. In his account of Peter Stuyvesant's first voyage up the Hudson, which employs sensuous landscape description for slyly comic ends, Irving seems to be having at the tradition of erotic embassies. Thus, at first glance, the episode appears to be an extended evocation of the "fresh green breast" school of landscape description associated with first encounters of Europeans and the New World: "Wildness and savage majesty reigned on the borders of this mighty river—the hand of cultivation had not as yet laid low the dark forest, and tamed the features of the landscape—nor had the frequent sail of commerce yet broken in upon the profound and awful solitude of ages" (622). But Irving's emphasis is not entirely on the element of sublime wildness that he later will associate with the Catskills:

> *Through such scenes did the stately vessel of Peter Stuyvesant pass. Now did they skirt the bases of the rocky heights of Jersey, which spring up like everlasting walls, reaching from the waves unto the heavens, and were fashioned, if tradition may be believed, in times long past, by the mighty spirit Manetho, to protect his favorite abodes from the unhallowed eyes of mortals. Now did they career it gaily across the vast expanse of Tappan*

> *Bay, whose wide-extended shores present a variety of delectable scenery—here the bold promontory, crowned with embowering trees advancing into the bay—there the long woodland slope, sweeping up from the shore in rich luxuriance, and terminating in the rude upland precipice—while at the distance a long waving line of rocky heights threw their gigantic shades across the water. Now would they pass where some modest little interval, opening among these stupendous scenes, yet retreating as it were for protection into the embraces of the neighboring mountains, displayed a rural paradise, fraught with sweet and pastoral beauties; the velvet-tufted lawn—the bushy copse—the tinkling rivulet, stealing through the fresh and vivid verdure—on whose banks was situated some little Indian village, or, peradventure, the rude cabin of some solitary hunter.* (633)

There is, as in Governor Pownall's description of this same scene, a mingling and opposition of masculine, mountainous terrain and the feminine aspects of nature, associated with signs of domestic order or the advances of civilization; but to all of this Governor Stuyvesant is oblivious.

Stuyvesant's boat pushes on upstream through paradisiac surroundings, its bow decorated with "figures of little pursy Cupids with periwigs on their heads, and bearing in their hands garland of flowers . . . the matchless flowers which flourished in the golden age, and exist no longer"; but the Governor-General of New Amsterdam, "that miracle of hardihood and chivalric virtue," is indifferent to the sensuous pageant of virgin territory through which he passes. "All these fair and glorious scenes were lost upon the gallant Stuyvesant; naught occupied his active mind but thoughts of iron war, and proud anticipations of hardy deeds of arms" (622, 625). Though his boat bears all the symbols of an erotic errand, and though the Hudson puts forth the sensual signals of Imlay's Ohio River, Stuyvesant is on a martial errand, purely, his stance like that of Hawthorne's Endicott in the midst of the maypole revelers.

If we can accept the possibility that Irving saw in Stuyvesant Dutch equivalents to the martial qualities of George Washington ("hardihood and chivalric virtue"), then an added measure of irony may be found in the governor's passage up the Hudson, for Stuyvesant shows the same determined, iron-jawed profile found in depictions of Washington crossing the Delaware. Finally, in Irving's description of Stuyvesant's armor-clad indifference to the natural beauty around him, we have not only a suggestion of Hawthorne's Endicott but a foreshadowing of

Captain Ahab's maniacal obliviousness to the harmonies of nature in an equivalent scene in *Moby-Dick*. It is a ligature that connects Melville's mad champion of technology as an iron way to Truth with the Enlightenment zeal for improvements that was exemplified by the foremost Forefather of them all.

These are the connections that give ideological heft to the otherwise lighthearted works of a man whose very name expressed the high regard with which Washington was regarded in 1783 (the year in which Irving was born), though the name was otherwise unsuited to the character of the author himself. Doubtless a figure of Duty as he labored at his monumental biography of that monumental man, Irving as a literary artist was generally otherwise disposed, and in his best-known writings expresses values quite different from those associated with the accomplishments of the great general, for whom rivers were chiefly a means of penetrating mountains toward valleys beyond. Worshipped by his contemporaries as a genial presiding spirit of American letters, Irving was the indolent progenitor of the Romantic response to the Enlightenment, a reaction presaged both by Brockden Brown's dark anti-Jacobin fiction and John Neal's poetic celebration of wildness undefiled. His was a radical conservatism that resisted the blandishments of progress, especially when figured as commerce and invention (and their avatar, the Yankee), and cherished rather the preserves of untouched nature symbolized by the impenetrable fastness of the Catskills, that easternmost extension of Lewis Evans' Endless Mountains, the passage through which was associated with Washington so early in his career and with the Erie Canal in 1825.

Irving's four-volume biography of Washington was an estimable accomplishment, as were his accounts of the lives of other men associated with the discovery, exploration, and exploitation of North America, like Columbus, Captain Bonneville, and John Jacob Astor; but it was his *Sketch Book* that chiefly guaranteed his fame, both as a literary artist and an influence on the emerging shape and emphasis of American letters. His greatest creation, finally, was the very antithesis of the virtues celebrated in George Washington—being a lazy, affable ne'er-do-well whose favorite activity is avoiding work, mostly by going hunting, an activity that puts him into contact, not with a Kentucky of boundless opportunity, but with the mountain spirit of Liberty. Irving thereby not only gave literary expression to his reservations about progress in America, but ironically provided the champions of internal improvements like Henry S. L. Dearborn with a suitable epithet when, in calling in 1839 for a regional system of railroads in New England, he tried to awaken "the Rip Van Winkles" of backwoods Maine before it was too late (88).

Epilogue

Ceaseless Turmoil

i

John Quincy Adams, whose administration inherited the Era of Good Feelings from Monroe only to founder in the party quarrels that followed, was the last president of the United States to identify his administration with a program of internal improvements. Indeed, it was his advocacy of public works that helped assure his defeat, for not only did it arouse Republican suspicions that Adams had not put his Federalist faith in strong central powers sufficiently behind him, but it also gave additional political force to Andrew Jackson's candidacy. Jackson's election put an end to any hope for direct government support of a national system of internal improvements, though it hardly put an end to public discussion of the issue. Jackson himself was not an inveterate enemy of public works, but their championship was generally identified during his administration and afterwards with Whigs who, like Adams, could never quite rid themselves of the trailing shadow of Federalism.

Thus, if no American president after Adams spoke out for a national program of internal improvements, it was because the Whig Party was chronically unable to mount a successful candidacy for that high office. Moreover, when they were finally victorious in that regard, the Whigs had not been sufficiently careful in considering the age and health of that candidate, or the depth of commitment to the party platform of their candidate for the second highest office. In sum, Tyler was not sufficiently "too." When the Whigs again succeeded in electing one of their own to the White House, the issue of internal improvements as the means of cementing a more perfect union had lost its savor. Slavery, not Space, was the problem; Compromise, not Canals, the solution. Zachary Taylor conceived of himself as another Washington in office, being a former Army general and a champion of the Union, and he certainly followed Washington's example by dying conveniently (in terms of chronology) at mid-century. For after 1850, it must be said, much of Washington's farewell address seemed no longer relevant. The Whig Party soon followed suit, literally dying, not with Taylor, but with the man whose public career seems an epitome of the Whig faith in keeping alive the Forefathers' vision of a more perfect union realized by roads and canals: Daniel Webster.

ii

Even to themselves, the Whigs (like Webster) were hard to define, being less a political party than a loose aggregate of interests with about as much stability as the coalition cabinets they tended to form. But in certain, strategic ways, the Whigs preserved the vision associated with what can (once again) be called the Jacobin strain of early Republicanism, which was radical in its faith in progress. The Whigs were born like Venus on the rising wave of national optimism in the 1820s, a political expression of the tidings of cheer and good will toward all sections that so briefly characterized the period and the celebration of the opening of the Erie Canal. But in many salient aspects, the Whigs were also a resurrection of the equivalent optimism so briefly enjoyed by Americans during the first decade of the century, the mood that warranted that thoroughly Whiggish document, Gallatin's report.

To return to that period is to acknowledge the omnipresence of Joel Barlow as an apostle of progress, and in a Fourth of July address Barlow delivered in 1809, the Republican Bard used the occasion of the publication of Gallatin's report to enlarge upon themes that would later appear in the speeches of Webster and Clay: "Public improvements, such as roads, bridges, and canals," said Barlow, "are usually considered only in a commercial and economical point of light; they ought likewise to be regarded in a moral and political light" (*Works*: I: 529). The public funding of "progressive improvements" such as Gallatin had called for, would make a garden of the United States and people it with a race worthy to enjoy it: "a garden extending over a continent; giving a glorious example to mankind of the operation of the true principles of society, the principles recognized in your government. Many persons now in being might live to see this change effected; and most of us might live to enjoy it in anticipation by seeing it begun" (535).

In Barlow's Fourth of July speech, as in his celebration of technological progress in the unfinished *Canal,* we can see the dim outlines of much that was to be championed, if not realized, by politicians and platforms in years to come. Barlow's coupling of "public improvements" with "moral and political" considerations established twin parallels with a considerable trajectory into the future, the first of which linked *"public improvements* and *public instruction,"* while the second associated the first with the means by which "this beneficient union" would be held together (529). To attempt to preserve the union by force, said Barlow, would be impossible, for no standing army or legislative act could hold back "a whole geographical district of rebels" (530). The only possible means was a direct appeal "to the interest and convenience of the people," who must "become habituated to enjoy a visible, palpable, incontestible good: a greater good than they could promise themselves from any change" (530). Education would serve to instruct citizens "as members of the great community," but a national system of public works was the best teacher, for it would "bind the states together in a band of union that every one could perceive, that every one must cherish, and nothing could destroy" (533).

John Quincy Adams retained enough of his Federalist scorn of novelty in 1807 to parody Barlow's poem written that year in praise of Meriwether Lewis, but he was sufficiently committed to his newfound Republicanism to sponsor legislation similar to that which finally authorized the writing of Gallatin's report. And, by the time of his election to the presidency of the United States, he was capable of sentiments remarkably similar to those in Barlow's Fourth of July speech. In his first annual message, Adams stated the principle that "the great object of the institution of civil government is the improvement of [its citizens'] condition," then pointed out that "the most important means of improvement" are roads and canals, which make possible "the communications and intercourse between distant regions and multitudes of men" (*Messages*: II: 311). Like Barlow, moreover, Adams reminded his listeners that improvements of a "moral, political, and intellectual" nature are equally essential, and that among George Washington's favorite projects, as yet unrealized, was a national university in the city bearing his name.

Adams likewise urged the passage of laws "promoting the improvement of agriculture, commerce, and manufactures, the cultivation and encouragement of the mechanic and of the elegant arts, the advancement of literature, and the progress of sciences, ornamental and profound" (316). He even provided a paragraph in praise of the spirit of exploration, citing in particular "The River of the West, first fully discovered and navigated by a countryman of our own," by which he probably meant Captain Gray of the *Columbia,* not Captain Lewis of the *Experiment,* but the phrasing is sufficiently ambiguous to satisfy the casual auditor (313). To place Barlow's and Adams' speeches side by side is to understand the extent to which the Whigs borrowed from the earlier Republican litany of Enlightenment ideals, a belief in economic and technological progress coupled with a deep and abiding faith in morality and education as twin vectors essential to an enduring republic and an indestructible Union. The emergence of the Whig Party is generally identified with Henry Clay's American System, outlined in his speech on the tariff in 1824; but in terms of principles, as opposed to particulars, there is a consistency and a continuity between the Jacobin idealism of Barlow and the consolidationalist zeal of Calhoun, Adams, Clay, and Webster.

During Adams' first year in office, Jefferson and John Adams died, the old Republican and his Federalist adversary having long since buried their differences and formed a committee of correspondence of two. Out of that symbolic union emerged the man who inherited the ideals of both the Republican and the Federalist parties, and who, as the literal son of a Founding Father, completed the transition from one generation to another: "To preserve, to improve, and to perpetuate the sources and to direct in their most effective channels the streams which contribute to the public weal is the purpose for which Government was instituted" (379). This strong, declarative phrase appears in John Quincy Adams' third annual message to Congress, and it combines a Federalist faith in a strong central government with a Republican faith in universal progress, in effect espousing the main point of Monroe's "View" while disregarding the heavy load of Constitutional argumentation. It should remind us also of James Wilson's metaphor in defending the Constitution as an instrument of compromise: the "channels" and "streams" now issue forth *from* the government *to* the people, benefits returning to the source of power. A new age had clearly begun.

John Quincy Adams, the first of the new generation of statesmen to hold the office of President, was therefore the first to labor under the weight of the Founding Fathers: "We are their monuments," declared the adjutant-general of the United States upon the deaths of Adams and Jefferson; yet at times the Sons felt more like pedestals (*Messages*: II: 349). Still, they took their responsibilities seriously, and if that generation may be called "Inventors of the Promised Land," the promise was more often than not fulfilled by invention, most especially the kinds of public works that approximate monumental architecture: "The dominion of man over physical nature," said Adams in his inaugural address, "has been extended by the invention of our artists," the "magnificence and splendor" of whose works were comparable to those of ancient empires, and which promised likewise to remain as monuments to enterprise, "imperishable glories" (295, 298).

John Quincy continued the great Adams tradition of one term in office, yet he did not follow his father into retirement, but continued as a member of Congress to work for internal improvements, which increasingly became (like Adams himself) a cause without real party affiliation, attacked (while often funded) by Democrats and neglected (while publicly espoused) by the Whigs. Like the Adamses, the use of internal improvements tended to precipitate divisiveness; its stated ideal of perfecting the national union acting chiefly to arouse sectional jealousies. Before the issue of abolition became acrimonious (thanks in no small part to John Quincy's efforts on its behalf), no single cause so dramatically demonstrated the essential disjunctiveness of the national fabric, and as the Era of Good Feelings gave way to mounting tensions between North and South, the gradual withdrawal of Calhoun's support of internal improvements, accompanied by Daniel Webster's increasing identification with them, provided a round dance of sorts, exemplifying the perilous balance between sectional interests and the national good.

At the outset of John Quincy Adams' administration, however, both his father and Jefferson still lived, good feelings prevailed, and unanimity of sectional interests seemed possible. Toward the end of his first annual message, Adams exulted that "the spirit of improvement, is abroad upon the earth," and reminded the Congress that "in the course of the year now drawing to its close we have beheld . . . under the perservering and enlightened enterprise of . . . one State of this Union . . . the waters of our Western lakes mingle with those of the ocean," an allusion to the Erie Canal, completed in 1825 (*Messages*: II: 316). More than three hundred miles of artificial river that linked the Hudson to the Great Lakes, the Erie Canal was seen by Adams and his contemporaries as a harbinger of even greater "improvements" to come. It was certainly the apogee and culmination of more than a half-century of rhetorical efforts, and though it might be duplicated and eventually transcended by other improvements, it was not equalled in its geopolitical impact until the completion of the transcontinental railroad a half-century later.

Coeval with the Erie Canal was Henry Clay's "American System," a three-part program of protective tariffs, internal improvements, and a national bank that was designed to provide a commercial mechanism equivalent to the tripartite Constitution; and both the canal and the system were intended to become beautiful machines that would contribute to the stability and orderly expansion of the republic. In defending his Tariff Bill, Clay strove to convince the Congress that "the interests of these great sections of our country are [not] irreconcilable with each other" but could be made to harmonize by "mutual concession" and "fair compromise" (*Papers*: III: 723–24). He invoked "that saving spirit of mutual concession under which our blessed Constitution was formed, and under which alone it can be happily administered," and in doing so evoked, without explicit reference, the main burden of Washington's farewell address, that "the stability of our Union" was the "paramount and greatest of all our interests" (726, 724).

Clay's repeated emphasis on mutual concessions and compromise defined the key terms for a position that would take on connotations, not of statesmanlike elevation above particulars, but of the surrender of moral purity for the sake of commercial considerations. By that time, also, so closely had the catchwords "internal improvements" become associated with political expediency that they would inspire the Transcendentalists' suggestion that other kinds of internal improvements should be given priority—namely, those aimed at reforming the essential spirit of the American people. In the speeches of Barlow and Adams the two kinds of "improvements" were virtually one and the same, but in time the unanimity eventually ceased—as did the Whig Party.

And yet, during the first quarter of the nineteenth century, president after American president called out for the implementation of a national system of roads and canals such as Washington had envisioned, a system whose champions included Robert Fulton, who to his contemporaries was "improvement" personified, and Joel Barlow, surely the most important poet of the post-Revolutionary era. Associated by proponents with the desirable and even urgent goal of strengthening the Union, and by Monroe with mechanisms of expansion, the idea of internal improvements was central to Henry Clay's American System and became a cause with which John Quincy Adams associated not only his presidency but his subsequent political career.

If, as it has been claimed, the Constitution was a creation of the European political phase of the Enlightenment, then the notion of internal improvements gave physical form in America to the Enlightenment faith in the capacity of people not only to lend definitive shape to their environment but to accelerate the kinds of exchange—both of commodities and culture—with which progress was invariably associated. Where, during the years immediately before and for some time after the French and Indian War, North American rivers were seen as natural conduits of advancing and cohering empire, the post-Revolutionary years witnessed a new direction, an emphasis on the necessity of creating artificial equivalents—roads and canals—that would complete the great, God-given diagram.

In sum, Union having been accomplished, a Nation had yet to be created, and rivers having been distributed by an all-wise Providence, canals were yet to be dug. Of all the planned, projected, and putative canals, the Erie, whose completion John Quincy Adams identified with "the spirit of improvement," would emerge as *the* canal in America, not only because it was the first of such size that had been completed, but because it had a massive impact on the future course of the United States. The completion of the Erie Canal both effected the link between the East and West that Washington and Jefferson had long argued for, and located it in the place they had long feared. In providing an important adjunct to a more perfect union, the Erie Canal set in motion geopolitical forces that would, by the middle of the nineteenth century, threaten the cohesiveness of that union with which George Washington identified himself and to whose sanctity the Whig Party devoted so much magnificent rhetoric. The Erie Canal was hardly *the* cause of the Civil War, but had Providence disposed the geography of the North American continent in such a way that Washington's Potomac route had been more feasible, then the shape and even the mission of the nation that emerged during the first half of the nineteenth century would surely have been much different from what they became.

iii

In 1804, the inveterate New England traveler and diarist Timothy Dwight visited Buffalo on Lake Erie, and without the help of any angel was able to ascertain the commercial potential of the scene spread out before him. It was, he reflected, a "boundless view" in a double sense, the immensity of the prospect being given particular point by seven ships riding in the harbor, which presented "an image of business and activity which, distant as we were from the ocean, was scarcely less impressive than that presented by the harbor of New York when crowded with almost as many hundred" (*Travels*: IV: 45). Though struck by the placid beauty of the lake, Dwight recognized that "the period is not distant when the commerce of this neighborhood will become a great national object, and involve no small part of the interests and happiness of millions" (44). To Dwight's way of thinking (which depends on Jonathan Carver's account of the Great Lakes), Erie was but part of a gigantic river, a system of interlocking waters that started with Lake Superior, a "vast reservoir" that empties through the "channel" of the St. Mary's River into Lake Huron, which also "received the water of Michigan," and "the river Huron is the channel through which this accumulated mass flows into Lake Erie," and from thence into "that part of the St. Lawrence which is termed the river Niagara, a stream inferior in splendor to none perhaps in the world" (46–52).

This eastward flow of great waters is a conceit not far different from Gouverneur Morris' vision of releasing Lake Erie by means of a canal, a Federalist notion that has its geopolitical counterpart in James Wilson's figure of confederated compromise. Thus Dwight could point out that what was thought of as "numerous, disjointed parts" was actually "one vast, continued stream"; the Niagara River being the "great outlet of this world of waters, covering about ninety-six million acres, or one hundred and fifty thousand square miles," an "amazing mass of waters" that rolled over the Falls "with a force and grandeur of which my mind had never before formed a conception. The torrent is thrown up with immeasurable violence, as it rushes down the vast declivity, between two and three miles in breadth, into a thousand eminences of foam" (57). As we have seen, from Pownalls' Cohoes to Jefferson's confluence of the Potomac and Shenandoah, American falls and rapids in topographical descriptions are an indirect means of conveying the energies of the opening and expanding republic.

During his 1812 tour, De Witt Clinton paid a visit to Niagara Falls, and though he ritualistically observed "the sublimity of the spectacle," he was mostly interested in the survival rate of living creatures (including an Indian) that had floated over the cataract, and observed matter-of-factly (if prophetically) that "Goat Island belongs to the State, and must be extremely valuable for hydraulic works" (*Life*: 130). Dwight's description, by contrast, is an amazing tangle of Romantic wonder and Enlightenment assessment, and bears comparison with Meriwether Lewis' aesthetic diagram of the falls at the head of the Missouri.

It is as if the New England preacher were attempting to take the measure of the universe with a carpenter's plumb-bob, an inadequate endeavor best demonstrated by a paragraph that begins, "the cause of this singular phenomenon may be . . . understood," and ends:

> *This was a scene which I was unprepared to expect, and an exhibition of the force of water I had never before imagined* (60). *The emotions excited by the view of this stupendous scene are unutterable . . .* [*and*] *are heightened to a degree which cannot*

be conjectured by the slowly ascending volumes of mist, rolled and tossed into a thousand forms by the varying blast, and by the splendor of the rainbow, successively illuminating their bosom. At the same time, the spectator cannot but reflect that he is surveying the most remarkable object on the globe. Nor will he fail to remember that he stands upon a river, in most respects equal, and in several of high distinction superior, to every other; or that the inland seas which it empties, the mass of water which it conveys, the commercial advantages which it furnishes, and the grandeur of its disruption in the spring are all suitable accompaniments of so sublime and glorious a scene. (62–63)

Where many visitors would invariably be reminded of God's omnipotence by the sight of the great falls, the Reverend Timothy Dwight's reaction was closer to that of Henry Adams when confronting the Dynamo: The "vast thunder" of the falls, along with an attendant vapor, which forms "a large, majestic cloud, visible from an advantageous position for a great number of miles," when coupled with the idea of "commercial advantages" inspired by the situation, graces the notion of power with a Mosaic analogy, conveying the idea of pure energy in the abstract; not only the terrific spectacle produced by a combination of natural forces, but the kind of power that "commercial advantage" on an epic scale will endow the nation that can harness them.

Had Gouverneur Morris' idea of releasing the waters of the Erie been realizable, the effect would have been to channel the power described by Dwight, harnessing it for "commercial advantages" entirely. Like Morris, Dwight died as the work was begun on the Erie Canal, and though he would surely have approved of the project in principle, its scale of flow would not have interested him, for, like the British aesthete, William Gilpin, he was always put off aesthetically by "that dull, canal-like appearance which frequently lessens the beauty of . . . streams" (54). Yet the Erie would throw the current of Dwight's great fountain in a direction that would effect geopolitical changes with an impact commensurate with the geophysical force of Niagara Falls. "The motion," Dwight observed of the descending cataract, "remarkably resembles that of a wheel rolling toward the spectator. . . . The effort of this motion of so vast a body of water, equally novel and singular, was exquisitely delightful. It was an object of inexpressible grandeur, united with intense beauty of figure" (62). The Niagara, therefore, combines both sublimity *and* beauty, aesthetic ideals difficult to apply to the Erie Canal, that dull line of unrelieved level, millennial in concept but boring to man's sensibilities upon actual encounter. Yet the canal was also in figurative terms very much like a wheel, counterpart to the revolving fleet that Colonel Stone described at the mouth of the Hudson in 1825, tremendous in implication, being an engine affecting progressive motion while apparently standing still. In general appearance it resembled the track made by a tire of the Reverend Dwight's sulky as he drove along the muddy roads of the Mohawk Valley, slowly filling up with the creep of dull water; yet it was a vast rondure upon which the geopolitical balance of power in the whole United States slowly turned until it was aligned east and west.

iv

In 1822, as the Erie Canal was pushed toward completion, the aging Thomas Jefferson wrote a letter to Governor De Witt Clinton in which he maintained he could not remember having said to Joshua Forman in 1809 that the scheme of digging a canal from Lake Erie to the Hudson was "little short of madness," while admitting that he may very well have done so, for even in 1822, "many, I dare say, still think with me that New York has anticipated, by a full century, the ordinary progress of improvement"; from which observation Jefferson drifted off into a complex sequence of perplexing questions:

> *This great work* [*the Erie Canal*] *suggests a question, both curious and difficult, as to the comparative capability of nations to execute great enterprises. It is not from greater surpluses of produce, after applying their own wants, for in this New-York is not beyond some other states; is it from other sources of industry additional to her produce? This may be; —or is it a moral superiority? a sound calculating mind, as to the most profitable employment of surplus, by improvement of capital, instead of useless consumption? I should lean to this latter hypothesis, were I disposed to puzzle myself with such investigations; but at the age of 80, it would be an idle labour, which I leave to the generation which is to see and feel its effects, and add therefore only, the assurance of my great esteem and respect.* (Hosack: 340n)

Was Jefferson pondering not only the success of the Erie Canal but the slow progress of his own pet Potomac system when he entered the factor of "moral superiority"? Was he thinking of the heritage of the Puritans as opposed to that of the Cavaliers when he spoke of a "sound calculating mind" that preferred "profitable employment of surplus" to the "useless consumption" of capital? True prophecy is always dark, and Jefferson could become very vague as he ventured into the proximity of shadowy truths.

"A few American steam boats," wrote John Adams to Jefferson early the next year, "and our Quincy Stone Cutters would soon make the Nile as navigable as our Hudson [,] Potomac or Mississippi" (*Adams-Jefferson*: II: 591). Jefferson's aged friend and old antagonist had clearly come a long way since making his pilgrimage to Portsmouth three-score and ten years earlier; yet like his republicanized son, John Adams always placed any lever, however long, upon the obdurate point and fulcrum of his beloved New England. Against the expansionist schemes of Jefferson and Washington it was, finally, the consolidationist instincts of the Adamses, with their virtually congenital deference to the law, that triumphed. Where Jefferson had viewed the world as from a hilltop, Adams had a more earthbound view, revealed throughout his diary but no more so than in its penultimate pages as he went about managing his paternal acreage at Quincy, preparatory to assuming the highest office of the land.

Quoting Homer's Ulysses (through the rhymed agency of Pope), "Sure fix'd on Virtue may your nation stand and public Evil never touch the Land," which he regarded as "the Essence and Summary of Politicks," Adams concluded that "a nation can stand on no other Basis, and standing on this it is founded on a Rock," then turned to more mundane but not unrelated matters (*D&A*: III: 244). Building walls, widening a brook, enriching his manures and land with mud and seaweed, Adams also did what he could to widen his domain, bargaining with a neighbor for "Mount Arrarat," and having bought it, indulging the countryman's pleasure of walking his expanded bounds: "Walked, with my Brother to Mount Arrarat, and find upon Inquiry that Jo. Arnold's Fence against the New Lane begins at the Road by the Nine mile Stone. My half is towards Neddy Curtis's Land lately Wm. Fields. The Western Half of the Fence against Josiah Bass, or in other Words that Part nearest to Neddy Curtis's is mine. Against Dr. Greenleaf my half is nearest to Josiah Bass's Land" (248). This is a far different Ararat from the mountain on the far western reach of Peter Fontaine's map, yet it is also an imperial point, and no ark was more securely planted upon it than that of the Adams family. From that point, as from Forefathers Rock, the Adamses and much of New England expanded their perimeters, slowly but inexorably moving out along roads across bridges solidly built, traveling in the tracks laid down by Manasseh Cutler's shay, evincing the turtle-like creep that fabulously wins the race.

In responding to Adams' boast about steamboats and stonecutters from Quincy, Jefferson in his next letter chose to change the subject by seizing upon another remark in order to attack Calvin as "an Atheist," whose "religion was Daemonism," something very much like the singular force that found itself directed and defined by the westward line of the Erie Canal (591). Along that manmade river in days to come would travel a number of American messiahs driven by various versions of Calvinistic daemonism, including the American Mahomet, Joseph Smith, and the American Cromwell, Captain John Brown: New England men whose very names suggest democratic anonymity but whose careers demonstrate the radical changes in American life during the epoch immediately following the Era of Good Feelings, many of which were dictated by the fact and direction of the Erie Canal. Though dug by New Yorkers, it stretched out like a long thin arm of New England pointing west, much as the finger of John Quincy Adams had delineated the boundary of Oregon (and hence the Oregon Trail) not long after his father had paced out the limits of his Quincy farm.

V

Among the signal events of that year of wonders, 1807, was the birth of a child who could bear with him the fate of a nation even then being sealed. A Virginian by heritage, the child would choose a kinsman, George Washington, as his model and would strive to shape his conduct as he imagined the greatest of the Founding Fathers would have done. A Whig in sympathies, he effectively removed himself from active politics by electing a military career, and, having graduated from West Point, became an Army engineer, devoting the first twenty years of his public service to realizing Washington's old dream of improved waterways that would open the way westward. He first saw armed combat during the Mexican War, showed much promise on the field, and assisted in securing while widening his nation's imperial bounds. Chance and fate determined that it would be this Army engineer who would apprehend John Brown at Harper's Ferry, on the very threshold to soil made sacred by Washington's memory, and it was his Virginian heritage also that overrode his Unionist ideology, so that the most brilliant strategist the United States had yet produced would lead the secessionist forces in the struggle that followed.

Robert E. Lee was in truth a Washington for the nineteenth century, with all that the complex phrase suggests, and given the subsequent history and tragedy of the man and his region, he was perhaps the most important of the thousands who gathered in 1825 to view the grand procession of canal barges and steamboats down the Hudson River to New York City. For in the fall of that year, Lee had entered West Point, in time to hear the thunderous salute of guns that signalled not only the passage of the ceremonial fleet downriver but the opening of the eastern portals to the West, a geopolitical shift of power that would determine Lee's career and the fate of his region and nation thereafter. Like the rifle fire exchanged between Washington's men and Jumonville's party at the headwaters of the Monongahela seventy-five years earlier, that cannonade was thunder anticipating a conflict still decades away. Though purely celebratory, the firing of cannon not ten years released from duty during the War of 1812 may be accounted the first shots of the Civil War, of which the firing upon Fort Sumter was merely a rolling echo and response.

"It Is Done!" was the message sent by those cannon, with the double-edged implication of all prophetic pronouncements, for the completion of the canal set in motion a future that in so many ways the direction and location of the canal had sealed.

Textus: A Note

The present volume is intended to be read as an independent study, but is a sequel nonetheless to *Prophetic Waters: The River in Early American Life and Literature,* which appeared in 1977. That book contained a final section, a bibliographic essay in which I attempted to place my own angle of approach within the spectrum of previous studies of the colonial period, both historical and literary. That section still stands as definitive of my purpose here. What follows is somewhat less ambitious, a summary of the books that have proven especially useful in providing guidelines and material for the present volume; a listing that, with the alphabetized bibliographic section that follows, is intended as a substitute for a lengthy apparatus of footnotes and interruptive intertextual references.

We may start at the beginning, in terms of alphabet and chronology, with Henry Adams' monumental *History of the United States During the Administration of Thomas Jefferson and James Madison* (1889). History is no longer written with such grace and irony, and the literary possibilities of American chronicles of folly during the early years of the republic—especially the high farce known as the Burr conspiracy—have not since been so effectively set forth. By contrast, yet with no intention of diminution, is the tone and emphasis of Lawrence Henry Gipson's magnum opus, *The British Empire Before the American Revolution* (1936–1970), a massively detailed chronicle of the impress of British empire on an emerging American map, an extended diagram of the process by which a new nation was conceived, not so much in liberty, as within the containing womb of imperial ambition. Where for Francis Parkman the same period was something of a field of honor on which two empires contended for the prize of continental domain, for Gipson the thirty years of war and intrigue immediately preceding the outbreak of the war for independence was an extended demonstration of the underlying complexity of conflicting interests that were inherited by the original confederacy of seldom-united states. Gipson also edited the *Essays* and wrote a life of Lewis Evans, in a single eponymous volume (1939), as well as an introduction to an edition of Thomas Pownall's intended enlargement of his (and Lewis') *A Topographical Description of the Dominions of the United States of America* (1949)—both volumes proving invaluable to my understanding of the role played by the pioneering geographer and his visionary patron to the unrolling of the (literal) American map. Instrumental in that process also was John A. Schutz's *Thomas Pownall: British Defender of American Liberty* (1951).

The library of works on George Washington, covering all aspects of his life from farming to Freemasonry, is immense. Of particular use to my own study, aside from Garry Wills' *Cincinnatus: George Washington and the Enlightenment* (1984), which appeared at a critical moment in the revision of this book and further enhanced the possibilities of iconographic "readings" of the most-pictured of our founding fathers, were Mark E. Thistlethwaite, *The Image of George Washington* (1979), and Hugh Cleland, *George Washington in the Ohio Valley* (1955). The signal importance of the 1750s as the gestatory decade of the republic, in terms of emerging ideology as well as events, is reinforced by Richard Hofstadter's posthumous *America at 1750: A Social Portrait* (1972), and, most especially in terms of literature, Lawrence C. Wroth's *An American Bookshelf: 1755* (1934). I first came across this personable ramble while working on what became the initial volume of this study at the American Antiquarian Society, nearly twenty years ago. It was Wroth who first called my attention to the work of Lewis Evans, and I shall never forget the moment when I spread out on a table the 1755 printing of Evans' great map of the Middle British Colonies, across which ran the wrinkled contours of the Allegheny range, that massive barrier that for so long determined, even as it vexed, succeeding projects of colonial and then national expansion and unity. There is no more effective iconographic display delineating the geographical destiny of the emerging United States, and if this present study has a starting point, it was that moment.

Equally influential, if of less immediate impact, was John Logan Allen's *Passage Through the Garden: Lewis and Clark and the Image of the American Northwest* (1975), which sets out in clear, definitive terms the contemporary (and erroneous) notion of pyramidal geography so central to the great fiction upon which Jefferson's favorite project was launched. Influential also was Elliot Coues' opinionated and often chauvinistic, but highly readable, commentary attached to his edition of Biddle and Allen's *History of the Expedition of Lewis and*

Clark (1893). Any discussion of the prolonged search for the fabled Passage to India is of course indebted to Bernard DeVoto's *The Course of Empire* (1952); and let me here add an acknowledgement of gratitude to Donald Jackson also, whose *Letters of the Lewis and Clark Expedition* (1962) in effect provides a comprehensive history of the ideologies and logistics involved in the expedition. Here, too, belongs the name of Douglass Adair, with whom I studied at Claremont Graduate School, and who, using the problematic death of Meriwether Lewis as an exercise central to his course in American historiography, first awakened me to the essential drama of historical discourse.

Aside from the initial inspiration provided by the relevant chapters in Adams' *History* (already mentioned), I was aided in mapping the intricacies of a central episode in this book by Thomas Perkins Abernethy's *The Burr Conspiracy* (1954), to which needs to be added Charles F. Nolan, Jr., *Aaron Burr and the American Literary Imagination* (1980). Adams' Falstaffian profile of General Wilkinson should be supplemented by Royal Ornan Shreve, *The Finished Scoundrel* (1933), and Thomas Robson Hay and M. R. Werner, *The Admirable Trumpeter* (1941). For Harmon Blennerhassett, we must revert to the memoir by William H. Safford in his edition of *The Blennerhassett Papers* (1861).

Like Washington himself, the Erie Canal, which realized his fears of the New York contingency with western waters, has launched considerable historical cargo. Here belongs George Roger Taylor, *The Transportation Revolution, 1815–1860* (1951), which places the early enthusiasm for canals within the larger context of internal improvements; on which the definitive work is still the dissertation of Joseph H. Harrison, Jr., "The Internal Improvement Issue in the Politics of the Union" (1954), fast fading from the stored reels of University Microfilms International. Unsurpassed for detail and documentation is Noble E. Whitford, *History of the Canal System of the State of New York*, two volumes, issued as the *Supplement to the Annual Report of the State Engineer and Surveyor of the State of New York for . . . 1905* (1906). On the extent to which the canal made New York the Empire State, despite its titular emphasis, see Robert G. Albion, *The Rise of New York Port: 1815–1860* (1939). Informative in terms of iconography, though posterior to the half-century covered here, is the catalogue for the exhibit, *The Course of Empire: The Canal and the New York Landscape: 1825–1875*, by Patricia Anderson (1984).

The historical literature on steamboating in the United States is also awesomely large, inspired often by nostalgia rather than a desire for accuracy and detail, but out of the pile, two volumes were of especial value to me: James T. Flexner's vivid and sinewy *Steamboats Come*

True (1944) and Louis C. Hunter's detailed and documented *Steamboats on the Western Rivers* (1949), which covers the subject from gauge cocks to the origin of the Texas deck. On the earlier, "heroic" age of navigation beyond the Alleghenies, Leland D. Baldwin's *The Keelboat Age on Western Waters* (1941) provides a stylish, conversational overview.

Essential to an understanding of the republican penchant for celebrations, with their intensely self-conscious iconographic displays, is Kenneth Silverman's monumental and elegant *A Cultural History of the American Revolution* (1976), to which my own debt is embarrassingly unfunded. Let me add to that great resource Michael Kammen's *A Season of Youth: The American Revolution and the Historical Imagination* (1978), the which, although it treats the nineteenth century entire, is a valuable lesson in how to "read" the subtext in both books and works of art. Also helpful in that regard is Clive Bush's *The Dream of Reason* (1977), another exercise in iconographic readings of popular images. On Lafayette's tour, we can start with Anne C. Loveland, *Emblem of Liberty: The Image of Lafayette in the American Mind* (1971), but essential to an understanding of the Lafayette phenomenon is the monumental collection of clippings assembled by Edgar Ewing Brandon, *Lafayette, Guest of the Nation: A Contemporary Account of the Triumphal Tour of General Lafayette Through the United States in 1824–1825*, in three volumes (1950–1957). To these let me add Stanley J. Idzerda et al., *Lafayette, Hero of Two Worlds: The Art of Pageantry of His Farewell Tour of America, 1824–1835* (1989), which arrived too late to be of help here.

On the perplexities of the post-1815 period, there is no more helpful text than George Dangerfield's *The Era of Good Feelings* (1952), although it must be supplemented by two highly influential studies with an ideological emphasis, which explore the darkness beneath the sparkling surface: Lawrence J. Friedman, *Inventors of the Promised Land* (1975), and Fred Somkin, *Unquiet Eagle: Memory and Desire in the Idea of American Freedom, 1815–1860* (1967). The idea of union, which underlay so much political and projectoring rhetoric of the period, is treated by Paul C. Nagel in *One Nation Indivisible: The Union in American Thought: 1776–1861* (1964), while a balanced, comprehensive account of the chief representatives of the Whig Party, chief champions of the idea of union and the internal improvements that would realize it, is Thomas Brown, *Politics and Statesmanship: Essays on the American Whig Party* (1985). A detailed study of Whig ideology, which has helped offset the imbalance of attention paid hitherto to that loathed party of commerce and compromise, is Daniel Walker Howe, *The Political Culture of American Whigs* (1979). Merrill Peterson's *The Great Triumvirate: Webster, Clay, Calhoun*

(1987) appeared too late to be influential here, but will be accounted for in a study now in progress.

Let me now, in concluding this section, list several works that have carried forward the tradition in American Studies initiated by Henry Nash Smith and Leo Marx, along lines I should hope my own work may be considered parallel to. I start with John F. Kasson, *Civilizing the Machine: Technology and Republican Values in America, 1776–1900* (1976), which, while emphasizing the impact of the steam-powered factory as an essential *figura,* has contributed much to my understanding of the complex of ideas that, having originated during the early years of the republic—chiefly identified with Alexander Hamilton and Tench Coxe—became a presiding ideology in the speeches and platforms of the Whig Party. Perhaps closer to my own purview is Cecilia Tichi's *New World, New Earth: Environmental Reform in American Literature from the Puritans Through Whitman* (1979), the first study of the conflict between technology and pastoralism that emphasized the importance of Joel Barlow and Jedidiah Morse to the advocacy of progress as essential to republican ideology.

The contents of the present volume were essentially complete by 1981, but during the hiatus necessitated by several revisions—of text and career—there appeared two studies that must be listed here. The first is Myra Jehlen's *American Incarnation: The Individual, the Nation and the Continent* (1986), which shares my interest in the reciprocity between ideology and the emerging shape of the American landscape, but without bending to the particulars of canals and corporations (perhaps the truest meaning of "incarnation" in the United States); taking instead the high, intellectual road, lining the route between Jefferson's busy hive of a head and Emerson's honeyed cells. For Jehlen, the shape of the landscape is largely a matter of interior landscape gardening. The second text, Robert Lawson-Peebles, *Landscape and Written Expression in Revolutionary America: The World Turned Upside Down* (1988), is much closer in purpose to my own study. Though Lawson-Peebles' purview is almost exclusively literary, and the burden of his argument rests on the perceived "quarrel" between the Old World and the New, he comes to conclusions regarding republican disarray reassuringly close to mine, albeit too late to affect my own study.

Finally, let me here list the twenty volumes of the *Dictionary of American Biography* (1943 edition), my constant companion and perennial resource, without which this book could not have been written; and acknowledge as well the indispensability to anyone lost in the thickets of late–eighteenth-century bibliography of R. G. Vail's *The Voice of the Old Frontier* (1949).

Bibliography of Works Quoted

Adams, Henry. *History of the United States During the Administration of Thomas Jefferson and James Madison* [1889–1891]. Intro. by Henry Steele Commager. 9 vols. New York: 1930.

Adams, John. *The Adams–Jefferson Letters. . . .* 2 vols. Ed. Lester J. Cappon. Chapel Hill, N.C.: 1959.

———. *Diary and Autobiography*. 4 vols. Ed. Lyman H. Butterfield. The Adams Papers. Cambridge, Mass.: 1962. (*D&A*)

———. *The Earliest Diary of John Adams*. Ed. Lyman H. Butterfield. The Adams Papers. Cambridge, Mass.: 1966. (*ED*)

Adams, John Quincy. "An Oration, pronounced July 4th, 1793, at the Request of the Inhabitants of the Town of Boston. . . ." *Eloquence of the United States*, 5 vols. (1827). Vol. 5.

———. *Report on Weights and Measures*. Washington, D.C.: 1821.

———. *Writings. . . .* 7 vols. Ed. Worthington C. Ford (1913). New York: 1968.

———. See also *Messages and Papers of the Presidents.*

Ames, Nathaniel. *The Essays, Humor, and Poems. . . .* Ed. Sam Briggs. Cleveland, Ohio: 1891.

Ashe, Thomas. *Travels in America*. London, 1809.

Bacon-Foster, Mrs. Corra. "Early Chapter in the Development of the Potomac Route to the West." *Records of the Columbia Historical Society*, vol. 15. Washington, D.C.: 1912. (B-F)

Barlow, Joel. *The Canal*. Kenneth R. Ball, "Joel Barlow's Canal and National Religion," *Eighteenth-Century Studies*, vol. 2. Berkeley, 1969.

———. *The Works. . . .* 2 vols. Facsimile ed. Intro. by William K. Bottorff and Arthur L. Ford. Gainesville, Florida: 1970.

Bartram, John. William Storck, *A Description of East-Florida, with a Journal Kept by John Bartram. . . . The Third Edition*. London, 1769.

———. *A Journey from Pennsylvania to Onandega in 1743*. Intro. by Whitfield J. Bell, Jr. Barre, Mass.: 1973.

Bartram, William. *The Travels . . .* [1791]. Naturalist's Edition. Ed. Francis Harper. New Haven, Conn.: 1958, 1967.

Benton, Thomas Hart. *Thirty Years' View. . . .* 2 vols. New York: 1856.
Blodgett, Samuel. *Facts and Arguments Respecting . . . an Extensive Plan of Inland Navigation. . . .* Philadelphia, Pa.: 1805.
Brackenridge, Hugh H. *Modern Chivalry* [1792–1815]. Ed. Claude M. Newlin. New York: 1937.
———. *A Poem on the Rising Glory of America* [1772]. See entry under Freneau.
Brandon, Edgar Ewing. *Lafayette, Guest of the Nation: A Contemporary Account . . . as Reported by the Local Newspapers.* 3 vols. Oxford, Ohio: 1950–1957.
Brown, Charles Brockden. "Abstract of a Report on American Roads," *Literary Magazine* 5 (1806).
———. *Address to the Government of the United States on the Cession of Louisiana to the French. . .* (1803).
———. *Edgar Huntly; or, Memoirs of a Sleep Walker* [1799]. Ed. Sydney J. Krause *et al.* Kent, Ohio: 1984.
———. *Wieland; or, The Transformation . . .* [1798]. Ed. William S. Koble. Columbus, Ohio: 1969.
Brown, William Hill. *The Power of Sympathy.* 2 vols. [1789]. Facsimile edition. New York: 1937.
Bryan, Daniel. *The Mountain Muse, Comprising the Adventures of Daniel Boone. . . .* Harrisonburg [Va.], 1813.
Byles, Mather. *Poems on Several Occasions* [1744]. Facsimile ed. Intro. C. Lennart Carlson. New York: 1940.
Calhoun, John C. *Works. . . .* 9 vols. New York: 1851–1856.
Carver, Jonathan. *Travels Through the Interior Parts of North America. . . . Third Edition* [1781] Minneapolis: 1956.
Chastellux, Marquis de. *Travels in North-America. . . .* New York: 1827.
Chateaubriand, Françoise Rene de. *Atala; or, The Love and Constancy of Two Savages. . . .* Trans. Caleb Bingham. Second ed. Boston: 1814.
Clark, David Lee. *Charles Brockden Brown. . . .* Chapel Hill, N.C.: 1952.
Clay, Henry. *The Papers. . . .* 5 vols. (continuing). Ed. James F. Hopkins. Lexington, Ky.: 1959– .
Clinton, De Witt. *Life and Writings. . . .* Ed. William W. Campbell. New York: 1849.
Colden, Cadwallader C. *Memoir Prepared at the Request of a Committee of the City of New York . . . of the Completion of the New York Canals.* New York: 1825.
Colles, Christopher. *Proposals for the Speedy Settlement of the . . . Lands in the Western Frontier of . . . New York, and for the Improvement of the Inland Navigation Between Albany and Oswego.* New York: 1785.
Cooper, James Fenimore. *Correspondence. . . .* 2 vols. Ed. James Fenimore Cooper. New Haven, Conn.: 1923.
———. *The Last of the Mohicans: A Narrative of 1757* [1826]. Darley edition. New York: 1859.
———. *Notions of the Americans Picked Up by a Traveling Bachelor* [1828]. 2 vols. Ed. Robert E. Spiller. New York: 1963.

———. *The Pioneers; or, The Sources of the Susquehanna* [1823]. Darley edition. New York: 1861.

Cramer, Zadoc. *The Navigator. . . . Eighth Edition* [1814]. Facsimile ed. Readex (N.P.): 1966.

Crèvecoeur, J. Hector St. John de. *Letters from an American Farmer* [1782], and *Sketches of 18th-century America* [1925]. Ed. Albert E. Stone. New York: 1981.

Cushman, Robert. *Self-Love* [1621]. New York: 1857.

Cutler, Manasseh. *An Explanation of the Map which delineates that part of The Federal Lands . . . between Pennsylvania West Line, The Rivers Ohio and Siotu, and Lake Erie* [1787]. As *The First Map and Description of Ohio* [1787]. Ed. P. Lee Phillips. Washington, D.C.: 1918.

———. "Sermon Preached at Campus Martius, Marietta . . . 1788." *Life Journals and Correspondence. . . .* 2 vols. Ed. William Parker Cutler and Julia Perkins Cutler. Cincinnati: 1888. Vol. 2, Appendix E.

Dearborn, Henry A. S. *Letters on the Internal Improvements and Commerce of the West.* Boston: 1839.

Dwight, Timothy. *The Major Poems. . . .* Facsimile ed. Intro. by William J. McTaggert and William K. Bottorff. Gainesville. Fla.: 1969.

———. *Travels in New England and New York* [1821–22].4 vols. Ed. Barbara Miller Solomon. Cambridge, Mass.: 1969.

Eloquence of the United States. 5 vols. Compiled by E. B. Williston. Middletown, Conn.: 1827.

Evans, Lewis. *Geographical, Historical, Political, Philosophical and Mechanical Essays. The First . . .* [1754]. *Number 11 . . .* L. H. Gipson, *Lewis Evans.* Philadelphia: 1939.

Farrand, Max, ed. *The Records of the Federal Conventions of 1787.* 4 vols. Revised ed. New Haven, Conn.: 1937, 1966.

The Federalist [1787–88]. Ed. Jacob E. Cooke. New York: 1961.

Filson, John. *The Discovery and Settlement of Kentucke* [1784]. Facsimile edition. Ann Arbor, Mich.: 1966.

Fontaine, James, *et al. Memoirs of a Huguenot Family.* Translated and compiled by Ann Maury. New York: 1852.

Franklin, Benjamin. *Autobiography. . . .* Ed. Leonard W. Laboree *et al.* New Haven, Conn.: 1964.

———. *The Papers. . . .* 20 vols. (continuing). Ed. Leonard W. Labaree *et al.* New Haven, Conn.: 1959– .

Freneau, Philip. *The Last Poems. . . .* Ed. Lewis Leary. New Brunswick, N.J.: 1945.

———, and Hugh Henry Brackenridge. *A Poem on the Rising Glory of America* [1772]. *The Poems of Philip Freneau.* 3 vols. Ed. Fred Lewis Pattee. Princeton, N.J.: 1902–1907. Vol. 1.

Fulton, Robert. *A Treatise on the Improvement of Canal Navigation.* London: 1796.

Gallatin, Albert. *Roads and Canals. American State Papers, Miscellaneous.* Vol. 1. Washington, D.C.: 1834.

Hamilton, Alexander. *Report on the Subject of Manufactures.* [1791]. *The*

Papers of Alexander Hamilton. Vol. X. Ed. Harold C. Syrett. New York: 1966.

Hosack, David. *Memoir of De Witt Clinton*. New York: 1829.

Hutchins, Thomas. *An Historical Narrative and Topographical Description of Louisiana and West Florida. . . .* [1784]. Facsimile edition. Intro. Joseph G. Tregle, Jr. Gainesville, Florida: 1968.

———. *A Topographical description of Virginia, Pennsylvania, Maryland, and North Carolina. . .* [1778]. In Imlay, *A Topographical Description of the Western Territory of North America. . . . The third edition* [1797]. (*TD*)

Imlay, Gilbert. *The Emigrants* [1793]. Facsimile edition. Intro. by Robert R. Hare. Gainesville, Florida: 1964.

———. *A Topographical Description of the Western Territory of North America. . . .* London: 1792.

———. *A Topographical Description of the Western Territory of North America. . . . The third edition, with great additions* [1797]. Facsimile edition. New York: 1968. (*TD*)

Irving, Washington. *History, Tales and Sketches*. Ed. James W. Tuttleton. New York: 1983.

Jefferson, Thomas. *The Adams-Jefferson Letters. . . .* See entry under John Adams.

———. Letters of the Lewis and Clark Expedition. . . . Ed. Donald Jackson. Urbana, Ill.: 1962. (*LL&CE*)

———. *Notes on the State of Virginia*. Ed. William Peden. Chapel Hill, N.C.: 1955.

———. *Report on the Navigation of the Mississippi. Class I, Foreign Relations. American State Papers*, Vol. 1. Washington, D.C.: 1833.

———. See also *Messages and Papers of the Presidents.*

Johnson, Edward. *A History of New England* [1654]. Reprinted as *Johnson's Wonder Working Providence*. Ed. J. Franklin Jameson. New York: 1910.

Lewis, Meriwether. *History of the Expedition Under the Command of Lewis and Clark. . . .* [1814]. 4 vols. Ed. Elliott Coues. New York: 1893.

Longfellow, Henry W. *Final Memorials*. Ed. Samuel Longfellow. Boston, Mass.: 1887.

Madison, James. See *Messages and Papers of the Presidents.*

Melville, Herman. *Battle-Pieces and Aspects of the War* [1866]. Ed. Hennig Cohen. New York: 1963. (BP)

Messages and Papers of the Presidents: 1789–1897. 10 vols. Ed. James D. Richardson. N.p.

Monroe, James. *Writings. . . .* 7 vols. Ed. Stanislaus M. Hamilton. New York: 1898–1903.

———. See also *Messages and Papers of the Presidents; A Narrative of a Tour. . . .*

Morris, George P. The Croton Ode. Sung . . . on the Completion of the Croton Aqueduct. . . . New York: 1842 [Broadside.]

Morse, Jedidiah. *The American Geography. . . .* Elizabethtown, N.J.: 1789.

———. *The American Universal Geography. . . . Part 1. Being a New Edition of the American Geography. . . .* Boston: 1793.

A Narrative of a Tour of Observation, Made During the Summer of 1817 by James Monroe. . . . Philadelphia: 1818.

Neal, John. *Battle of Niagara. A Poem Without Notes. . . . By Jehu O'Cataract* [pseud.]. Baltimore: 1818.

———. *Wandering Recollections of a Somewhat Busy Life.* Boston, Mass.: 1869.

Newlin, Claude Milton. *The Life and Writings of Hugh Henry Brackenridge.* Princeton, N.J.: 1932.

Niles, Hezikiah. *Niles' Weekly Register. General Index to the First Twelve Volumes. . . .* Baltimore: 1818.

Ogg, F. A. D. *The Opening of the Mississippi. . . .* New York: 1904.

Pike, Zebulon M. *An Account of Expeditions to the Sources of the Mississippi and Through the Western Parts of Louisiana. . . .* [1808]. Facsimile edition. Readex (n.p.): 1966.

Pownall, Thomas. *Scenographia Americana.* London: 1768.

———. *A Topographical Description of the Dominion of the United States of America.* [1776] [*Being a Revised and Enlarged Edition of*] *A Topographical Description of Such Parts of North America as Are Contained in the (Annexed) Map of the Middle Colonies &c. in North America.* Ed. Lois Mulkearn. Pittsburgh: 1949.

Silverman, Kenneth. *A Cultural History of the American Revolution.* New York: 1976.

Stone, William Leete. "From New York to Niagara: Journal of a Tour . . . in the Year 1829." *Publications of the Buffalo Historical Society* 14 (1910).

———. *Narrative of the Festivities Observed in Honor of the Completion of the Erie Canal.* New York: 1825. (Bound and paged consecutively with Colden, *Memoir.*)

Todd, Charles B. *Life and Letters of Joel Barlow. . . .* New York: 1886.

Tuthill, Mrs. L. C. *Success in Life. The Mechanic.* New York: 1852.

Waldo, S. Putnam. *The Tour of James Monroe . . . in the Year 1817.* Hartford, Conn.: 1818.

Warville, J. P. Brissot de. *New Travels in the United States of America* [1788]. Ed. Durand Echeverria. Cambridge, Mass.: 1964.

Washington, George. *Diaries. . . .* 6 vols. Ed. Donald Jackson. Charlottesville, Va.: 1976–79.

———. *The Journal of Major George Washington. . . .* [1754]. Facsimile edition. New York: 1959.

———. See also *Messages and Papers of the Presidents.*

Watson, Elkanah. *History of the . . . Western Canals . . . Together with the . . . Modern Agricultural Societies. . . .* New York: 1820.

———. *Men and Times of the Revolution. . . .* Ed. Winslow C. Watson. Second edition. New York: 1861.

Wilson, James. *Selected Political Essays.* Ed. Randoph G. Adams. New York: 1930.

Wirt, William. "Speech . . . in the trial of Aaron Burr, for treason. . . ." *Eloquence of the United States,* 5 vols. (1827). Vol 5.

Index